Current Topics in Behavioral Neurosciences

Volume 66

Current Topics in Behavioral Neurosciences provides critical and comprehensive discussions of the most significant areas of behavioral neuroscience research, written by leading international authorities. Each volume in the series represents the most informative and contemporary account of its subject available, making it an unrivalled reference source. Each volume will be made available in both print and electronic form.

With the development of new methodologies for brain imaging, genetic and genomic analyses, molecular engineering of mutant animals, novel routes for drug delivery, and sophisticated cross-species behavioral assessments, it is now possible to study behavior relevant to psychiatric and neurological diseases and disorders on the physiological level. The Behavioral Neurosciences series focuses on translational medicine and cutting-edge technologies. Preclinical and clinical trials for the development of new diagnostics and therapeutics as well as prevention efforts are covered whenever possible. Special attention is also drawn on epigenetical aspects, especially in psychiatric disorders.

CTBN series is indexed in PubMed and Scopus.

Founding Editors:

Michael Browning • Philip J. Cowen •
Trevor Sharp
Editors

Emerging Neurobiology of Antidepressant Treatments

Springer

Editors
Michael Browning
Department of Psychiatry
University of Oxford; Oxford Health NHS Trust
Oxford, UK

Philip J. Cowen
Department of Psychiatry
University of Oxford
Oxford, UK

Trevor Sharp
Department of Pharmacology
University of Oxford
Oxford, UK

ISSN 1866-3370 ISSN 1866-3389 (electronic)
Current Topics in Behavioral Neurosciences
ISBN 978-3-031-66355-0 ISBN 978-3-031-66356-7 (eBook)
https://doi.org/10.1007/978-3-031-66356-7

This Springer imprint is published by the registered company Springer Nature Switzerland AG
The registered company address is: Gewerbestrasse 11, 6330 Cham, Switzerland

Preface

Living in exciting times is regarded as a mixed blessing but for antidepressant treatment development, it is something that we welcome. Novel approaches involving both pharmacological agents and other therapeutic modalities are set to move beyond traditional monoamine potentiation and transform the medical treatment of clinical depression. The aim of this volume is to cover the key neurobiological mechanisms involved in both established antidepressant treatments and the most promising newer approaches.

For the last 30 years antidepressant prescribing has been dominated by selective serotonin reuptake inhibitors (SSRIs). Cowen outlines the benefits and problems of these agents and finds SSRIs lacking in ability to ameliorate the anhedonia and cognitive impairments associated with severe depression. An increasingly recognised problem of SSRIs is the experience of unpleasant withdrawal symptoms manifesting when treatment is discontinued. These important phenomena require better scientific understanding. Sharp and Collins assess the likely mechanisms and demonstrate that excessive 'rebound' firing of serotonin neurons upon SSRI withdrawal is a significant factor.

Dopamine pathways have long been implicated in the anhedonia and loss of motivation experienced by depressed patients. However, with the exception of bupropion, current antidepressant treatments have little direct impact on dopaminergic mechanisms. Halahakoon and Browning summarise evidence for the antidepressant efficacy of the dopamine receptor agonist, pramipexole. They suggest that an action of pramipexole to inhibit decay of reward value could underlie its therapeutic effects in treatment-resistant depression.

Opioids have powerful effects on mood and historically were used to treat people with severe melancholia. Jelen and colleagues describe the role that opioid neurotransmission plays in the manifestations of depression and the interactions of opioids with other important systems relevant to depressive states such as the HPA axis and the immune network. In the development of opioid-modulating drugs as potential antidepressants it will be important to avoid the dependence-producing properties of mu receptor agonists. Selective opioid kappa receptor antagonists provide an intriguing possibility in this respect.

A remarkable phenomenon in the psychopharmacology of depression has been the development of more rapidly acting and effective antidepressant treatments such as ketamine and psilocybin. Such agents produce immediate profound changes in mood and self-conscious state and lead to intriguing correlations between improvement in mood and the functioning of relevant neural networks.

Kamenish and Robinson explore the preclinical neuropsychological actions of these rapidly acting agents in comparison with more conventional monoamine reuptake blockers. They suggest an important role for differential effects on emotional memory. Whereas conventional antidepressants produce a positive bias in memories of new experiences, ketamine attenuates the negative emotional biases associated with aversive past experience. This may explain the rapid alleviation of the depressive state produced by ketamine in people with clinical depression.

In her review of the mechanism of action of ketamine in depressed patients, Costi describes changes in striatal connectivity which correlate with reductions in anhedonia. Tasked-based fMRI confirms that ketamine increases activity in the striatal and orbitofrontal regions involved in reward processes. This is consistent with the benefit ketamine produces in depressed patients in whom anhedonia has been unrelieved by conventional antidepressant agents. Costi points out that thus far, apart from ketamine, other NMDA receptor antagonists have not been developed successfully for clinical use. Whether an opioid action is a necessary part of ketamine's antidepressant effect remains an important question.

Nutt and colleagues explore the exciting effects of psychedelic drugs, such as psilocybin, in patients with depression. In contrast to other pharmacological treatments, the contention is that a single dose of a psychedelic drug, given in an appropriate setting, can produce enduring symptomatic benefit and improved psychosocial adjustment. The specific contribution of psychotherapy to this therapeutic effect is still to be defined. Neuroimaging studies indicate that psychedelic treatment is associated with a substantial increase in connectivity between brain regions that are not usually functionally linked. This offsets the functional segregation of networks associated with adult brain development and perhaps the entrenched negative thinking of the depressive state. This notion is similar to the idea that psychedelics increase brain entropy, thereby disrupting top-down control of cognition and facilitating reappraisal of negative self-referential themes.

Other novel approaches to the treatment of depression involve modifications to the gut microbiome and the immune system. Lachmansingh and colleagues explore the significance of what has come to be called, the 'gut-brain' axis. Associations have been discovered between compositional alterations of gut microbiota and clinical depressive symptomatology. An increasing number of studies are assessing the value of pre- and probiotics as augmentation treatments in depression. Future work in this area will be directed both to the role of specific gut microorganisms and using personalised approaches to match particular prebiotic fibres to individual patient characteristics.

Cavanagh describes the role that inflammatory mechanisms play in some patients with clinical depression. The evidence for antidepressant effects of anti-inflammatory drugs is not easy to interpret because of the wide variety of agents used and the heterogeneity of patient samples. However, there is evidence for antidepressant effects of some NSAIDS and cytokine inhibitors. Future work should involve phenotyping depressed patients for a range of inflammatory markers and better controlled clinical trials of anti-inflammatory agents combined with experimental medicine studies of mechanism.

Transcranial Magnetic Stimulation (TMS) is increasingly used as a neurostimulatory treatment for depression. A variety of coil positions and stimulus parameters have been employed, making it difficult to discern a common mode of action. Downar and colleagues provide a wide-ranging review of TMS mechanisms which discusses effects on neural networks, synaptic plasticity, and modification of dynamic synaptic range. Effects on the latter process could explain the time-course of TMS response and its ability to alleviate the symptomology of apparently diverse neuropsychiatric conditions. Increasing sophistication of neural network analysis with fMRI is offering the possibility of effective personalising of TMS treatment for individual patients.

Electroconvulsive therapy (ECT) continues to be regarded as the most effective of the biological antidepressant treatments. Belge and colleagues review the mechanisms involved in the antidepressant action of ECT focusing on the structural and functional changes revealed by MRI in depressed patients. There is reliable evidence that ECT produces increases in the volume of the hippocampus, reversing the changes seen in depressed patients prior to treatment. Whether this correlates with clinical outcome or cognitive side effects is uncertain. Volumetric changes in other brain regions during ECT, for example the striatum, might be more clinically relevant. FMRI studies suggest that ECT produces changes in functional connectivity in neural networks that generally reverse the abnormal connectivity patterns seen in the depressed state.

Theiss and colleagues discuss the potential neurobiological mechanisms associated with a range of more invasive neuromodulatory treatments that have been trialled in patients with refractory depressive states. These include vagal nerve stimulation and deep brain stimulation. With vagal nerve stimulation an important target is likely to be the nucleus of the solitary tract which sends projections to several brain regions involved in the pathophysiology of depression. Clinical results from studies of deep brain stimulation have been inconsistent. This is likely to reflect the difficulty in targeting the precise brain regions relevant to the expression of depression in individual patients.

The development of new treatments for depression recognises the importance of this clinical condition and the limitations of conventional therapies. The next few years are likely to see availability of a much wider range of antidepressant treatments that include manipulation of the gut microbiome, modification of immunological mechanisms, and the use of psychedelic-assisted psychotherapy. This enlarged scope of therapies will make it increasingly important to develop the so-called ‘precision’ approaches where clinical and biological phenotyping permit matching of depressed patients to the treatment most likely to help them.

Oxford, UK Michael Browning
Oxford, UK Philip J. Cowen
Oxford, UK Trevor Sharp

Contents

SSRIs in the Treatment of Depression: A Pharmacological CUL-DE-SAC?

Philip J. Cowen

Contents

Abstract The widespread adoption of selective serotonin reuptake inhibitors (SSRIs) as first-line pharmacological treatments in the management of clinical depression transformed the landscape of drug therapy for this condition. SSRIs are

P. J. Cowen (✉)
Department of Psychiatry, Warneford Hospital, University of Oxford, Oxford, UK
e-mail: phil.cowen@psych.ox.ac.uk

Curr Topics Behav Neurosci (2024) 66: 1–20
https://doi.org/10.1007/7854_2023_447
Published Online: 4 November 2023

safer and better tolerated than the tricyclic antidepressants (TCAs) that they replaced. However, they have limitations that may have placed a ceiling on the expectations of first-line pharmacological treatment. Notable problems with SSRIs include induction of anxiety on treatment initiation, delayed onset of significant therapeutic effect, sexual dysfunction, sleep disturbance and overall modest efficacy. The latter is linked with an inability of SSRIs to effectively treat syndromes of anhedonia and cognitive impairment. Combined serotonin and noradrenaline reuptake inhibitors (SNRIs), such as venlafaxine, have produced some limited improvements over SSRIs in efficacy, at the cost of a greater side-effect burden. Attempts to supplement serotonin reuptake activity with actions at serotonin receptor sub-types have not yet yielded substantial benefits; however, vortioxetine may provide more utility in the management of cognitive impairment. Future advances might come from the development of SNRIs, which more closely mimic the actions of effective TCAs. There may also be possible benefits to be derived from combining SSRIs with 5-HT_4 receptor agonists and 5-HT_7 receptor antagonists.

Keywords 5-HT_{1A} receptor · 5-$HT_{2A/2C}$ receptor · 5-HT_4 receptor · 5-HT_7 receptor · Antidepressants · Anxiety · Depression · Serotonin · Serotonin (5-HT) receptors · SNRIs · SSRIs · TCAs · Vilazodone · Vortioxetine

1 Depression and Antidepressant Treatment

Clinical depression is a common condition with a high morbidity. In practical terms, this means that the experience of depression causes much hardship for patients and their families. Depression often begins in early adult life and can persist with remissions and recurrences throughout the lifespan; this contributes to the substantial overall morbidity associated with the condition. In global terms, recurrent clinical depression causes significantly greater decrements in health than other major chronic diseases, such as asthma, angina, arthritis and diabetes (Moussavi et al. 2007).

While depression has been recognised and treated by physicians since antiquity, the use of specific antidepressant medication is a more recent development, dating back to the second half of the twentieth century when two classes of pharmacological agents, tricyclic antidepressants (TCAs) and monoamine oxidase inhibitors (MAOIs), were discovered by chance (see Healy 1997). Subsequently, it became apparent that MAOIs were associated with potentially dangerous food and drug interactions; also, a major trial conducted by the Medical Research Council questioned their efficacy (Medical Research Council 1965). TCAs therefore became the mainstay of treatment for clinical depression of any severity.

Investigations in experimental animals suggested that the therapeutic effect of TCAs was mediated by their ability to potentiate the activity of the brain neurotransmitters noradrenaline and serotonin (5-Hydroxytrytamine (5-HT)). This effect was achieved through a blockade of a reuptake pump located on the cell membrane of the neuron from which the transmitters were released. Therefore, TCAs and related drugs became known as 'reuptake inhibitors'. TCAs were regarded as useful drugs, and both general practitioners and psychiatrists became familiar with their clinical use (Harrison et al. 2018).

The adverse effects of TCAs were believed to be due to pharmacological actions other than noradrenaline and serotonin reuptake blockade. For example, the blockade by TCAs of acetylcholine receptors caused so-called anticholinergic effects, which were experienced as dry mouth, blurred vision, constipation and urinary hesitancy. The ability of TCAs to block noradrenaline alpha$_1$ adrenoreceptors was associated with postural hypotension. The histamine H1-receptor-blocking effects of TCAs caused sedation and drowsiness. Most seriously, the membrane-stabilising effect of TCAs predisposed to cardiac arrhythmias, which could have fatal consequences in even modest overdoses, for example the acute ingestion of only 1 week's supply of medication (Harrison et al. 2018).

Because of the problems associated with TCA treatment and the scientific perception that the therapeutic and adverse effects of TCAs were probably pharmacologically dissociable, there was a search by the pharmaceutical industry for antidepressant compounds that would be better tolerated, would be safer in overdose and hopefully would be more efficacious than TCAs. This led to the development of potent monoamine reuptake inhibitors that lacked the other pharmacological properties of TCAs. These new compounds included the selective serotonin reuptake inhibitors (SSRIs) which over time superseded the TCAs as first-line antidepressant treatments.

2 Development of SSRIs

Currently, in most clinical settings, the decision to employ a pharmacological treatment for depression results in the initial prescription of a selective serotonin reuptake inhibitor (SSRI). SSRIs are widely available in generic form and are safer and better tolerated than their 'parent' compounds, the TCAs. In clinical use, SSRIs have some drawbacks, which are described below. However, in the last decades, no new antidepressant medication has seriously challenged their dominant prescribing position. Is it possible that serotonin reuptake blockers have reached a point of development where further pharmacological modifications cannot lead to useful advances in treatment?

The first SSRI to be introduced in the UK was zimelidine, which was marketed in 1982. However, it was soon withdrawn because of a rare association with the Guillain-Barre syndrome. Subsequently, several other SSRIs became available: fluvoxamine (1983), fluoxetine (1987), sertraline (1990), paroxetine (1991), citalopram (1995) and escitalopram (2002). Initially, TCAs continued to be recommended as first-line antidepressants. However, in its 2004 Depression Guideline, the National Institute for Health and Care Excellence (NICE) recommended that SSRIs should become first-choice antidepressants because of their improved

Table 1 Benefits and disadvantages of SSRIs in the treatment of clinical depression

Benefits of SSRIs	Disadvantages of SSRIs
Simple dosing	Delayed onset of action
Effective in co-morbid anxiety states	Limited efficacy (especially in severe depression)
Relatively safe in overdose	Limited ability to relieve anhedonia and cognitive impairment
General cardiac safety (but citalopram and escitalopram increase QTc)	Induction of anxiety early in treatment (possible link to suicidality)
Reasonably well tolerated by most people	Lack of benefit from sleep
	Sexual dysfunction
	Emotional blunting
	Withdrawal symptoms

safety and tolerability relative to TCAs. Though some newer antidepressants (mirtazapine, venlafaxine, duloxetine, agomelatine, vilazodone and vortioxetine) have been marketed since the advent of SSRIs, the hegemony of SSRIs as customary first-line agents has remained unchallenged for 20 years (Harrison et al. 2018). What are the benefits and problems of SSRIs as antidepressant medications (see Table 1)?

3 Advantages of SSRIs

3.1 Simplicity of Dosing

With TCAs, gradual dose titration is necessary to improve tolerability. In addition, there is some uncertainty as to what the appropriate dose might be for an individual patient. With SSRIs, it is possible to prescribe a therapeutic dose from the beginning of treatment in a simple once-daily schedule. This dosing is supported by positron emission tomography (PET) imaging studies, which demonstrate that at their minimally effective doses, SSRIs occupy about 80% of serotonin transporters with little further increase occurring as the dose is raised (Meyer et al. 2004).

Whether the efficacy of SSRIs is improved with increased dosing has been controversial. The most recent meta-analyses suggest that for most depressed patients, increasing the SSRI dose above the minimally effective level does not bring about additional benefits (Furukawa et al. 2019, 2020). However, such analyses may not account for those individual patients who metabolise SSRIs to a greater extent than normal. It is also possible that at high doses, additional pharmacological properties of some SSRIs may supplement their effects on serotonin. For example, sertraline has an appreciable affinity for the dopamine transporter, which may become functionally relevant at higher doses (see Table 2).

Table 2 Binding affinity (nM) of selected antidepressants at key neurotransmitter targets (note that a lower number denotes more potent binding affinity)

Drug	SERT	NET	DAT	5-HT2A	5-HT2C	H-1	Musc
Amitriptyline	43.0	61.3	3,944	5.6	6.2	5.5	11
Nortriptyline	260	4.0	1,700	41	8.5	9.0	55
Clomipramine	0.7	45	1,800	13	37	22	37
Sertraline	0.2	679	25.5	5,353	2,298	5,810	10,000
Escitalopram	1.8	7,150	10,000	10,000	2,351	1970	1,242
Venlafaxine	7.7	630	8,473	10,000	10,000	10,000	10,000
Duloxetine	5.0	45	435	504	916	2,300	1,600

SERT serotonin transporter, *NET* noradrenaline transporter, *DAT* dopamine transporter, *H-1* H1-histamine receptor, *Musc* acetylcholine muscarinic receptor (estimates from https://pdsp.unc.edu/databases/kidb.php; Besnard et al. 2012)

3.2 *Efficacy Against Anxiety Symptoms*

Patients with major depression often present with co-morbid anxiety. These conditions can include generalised anxiety disorder, social anxiety disorder and panic disorder. Diagnosing the predominant condition to treat is sometimes difficult, and it is therefore helpful that in addition to their antidepressant properties, SSRIs have a broad range of therapeutic activity in anxiety disorders. This extends to patients with obsessive-compulsive disorder (OCD) and, to a somewhat lesser extent, post-traumatic stress disorder (PTSD), where concomitant diagnoses of major depression are frequent (Baldwin et al. 2014; Harrison et al. 2018).

3.3 *Relative Safety in Overdose*

With the exception of lofepramine, TCAs have significant cardiac toxicity in overdose. In contrast, fatal overdose with an SSRI as a single substance is rare. Of the SSRIs, citalopram and escitalopram may be more likely to be involved in serious overdose toxicity because of their effects on the QTc interval (Hawton et al. 2010).

4 Problems with SSRIs

4.1 *Delayed Onset of Action*

Like other drugs that achieve their antidepressant effects through monoamine reuptake blockade, SSRIs have a delayed onset of therapeutic effect. This is the case even though, unlike TCAs, SSRIs can be administered at a therapeutically effective dose from the beginning of treatment. Clinically, this means that significant subjective improvement in depressive symptomatology may not be apparent for a number of weeks.

This delay in therapeutic effect has given rise to the notion that the antidepressant action of SSRIs is due to neuroadaptive changes occurring as a consequence of the acute increase in serotonin neurotransmission. Previous theories have implicated adaptive changes in inhibitory $serotonin_{1A}$ ($5\text{-}HT_{1A}$) autoreceptors, which require desensitisation before the effects of the SSRI on serotonin neurotransmission can be fully expressed. More recent formulations have focused on changes in synaptic plasticity and neurogenesis as being critical to antidepressant action (Harmer et al. 2017).

In fact, a meta-analysis of the time course of the antidepressant effect of SSRIs showed that, relative to placebo, there is a significant benefit of SSRIs to ameliorate depressive symptomatology during the first week of treatment. There follows a gradual linear improvement in the antidepressant effect, which grows over the following weeks (Taylor et al. 2006). This suggests that SSRIs start to act on depressive symptoms quickly but that a small weekly benefit needs to grow over time before becoming apparent to patients and clinicians.

This profile of effect is supported by the neuropsychological actions of SSRIs. Such studies show that very early in treatment, SSRIs act to reverse the negative emotional biases that are regarded as a key maintaining factor in clinical depression. The action on emotional biases occurs at a non-conscious level, so it is plausible that time is needed for the effects of this transformed emotional environment to be experienced as an improvement in mood at a conscious level (Harmer and Cowen 2013). This notion also suggests that SSRIs do not work directly to improve mood but produce their antidepressant effects indirectly through changing emotional processing. If this is the mechanism of action, it may put an inevitable constraint on how quickly SSRIs (and other monoamine reuptake blockers) are able to improve the clinical symptoms of depression.

4.2 Limited Efficacy

Network meta-analyses show that SSRIs are superior in efficacy to placebo (Cipriani et al. 2018). However, in real-world studies, their overall efficiency appears modest. While some patients do well with first-line SSRI treatment, about half do not achieve a useful clinical response (operationally defined as a 50% decrease in baseline score on a standardised depression rating scale), with probably about half of these subjects experiencing very little improvement at all (Rush 2007; Cowen 2017).

There has been a long-standing debate as to whether SSRIs have lesser efficacy in more severe depressive states, such as might be seen in depressed inpatients. In patients with depressive psychosis, there is some evidence that conventional TCAs might be more effective than newer antidepressants (Wijkstra et al. 2010). A related issue is the proposed lack of efficacy of SSRIs in patients with prominent symptoms of anhedonia (Shelton and Tomarken 2001; McMakin et al. 2012; Cao et al. 2019).

Such symptoms are common in patients with resistant depression and those with more severe depressive symptoms (Kelly et al. 2022).

4.3 *Suicidality*

Like other antidepressants, SSRIs carry a warning of an increased risk of suicidal behaviour during the early stages of treatment. Relative to placebo, there may be a small risk that SSRIs can increase rates of self-harm, but in the largest meta-analysis of adult patients, involving over 700 studies, the number needed to harm with SSRIs (759) was much greater than the number needed to treat (estimated to be between four and seven) (Gunnell et al. 2005).

In adolescents and children, the risk of self-harm with SSRIs might be greater. In a meta-analysis of 27 placebo-controlled trials in children and adolescents with a variety of diagnoses, including depression, Bridge et al. (2007) found no completed suicides but a small, significant increase in suicidal ideation and self-harm attempts with SSRIs compared with placebo (number needed to harm = 143). However, a subsequent network meta-analysis of randomised controlled trials found an increased risk of suicidality in depressed young people only with venlafaxine and not with SSRIs (Cipriani et al. 2016).

4.4 *Increased Anxiety Early in Treatment*

While over time SSRIs have useful therapeutic effects in anxiety disorders, early in treatment, their initiation can be associated with worsening anxiety symptoms. This can make the management of SSRI treatment difficult and cause patients additional distress. Increased anxiety and agitation early in treatment have also been putatively linked to worsening suicidal ideation (Sinclair et al. 2009; Amitai et al. 2015).

4.5 *Sleep Problems*

Sleep problems are an important part of depressive conditions, and persistent insomnia increases the risk of future episodes of depression and suicidal behaviour (Taylor et al. 2003). Generally, SSRIs are not sedating compounds and do not assist sleep in those depressed patients who present with significant insomnia. Indeed, if anything, the effect of SSRIs is to disrupt sleep (Sharpley and Cowen 1995; Sharpley et al. 1996). Though patients may report improved subjective sleep following successful SSRI treatment, polysomnography continues to show sleep disruption even in those whose depressive symptoms have improved to some extent (Wilson et al. 2000).

4.6 Sexual Dysfunction

SSRIs are commonly associated with sexual dysfunction, particularly delayed orgasm and anorgasmia. However, problems with desire and arousal are also reported. The incidence of sexual dysfunction due specifically to SSRIs varies between studies. Where effects on the sexual function are specifically elicited, the rates are probably around 50% of patients. Of course, sexual dysfunction becomes particularly problematic when SSRI treatment needs to be continued for longer periods of time (Jacobsen et al. 2020).

4.7 Emotional Blunting

Patients taking SSRIs sometimes complain of emotional blunting. This is a different experience from that of anhedonia because it applies to both positive and negative emotions and is described as an inability to feel. It has been difficult to establish whether such reports are due to residual depressive symptoms or are directly attributable to SSRI treatment. Goodwin et al. (2017) found, using a specially developed questionnaire, that just under half of depressed patients treated with a variety of monoaminergic antidepressants reported emotional blunting. Generally, reports of blunting increased with depression scores, suggesting that depression contributes to this experience. However, emotional blunting seemed to be somewhat less in patients taking the noradrenaline and dopamine reuptake inhibitor bupropion, suggesting that serotonergic antidepressants are more likely to be associated with a feeling of blunted emotions.

4.8 Cognitive Impairment

It is increasingly recognised that clinical depression is associated with significant cognitive impairment. These impairments affect a broad range of cognitive processes, including attention, executive function, and learning and memory. Cognitive deficits often seem to persist in patients who otherwise have had a reasonable antidepressant response to treatment. TCAs may actually worsen cognition because of their anticholinergic actions. While SSRIs are not thought to impair cognition, they do not seem to produce a specific benefit of reversing cognitive deficits (Culpepper et al. 2017; Colwell et al. 2022).

4.9 Withdrawal Symptoms

A sudden cessation of SSRIs gives rise to a characteristic withdrawal syndrome. To describe the range of possible symptoms, Berber (1998) suggested the mnemonic 'FINISH' (flu-like symptoms, insomnia, nausea, imbalance, sensory disturbances, hyperarousal). The incidence of withdrawal symptoms varies between studies and depends on the pharmacokinetic profile of the SSRI concerned. For example, withdrawal symptoms are less common after the cessation of fluoxetine because of the long half-life of its active metabolite, norfluoxetine. In contrast, the withdrawal symptoms after stopping paroxetine treatment are regarded as particularly problematic; this may be related to the shorter half-life of this agent (Haddad 2001).

How far slow tapering of SSRI treatment avoids troublesome withdrawal symptoms is not clearly established. It has recently been suggested that because of the hyperbolic nature of the interaction of SSRIs with the serotonin transporter, the tapering of SSRI dosage should become much more cautious as lower doses are reached during the withdrawal process (Horowitz and Taylor 2019). This is because at lower doses, reductions in the SSRI dose produce proportionately much greater decreases in transporter occupancy.

5 Modifications of Selective Serotonin Reuptake Inhibition as a Therapeutic Mechanism

A number of pharmacological approaches have been adopted to improve the efficacy of selective serotonin reuptake blockade as a treatment for depression and also to ameliorate some of the common side effects of this mechanism. One strategy has been to add noradrenaline reuptake blockade to that of serotonin, so-called dual reuptake inhibitors. Another approach has been to develop drugs with additional activities at specific serotonin receptor sub-types. The aim here is to modify the effect of the broad enhancement of serotonin in the CNS and, in this way, produce more effective and better-tolerated drugs. How successful have these attempts been?

5.1 Dual Reuptake Inhibitors

In terms of efficacy in treating depression, it is accepted that SSRIs have not superseded TCAs and, in some situations, may be less effective (Wijkstra et al. 2010). TCAs have the ability to potentiate both serotonin and noradrenaline activity, and it is possible that this dual reuptake blockade provides increased therapeutic benefits. Venlafaxine and duloxetine were introduced as dual monoamine reuptake blockers or serotonin noradrenaline reuptake inhibitors (SNRIs) (Harrison et al. 2018).

In fact, both SNRIs have differences from standard TCAs in their profile of ability to block the reuptake of serotonin and noradrenaline. Tertiary TCAs, such as amitriptyline, have a roughly equal equivalence to inhibit serotonin and noradrenaline transporters. However, in the body, tertiary tricyclics are metabolised to the corresponding secondary compound in the case of amitriptyline, this is to nortriptyline. Secondary tricyclics, such as nortriptyline, are potent noradrenaline reuptake blockers. Therefore, during clinical treatment with tertiary tricyclics, there might, in fact, be a preponderance of noradrenaline reuptake blockade over that of serotonin (Table 2).

From this, it can be seen that the pharmacological effects of treatment with SNRIs are different from those of TCAs. SNRIs predominantly block the reuptake of serotonin and are less potent at blocking noradrenaline reuptake. This may account for the presence of a potential dose-response curve with venlafaxine (Furukawa et al. 2019). As the dose is increased, the noradrenaline transporter may begin to be occupied by venlafaxine. However, if we regard TCAs as true, 'dual uptake', inhibitors, we can see that SNRIs fall short in that respect (Table 2).

Overall, SNRIs have a similar adverse effect profile to that of SSRIs; however, at higher doses, increases in blood pressure can occur with SNRIs, presumably because of the growing potentiation of noradrenaline activity. Orthostatic hypotension can also be problematic (Harrison et al. 2018). Generally, SNRIs are thought to be a little more efficacious than SSRIs but are said to be less well tolerated (Cipriani et al. 2018). The toxicity of venlafaxine in overdose is intermediate between that of SSRIs and TCAs (Hawton et al. 2010).

5.2 Serotonin Reuptake Inhibition and Serotonin Receptor Activity

Two antidepressant drugs have been developed, which, as well as having the ability to block the serotonin transporter, act at some sub-types of serotonin receptors. The aim has been potentially to improve the efficacy and/or tolerability of serotonergic antidepressant agents. The first medication of this sort to be marketed was vilazodone, which has a high affinity for the serotonin transporter, as well as the 5-HT_{1A} receptor, where it acts as a partial agonist (Table 3).

Theoretically, partial agonism at 5-HT_{1A} receptors could provide benefits over standard SSRI treatment. For example, the activation of 5-HT_{1A} autoreceptors is

Table 3 Binding affinity (nM) of vilazodone and vortioxetine at serotonin receptors

Drug	SERT	5-HT1A	5-HT1B	5-HT1D	5-HT3	5-HT7
Vilazodone	0.5	0.3				
Vortioxetine	1.6	15	33	54	3.7	19

STT serotonin transporter (estimates from https://pdsp.unc.edu/databases/kidb.php; Besnard et al. 2012; Sanchez et al. 2015)

thought to delay the therapeutic onset of SSRI action, and this activation might be prevented by 5-HT_{1A} receptor partial agonism. This could lead to a quicker onset of antidepressant action in patients taking vilazodone (Stahl 2014). Such an effect has been reported with a combination of SSRI and the beta-adrenoceptor antagonist pindolol, which also has partial 5-HT_{1A} receptor agonist properties (Whale et al. 2010).

Also, the combination of the 5-HT_{1A} receptor partial agonist buspirone with SSRIs has been reported to be helpful in patients with resistant depression and perhaps in the treatment of SSRI-induced sexual dysfunction. Accordingly, other potential benefits of vilazodone might include superior efficacy and less sexual dysfunction compared to SSRIs (Stahl 2014).

In fact, network meta-analyses do not indicate that vilazodone is more effective than SSRIs (Wagner et al. 2018). Its side-effect profile is similar to that of SSRIs, with increased rates, relative to placebo, of nausea, diarrhoea, dizziness and headache. Whether its use is associated with a lower rate of sexual dysfunction than that seen with conventional SSRIs is unclear (Chauhan et al. 2022). The efficacy and acceptability of vilazodone in generalised anxiety disorder are not established (Slee et al. 2019).

Vortioxetine is also a serotonin reuptake inhibitor with actions at multiple serotonin receptor sub-types. Sanchez et al. (2015) summarised these as antagonism at the 5-HT_{1D}, 5-HT_3 and 5-HT_7 receptors; partial agonism at the 5-HT_{1B} receptor; and agonism at the 5-HT_{1A} receptor (Table 3).

At its maximum daily licensed dose (20 mg), vortioxetine demonstrates about 80% occupancy of brain serotonin transporters. The actions of vortioxetine at other serotonin receptor sub-types produce a complex neurochemical profile in animals with increases in cortical noradrenaline, acetylcholine, histamine and glutamate, as well as serotonin (Sanchez et al. 2015). It has been suggested that the effects of vortioxetine on acetylcholine, histamine and glutamate may allow some specific cognitive enhancing effects, not seen with standard SSRIs or SNRIs. How far such neurochemical effects might occur in depressed patients during the clinical use of vortioxetine is unclear. For example, a PET study in healthy volunteers who took 30 mg daily of vortioxetine for 9 days showed no significant occupancy of 5-HT_{1A} receptors (Stenkrona et al. 2013). This suggests that during clinical use, the main effects of vortioxetine are likely to be due to the occupation of the serotonin transporter and the blockade of the 5-HT_3 receptor (Table 3). Indeed, pre-clinical models have implicated 5-HT_3 receptor blockade in the therapeutic effects of vortioxetine (Riga et al. 2016).

In clinical trials of patients with major depression, vortioxetine was compared with placebo, as well as SNRIs such as duloxetine and venlafaxine. Overall, the efficacy of vortioxetine was greater than that of placebo and mostly equivalent to that of SNRIs. However, duloxetine had efficacy advantages over vortioxetine on some secondary outcome measures (Koesters et al. 2017).

Vortioxetine produced greater benefits than duloxetine on tests of cognition, particularly the digit symbol substitution test, which taps a number of cognitive domains. The path analysis of the effects of vortioxetine in depressed patients suggested that at least part of the cognitive benefit of vortioxetine was directly mediated by an action on cognition, rather than being secondary to mood improvement (Culpepper et al. 2017). However, subsequent studies in depressed patients have yielded less consistent results in terms of the specific cognitive benefit of vortioxetine (Baune et al. 2018; Vieta et al. 2018). It also is not clear how much 'real-world' improvement vortioxetine might provide in terms of better work functioning, for example.

The side-effect profile of vortioxetine is similar to that of SSRIs, with the most common adverse reactions being nausea, headache, dry mouth and dizziness. In this respect, it is of interest that the potent 5-HT_3 receptor blockade produced by vortioxetine does not apparently protect against the side effects of nausea. Vortioxetine at modest doses (10 mg or less) seems to produce less sexual dysfunction than standard SSRIs, such as paroxetine. However, this advantage may be diminished as the dose of vortioxetine increases towards its licensed maximum (Jacobsen et al. 2019).

How far vortioxetine is effective in the treatment of anxiety disorders is not clear. There have been a number of controlled studies on generalised anxiety disorder, but meta-analyses report that trial quality is low and the efficacy of vortioxetine in this condition is not clearly established (Qin et al. 2019; Slee et al. 2019). Vortioxetine has a long half-life (about 60 h), and therefore withdrawal symptoms would be expected to be mild. Placebo-controlled trials have supported this suggestion, though the literature contains cases where acute vortioxetine withdrawal has led to emotional lability, nervousness and agitation (Siwek et al. 2021).

6 Other Potential Serotonin Receptor Sub-type Mechanisms

6.1 5-$HT_{2A/2C}$ Receptor Blockade

There are currently no SSRI-like drugs, which also potently block 5-$HT_{2A/2C}$ receptors. Trazodone and nefazodone antagonise 5-HT_{2A} receptors, but their ability to inhibit serotonin reuptake seems weak, as judged by a lack of relevant changes in platelet serotonin levels (Mück-Šeler et al. 1991; Sargent et al. 1997) (Table 4).

However, there are reasons to believe that combining serotonin reuptake inhibition and 5-$HT_{2A/2C}$ receptor blockade could lead to therapeutic benefits. Pre-clinical studies have shown that both 5-HT_{2A} and 5-HT_{2C} receptors are involved in mediating 'long-loop' feedback inhibition to raphe neurons. The 5-HT_{2A} receptors involved in this effect are probably located on prefrontal cortical neurons, while 5-HT_{2C} receptor feedback may involve receptors expressed in the lateral habenula. Both 5-HT_{2C} and 5-HT_{2A} receptor antagonists enhance the ability of acute SSRI administration to increase 5-HT levels in vivo in animals (see Sharp et al. 2007).

Table 4 Binding affinity (nM) of selected antidepressants and pimavanserin at serotonin and alpha2-adrenoceptors

Drug	SERT	5-HT2A	5-HT2C	Alpha2	H-1
Trazodone	252	12.9	864	320	220
Nefazodone	220	8.5	72	640	370
Mirtazapine	10,000	36	40	20	1.6
Quetiapine	10,000	446	2,024	1,159	7.1
Pimavanserin		0.4	16		

(Estimates from https://pdsp.unc.edu/databases/kidb.php; Besnard et al. 2012; Hacksell et al. 2014)

Animal studies suggest that 5-HT$_{2C}$ receptor activation can inhibit the firing of dopamine neurons in the ventral tegmental area (VTA), an effect that is produced by some regimes of SSRI treatment (Dremencov et al. 2009). It is possible that this effect might explain the limited impact of SSRIs on the key clinical symptom of anhedonia, and indeed in healthy participants, short-term SSRI treatment produced a blunting of the neural response to a chocolate reward (McCabe et al. 2010). 5-HT$_{2C}$ receptor blockade may therefore have the ability to prevent the inhibitory effect of SSRIs on VTA activity and thereby improve their effect on symptoms related to reward and motivation.

In addition, the activation of 5-HT$_{2C}$ receptors has been implicated in some of the early adverse effects of SSRIs, including the exacerbation of anxiety and perhaps sleep disturbance (Lawlor et al. 1991; Burghardt et al. 2007). In this respect, it is of interest that tertiary tricyclic antidepressants, such as amitriptyline, also have quite potent 5-HT$_{2A/2C}$ antagonist effects (Table 2). Such TCAs are not associated with the induction of anxiety or sleep disturbance; however, it is likely that the sedative properties of their H1-receptor antihistaminic action play a role in this (Table 2). Whether TCAs are more effective than SSRIs in the treatment of anhedonia has not been established (Cao et al. 2019).

Several drugs with 5-HT$_{2A/2C}$ antagonist properties are used as augmentation (add-on) treatment in patients with inadequate response to SSRIs. These include mirtazapine and quetiapine (Zhou et al. 2015; Henssler et al. 2022). These drugs have many other pharmacological actions, so the specific benefit of 5-HT$_{2A/2C}$ receptor blockade in these circumstances is unclear (Table 4). Also, probably through H1-histamine receptor blockade, both mirtazapine and quetiapine cause sedation and weight gain, which can be problematic if treatment is sustained.

More recently, augmentation studies have been conducted with a selective 5-HT$_{2A/2C}$ receptor antagonist, pimavanserin, at doses that would be expected to produce a significant blockade of both 5-HT$_{2A}$ and 5-HT$_{2C}$ receptors (Stahl 2016) (Table 4). A phase 2 placebo-controlled study where pimavanserin was added to ineffective SSRI/SNRI treatment showed a robust antidepressant effect of this treatment with an effect size of 0.6 (Fava et al. 2019). However, subsequent phase 3 studies have failed to confirm the utility of add-on pimavanserin to SSRI non-responders (Soogrim et al. 2021).

It is also possible that combining an SSRI and a $5HT_{2A/2C}$ receptor antagonist from the start of treatment would lead to superior therapeutic outcomes. However, the largest trial of this effect did not show a positive benefit from a combination of mirtazapine and venlafaxine (Rush et al. 2011). Nevertheless, it does appear that the early addition of mirtazapine to potential SSRI non-responders may be worthwhile. Kato et al. (2017) randomised depressed patients who had not achieved remission after 3 weeks' treatment with sertraline to either a continuation of sertraline with placebo or the addition of mirtazapine. At 9 weeks, patients who had received mirtazapine addition had a significantly higher remission rate (36%) than those who had continued with sertraline (25%).

6.2 5-HT_4 Receptor Agonism

No currently available antidepressant medications have a significant affinity for the 5-HT_4 receptor. However, pre-clinical studies show the rapid and potent effects of $5HT_4$ receptor agonists in animal behavioural models of depression, as well as antidepressant-like changes in relevant neurobiological markers, for example synaptic plasticity. Also in animals, 5-HT_4 receptor agonists potentiate the effects of SSRIs to increase serotonin levels (Bockaert et al. 2004; Murphy et al. 2021).

5-HT_4 receptor agonists are active in animal models of cognition, where they improve performance on tasks of learning and memory. This effect could be related to the ability of 5-HT_4 receptor agonists to increase acetylcholine release. Consistent with this, 5-HT_4 receptor agonists reverse impairments in learning and memory produced by the muscarinic antagonist scopolamine (Lo et al. 2014).

The 5-HT_4 receptor agonist prucalopride is licensed for the treatment of refractory constipation. Studies and animals show that prucalopride has good brain entry and can produce behavioural effects. For example, in combination with donepezil, prucalopride attenuated the effects of scopolamine in the Morris water maze (Cachard-Chastel et al. 2008). Experimental medicine studies in healthy participants have shown that prucalopride improves episodic memory, an effect correlated with activity in hippocampal-related neural circuitry (De Cates et al. 2021). It is therefore possible that a combination of SSRIs and 5-HT_4 receptor agonists may produce benefits in the cognitive impairment apparent in many depressed patients, which does not seem to be well addressed by SSRIs when given as a sole treatment.

6.3 5-HT_7 Receptors

While amitriptyline and amisulpride have a significant affinity for 5-HT_7 receptors, SSRIs do not (Matthys et al. 2011). 5-HT_7 receptor knockout mice have an antidepressant profile in animal behavioural models, and a selective 5-HT_7 receptor antagonist produced antidepressant-like effects in the forced swim test and promoted hippocampal cell proliferation in rats (Mnie-Filali et al. 2011).

There is some evidence of synergy between 5-HT_7 receptor antagonists and SSRIs. Citalopram produced an antidepressant effect in the forced swim test in 5-HT_7 receptor knockout mice, showing that this effect was independent of 5-HT_7 receptor activation. However, a selective 5-HT_7 receptor antagonist potentiated the effects of citalopram in this test in wild-type mice when both drugs were combined (Sarkisyan et al. 2010).

In both animals and humans, the selective 5-HT_7 receptor antagonist JNJ-18038683 potentiated the ability of citalopram to decrease rapid eye movement sleep, an effect seen with many different kinds of antidepressant medication (Sharpley and Cowen 1995). In a subsequent clinical trial of depressed patients, JNJ-18038683 did not separate from placebo. However, in the same study, escitalopram was also without a specific benefit, suggesting that the study might represent a failed trial rather than evidence against an antidepressant effect of 5-HT_7 receptor blockade (Bonaventure et al. 2012). No clinical studies have yet reported the effects in depression of combined SSRI and 5-HT_7 receptor antagonist treatment.

7 Conclusions

In 1992, Guy Edwards described the advent of SSRIs as a ‘welcome though modest advance’ (Edwards 1992). This judgement still seems correct. The availability and tolerability of SSRIs have enabled many more depressed patients to benefit from antidepressant medication. However, with this acceptability has come the less welcome concern that antidepressants are prescribed too often for people with milder, depressive states and problems in living, for which psychosocial approaches are more appropriate. In more severely unwell, depressed patients, it is possible that SSRIs have somewhat less utility than TCAs. They seem relatively ineffective for the treatment of anhedonia, a key symptom domain of severe depression and treatment-resistant depression.

It seems unlikely that the further development of specific SSRIs will be of benefit in the management of depression. However, modifications of SSRIs might still provide treatment gains. SNRIs are probably somewhat more effective than SSRIs but still lack TCA levels of noradrenergic potentiation. A more truly acting, dual-reuptake inhibitor might produce further benefits.

In terms of combining an SSRI effect with additional actions at serotonin receptor sub-types, vilazodone does not seem to have provided a significantly increased benefit over SSRIs. Vortioxetine also does not seem any more effective as an antidepressant than SSRIs; however, it does have some measurable effects in improving cognition. It has yet to be demonstrated that these cognitive improvements provide ‘real-world’ benefits to patients.

Are there other serotonin receptor sub-types that might be able to enhance the action of SSRIs? Combining an SSRI with a $5HT_{2A/2C}$ receptor antagonist continues to be an interesting strategy and worth exploring. However, the failure of pimavanserin in phase 3 studies of depression is disappointing in this respect. The ability of 5-HT_4 receptor agonists to improve cognition is intriguing and suggests a possible combination of an SSRI with a 5-HT_4 receptor agonist. Similar comments apply to the possible combination of an SSRI and a 5-HT_7 receptor antagonist, where, based on animal studies, therapeutic potentiation looks possible.

References

Amitai M, Chen A, Weizman A et al (2015) SSRI-induced activation syndrome in children and adolescents – what is next? Curr Treat Opt Psychiatry 2:28–37

Baldwin DS, Anderson IM, Nutt DJ et al (2014) Evidence-based pharmacological treatment of anxiety disorders, post-traumatic stress disorder and obsessive-compulsive disorder: a revision of the 2005 guidelines from the British Association for Psychopharmacology. J Psychopharmacol 28:403–439

Baune BT, Sluth LB, Olsen CK (2018) The effects of vortioxetine on cognitive performance in working patients with major depressive disorder: a short-term, randomized, double-blind, exploratory study. J Affect Disord 229:421–428

Berber MJ (1998) Remembering the discontinuation syndrome. Flu-like symptoms, insomnia, nausea, imbalance, sensory disturbances, and hyperarousal (anxiety/agitation). J Clin Psychiatry 59:7275

Besnard J, Ruda GF, Setola V et al (2012) Automated design of ligands to polypharmacological profiles. Nature 492:215–220

Bockaert J, Claeysen S, Compan V et al (2004) 5-HT4 receptors. Curr Drug Targ-CNS Neuro Disord 3:39–51

Bonaventure P, Dugovic C, Kramer M et al (2012) Translational evaluation of JNJ-18038683, a 5-hydroxytryptamine type 7 receptor antagonist, on rapid eye movement sleep and in major depressive disorder. J Pharmacol Exper Ther 342:429–440

Bridge JA, Iyengar S, Salary CB et al (2007) Clinical response and risk for reported suicidal ideation and suicide attempts in pediatric antidepressant treatment: a meta-analysis of randomized controlled trials. JAMA 297:1683–1696

Burghardt NS, Bush DE, McEwen BS et al (2007) Acute selective serotonin reuptake inhibitors increase conditioned fear expression: blockade with a 5-HT2C receptor antagonist. Biol Psychiatry 62:1111–1118

Cachard-Chastel M, Devers S, Sicsic S et al (2008) Prucalopride and donepezil act synergistically to reverse scopolamine-induced memory deficit in C57Bl/6j mice. Behav Brain Res 187:455–461

Cao B, Zhu J, Zuckerman H et al (2019) Pharmacological interventions targeting anhedonia in patients with major depressive disorder: a systematic review. Prog Neuropsychopharmacol Biol Psychiatry 92:109–117

Chauhan M, Parry R, Bobo WV (2022) Vilazodone for major depression in adults: pharmacological profile and an updated review for clinical practice. Neuropsychiatr Dis Treat 18:1175–1193

Cipriani A, Zhou X, Del Giovane C et al (2016) Comparative efficacy and tolerability of antidepressants for major depressive disorder in children and adolescents: a network meta-analysis. Lancet 388:881–890

Cipriani A, Furukawa T, Salanti G (2018) Comparative efficacy and acceptability of 21 antidepressant drugs for the acute treatment of adults with major depressive disorder: a systematic review and network meta-analysis. Lancet 391:1357–1366

Colwell MJ, Tagomori H, Chapman S et al (2022) Pharmacological targeting of cognitive impairment in depression: recent developments and challenges in human clinical research. Transl Psychiatry 12:484. https://www.nature.com/articles/s41398-022-02249-6

Cowen PJ (2017) Backing into the future: pharmacological approaches to the management of resistant depression. Psychol Med 47:2569–2577

Culpepper L, Lam RW, McIntyre RS (2017) Cognitive impairment in patients with depression: awareness, assessment, and management. J Clin Psychiatry 78:1383–1394

De Cates AN, Wright LC, Martens MA et al (2021) Déjà-vu? Neural and behavioural effects of the 5-HT4 receptor agonist, prucalopride, in a hippocampal-dependent memory task. Transl Psychiatry 11:1–9

Dremencov E, El Mansari M, Blier P (2009) Effects of sustained serotonin reuptake inhibition on the firing of dopamine neurons in the rat ventral tegmental area. J Psychiatry Neurosci 34:223–229

Edwards JG (1992) Selective serotonin reuptake inhibitors. BMJ 304:1644

Fava M, Dirks B, Freeman MP et al (2019) A phase 2, randomized, double-blind, placebo-controlled study of adjunctive pimavanserin in patients with major depressive disorder and an inadequate response to therapy (CLARITY). J Clin Psychiatry 80:19m12928

Furukawa TA, Cipriani A, Cowen PJ et al (2019) Optimal dose of selective serotonin reuptake inhibitors, venlafaxine, and mirtazapine in major depression: a systematic review and dose-response meta-analysis. Lancet Psychiatry 6:601–609

Furukawa TA, Salanti G, Cowen PJ et al (2020) No benefit from flexible titration above minimum licensed dose in prescribing antidepressants for major depression: systematic review. Acta Psychiatr Scand 141:401–409

Goodwin GM, Price J, De Bodinat C et al (2017) Emotional blunting with antidepressant treatments: a survey among depressed patients. J Affect Disord 221:31–35

Gunnell D, Saperia J, Ashby D (2005) Selective serotonin reuptake inhibitors (SSRIs) and suicide in adults: meta-analysis of drug company data from placebo controlled, randomised controlled trials submitted to the MHRA's safety review. BMJ 330:385–392

Hacksell U, Burstein ES, McFarland K et al (2014) On the discovery and development of pimavanserin: a novel drug candidate for Parkinson's psychosis. Neurochem Res 39:2008–2017

Haddad PM (2001) Antidepressant discontinuation syndromes: clinical relevance, prevention and management. Drug Saf 24:183–197

Harmer CJ, Cowen PJ (2013) 'It's the way that you look at it' – a cognitive neuropsychological account of SSRI action in depression. Phil Trans Roy Soc B Biol Sci 368:20120407

Harmer CJ, Duman RS, Cowen PJ (2017) How do antidepressants work? New perspectives for refining future treatment approaches. Lancet Psychiatry 4:409–418

Harrison PJ, Cowen PJ, Burns T (2018) Shorter Oxford textbook of psychiatry. Oxford University Press, Oxford

Hawton K, Bergen H, Simkin S et al (2010) Toxicity of antidepressants: rates of suicide relative to prescribing and non-fatal overdose. Br J Psychiatry 196:354–358

Healy D (1997) The antidepressant era. Harvard University Press, Harvard

Henssler J, Alexander D, Schwarzer G et al (2022) Combining antidepressants vs antidepressant monotherapy for treatment of patients with acute depression: a systematic review and meta-analysis. JAMA Psych 79:300–312

Horowitz MA, Taylor D (2019) Tapering of SSRI treatment to mitigate withdrawal symptoms. Lancet Psychiatry 6:538–546

Jacobsen P, Zhong W, Nomikos G et al (2019) Paroxetine, but not vortioxetine, impairs sexual functioning compared with placebo in healthy adults: a randomized, controlled trial. J Sex Med 16:1638–1649

Jacobsen PL, Thorley EM, Curran C (2020) Real-world patient experience with sexual dysfunction and antidepressant use in patients with self-reported depression: a cross-sectional survey study. Neurol Psychiatry Brain Res 36:57–64

Kato M, Takekita Y, Koshikawa Y et al (2017) Non response at week 4 as clinically useful indicator for antidepressant combination in major depressive disorder. A sequential RCT. J Psychiatr Res 89:97–104
Kelly CA, Freeman KB, Schumacher JA (2022) Treatment-resistant depression with anhedonia: integrating clinical and preclinical approaches to investigate distinct phenotypes. Neurosci Biobehav Rev 14:104578
Koesters M, Ostuzzi G, Guaiana G et al (2017) Vortioxetine for depression in adults. Cochrane Database Syst Rev 7:CD011520
Lawlor BA, Newhouse PA, Balkin TJ et al (1991) A preliminary study of the effects of nighttime administration of the serotonin agonist, m-CPP, on sleep architecture and behavior in healthy volunteers. Biol Psychiatry 29:281–286
Lo AC, De Maeyer JH, Vermaercke B et al (2014) SSP-002392, a new 5-HT4 receptor agonist, dose-dependently reverses scopolamine-induced learning and memory impairments in C57Bl/6 mice. Neuropharmacology 85:178–189
Matthys A, Haegeman G, Van Craenenbroeck K et al (2011) Role of the 5-HT7 receptor in the central nervous system: from current status to future perspectives. Mol Neurobiol 43:228–253
McCabe C, Mishor Z, Cowen PJ et al (2010) Diminished neural processing of aversive and rewarding stimuli during selective serotonin reuptake inhibitor treatment. Biol Psychiatry 67: 439–445
McMakin DL, Olino TM, Porta G et al (2012) Anhedonia predicts poorer recovery among youth with selective serotonin reuptake inhibitor treatment–resistant depression. J Am Acad Child Adolesc Psychiatry 51:404–411
Medical Research Council (1965) Report by Clinical Psychiatry Committee. Clinical trial of the treatment of depressive illness. Br Med J i:881–886
Meyer JH, Wilson AA, Sagrati S et al (2004) Serotonin transporter occupancy of five selective serotonin reuptake inhibitors at different doses: an [11C] DASB positron emission tomography study. Am J Psychiatry 161:826–835
Mnie-Filali O, Faure C, Lambás-Señas L et al (2011) Pharmacological blockade of 5-HT7 receptors as a putative fast acting antidepressant strategy. Neuropsychopharmacology 36:1275–1288
Moussavi S, Chatterji S, Verdes E et al (2007) Depression, chronic diseases, and decrements in health: results from the world health surveys. Lancet 370:851–858
Mück-Šeler D, Jakovljević M, Deanović Ž (1991) Effect of antidepressant treatment on platelet 5-HT content and relation to therapeutic outcome in unipolar depressive patients. J Affect Disord 23:157–164
Murphy SE, De Cates AN, Gillespie AL et al (2021) Translating the promise of 5HT4 receptor agonists for the treatment of depression. Psychol Med 51:1111–1120
National Institute for Clinical Excellence (2004) Depression: management of depression in primary and secondary care. National Institute for Clinical Excellence. http://www.nice.org.uk/
Qin B, Huang G, Yang Q et al (2019) Vortioxetine treatment for generalised anxiety disorder: a meta-analysis of anxiety, quality of life and safety outcomes. BMJ Open 9:e033161
Riga MS, Sánchez C, Celada P et al (2016) Involvement of 5-HT3 receptors in the action of vortioxetine in rat brain: focus on glutamatergic and GABAergic neurotransmission. Neuropharmacology 108:73–81
Rush AJ (2007) Limitations in efficacy of antidepressant monotherapy. J Clin Psychiatry 68(suppl 10):8–10
Rush AJ, Trivedi MH, Stewart JW et al (2011) Combining medications to enhance depression outcomes (CO-MED): acute and long-term outcomes of a single-blind randomized study. Am J Psychiatry 168:689–701
Sanchez C, Asin KE, Artigas F (2015) Vortioxetine, a novel antidepressant with multimodal activity: review of preclinical and clinical data. Pharmacol Ther 145:43–57
Sargent P, Williamson DJ, Pearson G et al (1997) Effect of paroxetine and nefazodone on 5-HT1A receptor sensitivity. Psychopharmacology (Berl) 132:296–302

Sarkisyan G, Roberts AJ, Hedlund PB (2010) The 5-HT7 receptor as a mediator and modulator of antidepressant-like behavior. Behav Brain Res 209:99–108

Sharp T, Boothman L, Raley J et al (2007) Important messages in the 'post': recent discoveries in 5-HT neurone feedback control. Trends Pharmacol Sci 28:629–636

Sharpley AL, Cowen PJ (1995) Effect of pharmacologic treatments on the sleep of depressed patients. Biol Psychiatry 37:85–98

Sharpley AL, Williamson DJ, Attenburrow ME et al (1996) The effects of paroxetine and nefazodone on sleep: a placebo controlled trial. Psychopharmacology (Berl) 126:50–54

Shelton RC, Tomarken AJ (2001) Can recovery from depression be achieved? Psychiatr Serv 52: 1469–1478

Sinclair LI, Christmas DM, Hood SD et al (2009) Antidepressant-induced jitteriness/anxiety syndrome: systematic review. Br J Psychiatry 194:483–490

Siwek M, Chrobak AA, Gorostowicz A et al (2021) Withdrawal symptoms following discontinuation of vortioxetine – retrospective chart review. Pharmaceuticals 14:451

Slee A, Nazareth I, Bondaronek P et al (2019) Pharmacological treatments for generalised anxiety disorder: a systematic review and network meta-analysis. Lancet 393:768–777

Soogrim V, Ruberto VL, Murrough J et al (2021) Spotlight on pimavanserin tartrate and its therapeutic potential in the treatment of major depressive disorder: the evidence to date. Drug Des Develop Ther 13:151–157

Stahl SM (2014) Mechanism of action of the SPARI vilazodone: serotonin 1A partial agonist and reuptake inhibitor. CNS Spectr 19:105–109

Stahl SM (2016) Mechanism of action of pimavanserin in Parkinson's disease psychosis: targeting serotonin 5HT2A and 5HT2C receptors. CNS Spectr 2:271–275

Stenkrona P, Halldin C, Lundberg J (2013) 5-HTT and 5-HT1A receptor occupancy of the novel substance vortioxetine (Lu AA21004). A PET study in control subjects. Eur Neuropsychopharmacol 23:1190–1198

Taylor DJ, Lichstein KL, Durrence HH (2003) Insomnia as a health risk factor. Behav Sleep Med 1: 227–247

Taylor MJ, Freemantle N, Geddes JR et al (2006) Early onset of selective serotonin reuptake inhibitor antidepressant action: systematic review and meta-analysis. Arch Gen Psychiatry 63: 1217–1223

Vieta E, Sluth LB, Olsen CK (2018) The effects of vortioxetine on cognitive dysfunction in patients with inadequate response to current antidepressants in major depressive disorder: a short-term, randomized, double-blind, exploratory study versus escitalopram. J Affect Disord 227:803–809

Wagner G, Schultes MT, Titscher V et al (2018) Efficacy and safety of levomilnacipran, vilazodone and vortioxetine compared with other second-generation antidepressants for major depressive disorder in adults: a systematic review and network meta-analysis. J Affect Disord 228:1–12

Whale R, Terao T, Cowen P et al (2010) Pindolol augmentation of serotonin reuptake inhibitors for the treatment of depressive disorder: a systematic review. J Psychopharmacol 24:513–520

Wijkstra J, Burger H, Den Broek V et al (2010) Treatment of unipolar psychotic depression: a randomized, double-blind study comparing imipramine, venlafaxine, and venlafaxine plus quetiapine. Acta Psychiatr Scand 121:190–200

Wilson SJ, Bell C, Coupland NJ et al (2000) Sleep changes during long-term treatment of depression with fluvoxamine – a home-based study. Psychopharmacology (Berl) 149:360–365

Zhou X, Ravindran AV, Qin B et al (2015) Comparative efficacy, acceptability, and tolerability of augmentation agents in treatment-resistant depression: systematic review and network meta-analysis. J Clin Psychiatry 76:e487–e498

Mechanisms of SSRI Therapy and Discontinuation

Trevor Sharp and Helen Collins

Contents

Abstract SSRIs are one of the most widely used drug therapies in primary care and psychiatry, and central to the management of the most common mental health problems in today's society. Despite this, SSRIs suffer from a slow onset of therapeutic effect and relatively poor efficacy as well as adverse effects, with recent concerns being focused on a disabling SSRI discontinuation syndrome. The mechanism underpinning their therapeutic effect has long shifted away from thinking that SSRIs act simply by increasing 5-HT in the synapse. Rather, a current popular view is that increased 5-HT is just the beginning of a series of complex downstream

T. Sharp (✉) and H. Collins
Department of Pharmacology, University of Oxford, Oxford, UK
e-mail: trevor.sharp@pharm.ox.ac.uk

Curr Topics Behav Neurosci (2024) 66: 21–48
https://doi.org/10.1007/7854_2023_452
Published Online: 14 November 2023

signalling events, which trigger changes in neural plasticity at the functional and structural level. These changes in plasticity are then thought to interact with neuropsychological processes to enhance re-learning of emotional experiences that ultimately brings about changes in mood. This compelling view of SSRI action is underpinning attempts to understand fast-acting antidepressants, such as ketamine and psychedelic drugs, and aid the development of future therapies. An important gap in the theory is evidence that changes in plasticity are causally linked to relevant behavioural effects. Also, predictions that the SSRI-induced neural plasticity might have applicability in other areas of medicine have not yet been borne out. In contrast to the sophisticated view of the antidepressant action of SSRIs, the mechanism underpinning SSRI discontinuation is little explored. Nevertheless, evidence of rebound increases in 5-HT neuron excitability immediately on cessation of SSRI treatment provide a starting point for future investigation. Indeed, this evidence allows formulation of a mechanistic explanation of SSRI discontinuation which draws on parallels with the withdrawal states of other psychotropic drugs.

Keywords Antidepressants · Anxiety · Citalopram · Depression · Discontinuation · Fluoxetine · Ketamine · Neural plasticity · Paroxetine · Psychedelics · Serotonin · Serotonin (5-HT) receptors · SSRIs · Withdrawal

1 SSRI Benefits and Adverse Effects

Selective serotonin (5-hydroxytryptamine; 5-HT) reuptake inhibitors (SSRIs) are prescribed to many millions of people worldwide for major depression and anxiety, as well as other neuropsychiatric conditions including obsessive compulsive disorder, eating disorders and post-traumatic stress disorder. In England alone, it is estimated that each year about seven million adults (16% of the adult population) are prescribed an antidepressant, most frequently of the SSRI class (www.nhsbsa.nhs.uk). SSRIs are therefore one of the most widely used drug therapies in primary care and psychiatry, and central to the management of the most common mental health problems in modern-day society. The mental health benefits and general tolerability of SSRI therapy are evident in large scale meta-analyses (Cipriani et al. 2018).

Nonetheless, as noted in the previous chapter by Philip Cowen, not all patients experience an adequate level of therapeutic effect, and acute adverse effects such as anxiety, nausea, sexual dysfunction and GI disturbance are not uncommon and sometimes persist, resulting in non-compliance (Cartwright et al. 2016). Moreover, when SSRI therapy is abruptly stopped many patients immediately experience a constellation of disturbing psychological and somatic symptoms of unknown origin (Horowitz and Taylor 2019). In response to a recent UK Parliamentary review (APPG 2018) new evidence has emerged that the incidence, duration and severity of this discontinuation syndrome may be significantly greater than previously recognised, and potentially impacts on millions of patients in the UK alone (Davies and Read 2019). In this chapter, current knowledge of the molecular and cellular mechanisms underpinning these wanted and unwanted effects of a course of SSRI treatment are reviewed.

2 Mechanisms of SSRI Action

2.1 Importance of the 5-HT Transporter

As a drug class, SSRIs selectively and potently inhibit the 5-HT transporter (Owens et al. 1997), which plays a critical role in controlling the dynamics of transmission at 5-HT synapses. Although specific SSRIs have affinity for other neurotransmitter transporters and receptors, this is typically 10–100 times lower affinity than for the 5-HT transporter (Owens et al. 1997).

There is direct evidence that the 5-HT transporter is pivotal to SSRI action. In many preclinical experiments, genetic knockout of the 5-HT transporter phenocopies the effects of pharmacological inhibition using an SSRI. For example, similar to the effects of SSRI administration in wildtype mice, 5-HT transporter knockout mice demonstrate not only increased brain extracellular 5-HT and 5-HT release dynamics as well as adaptive changes in the expression/function of specific postsynaptic 5-HT receptors, but also similar behavioural changes such as increased fear learning (Jennings et al. 2010, 2012; Lima et al. 2019). In another important example, many preclinical measures of SSRI action are eliminated in mice with a genetic knock-in of a 5-HT transporter variant that is insensitive to SSRI binding (but otherwise functionally normal; Nackenoff et al. 2016). There are cases when genetic knockout of the 5-HT transporter does not phenocopy pharmacological inhibition of the 5-HT transporter, but these instances are likely explained in large part by life-long loss of the 5-HT transporter in the genetic model (Ansorge et al. 2004).

In humans, PET studies demonstrate high occupancy of the transporter at clinical SSRI doses, and there is evidence that forebrain occupancy and clinical response may be correlated (Spies et al. 2015). Moreover, earlier findings that acute depletion of the 5-HT precursor tryptophan causes relapse in patients treated with an SSRI (Delgado et al. 1990; Ruhe et al. 2007) is direct evidence that these drugs rely on increased 5-HT to produce their therapeutic effect.

Other candidate mediators of the therapeutic effects of SSRIs have been proposed including the interesting recent idea that they enhance the function of brain-derived neurotropic factor (BDNF) by directly targeting its receptor, TrkB (Casarotto et al. 2021). However, thus far the latter action has only been demonstrated for fluoxetine and evidence is awaited that it generalises to other members of the structurally diverse SSRI class of drugs (or indeed the many other types of antidepressant drugs, more generally). Moreover, fluoxetine's affinity for TrkB is low compared to the 5-HT transporter (mM vs. nM), which raises the question of whether the former target is sufficiently engaged at clinical doses to generate meaningful pharmacological, and hence therapeutic, effects.

2.2 Neuroadaptive Changes in 5-HT Neuron Function

The concept that blockade of the 5-HT transporter mediates the therapeutic effect of SSRIs seems at odds with the need to administer the drugs for several weeks for patients to experience the greatest benefit (Gelenberg and Chesen 2000). It is clear from PET studies that SSRIs achieve high occupancy of the 5-HT transporter within hours of administration of a single clinical dose (Meyer et al. 2004; Parsey et al. 2006). Thus, there is little doubt that SSRIs rapidly enter the brain and block the 5-HT transporter on acute administration. Therefore, since a course of repeated SSRI treatment is required to obtain the full antidepressant response, it seems likely that neuroadaptive changes play a critical role in the therapeutic effect.

Various neuroadaptive changes have been proposed to account for the paradoxical lag in onset of therapeutic effect of SSRIs and other antidepressants [e.g. (Sharp 2013)]. One popular idea is that this time lag relates to a gradual desensitisation of 5-HT$_{1A}$ autoreceptors located on the soma and dendrites of midbrain 5-HT neurons (Artigas et al. 1996; Blier and de Montigny 1994). This theory posits that blockade of 5-HT reuptake initially elevates extracellular 5-HT which then activates somatodendritic 5-HT$_{1A}$ autoreceptors to cause inhibition of 5-HT cell firing and release. This feedback response then suppresses transmission at the 5-HT synapse (Gartside et al. 1995). Evidence suggests that with continued SSRI treatment these autoreceptors desensitise leading to loss of inhibitory feedback and ultimately a greater increase in 5-HT transmission (Artigas et al. 1996; Blier and de Montigny 1994). The mechanism underlying 5-HT$_{1A}$ autoreceptor desensitisation remains surprisingly uncertain with decreased expression of the receptors being difficult to consistently demonstrate, although increased internalisation of the receptors is an interesting possibility (Riad et al. 2004).

Continued SSRI treatment likely impacts not only on 5-HT released at the 5-HT synapse but also co-released transmitters. A recent breakthrough in the 5-HT field is the discovery that 5-HT neurons release not only 5-HT but also glutamate. Across species the majority of midbrain 5-HT neurons express the vesicular glutamate transporter 3 (VGLUT3; Gras et al. 2002; Vigneault et al. 2015) which stores glutamate in presynaptic vesicles and likely co-exists with the vesicular monoamine transporter (VMAT2). Studies of the neurophysiological effects of optogenetically activated 5-HT neurons estimate that co-released glutamate mediates around half of all responses at 5-HT synapses (Sengupta et al. 2017). However, during repeated SSRI administration inhibition of the 5-HT transporter will likely shift the balance of 5-HT-glutamate co-release away from glutamate and in favour of 5-HT (Fischer et al. 2014). Thus, under conditions of chronic SSRI exposure 5-HT may preferentially mediate downstream signalling events versus co-released glutamate (see later).

2.3 *Role of Adaptive Changes in Neural Plasticity*

The 5-HT_{1A} autoreceptor desensitisation theory continues to be relevant although current thinking is that adjustments in the control of 5-HT neurons are just part of a complex sequence of neuroadaptive changes leading up to the antidepressant effect. An influential current theory is that the therapeutic effect is underpinned by an activated programme of gene expression that leads to a diversity of plasticity changes at the cellular level. These changes include strengthened synaptic efficacy and connectivity, increased neurogenesis, and even the switching of neural networks into a more immature and plastic developmental state (Sharp 2013; Castren 2005; Umemori et al. 2018; Harmer et al. 2017; Tartt et al. 2022). It is thought that these changes occur in key neural circuits within the limbic system such that these circuits become more resilient to adverse environmental changes (e.g. stress) and are more easily remodelled by incoming emotionally relevant stimuli.

The origin of this plasticity hypothesis is earlier preclinical studies demonstrating that repeated treatment with antidepressants including SSRIs, increased expression of BDNF and TrkB in hippocampus (Lindefors et al. 1995; Nibuya et al. 1995). This finding subsequently extended to other neurotrophic factors (Warner-Schmidt and Duman 2008; Elsayed et al. 2012) as well as a host of other genes important for processes linked to neural plasticity. This includes genes encoding transcription factors, vesicular glutamate transporters, postsynaptic density proteins, receptor chaperones as well as other transmitter signalling molecules and enzymes (for review see Sharp 2013; Harmer et al. 2017).

Although a clear molecular ‘fingerprint’ of antidepressant action has yet to emerge, the diversity of genes that change suggests the triggering of a wide range of adaptive cellular processes. Key cellular changes likely include the formation and strengthening of synapses as well as increased neuronal survival and even neurovascular support. Commonly used cellular markers of an antidepressant effect include enhanced formation of new neurons in hippocampus (Tartt et al. 2022) as well as increased dendritic arborisation and spine density of pyramidal neurons in the prefrontal cortex (Licznerski and Duman 2013).

The mechanisms underlying the alteration in gene expression induced by SSRIs and other antidepressants is not certain but candidates include activation of specific 5-HT receptor subtypes and their downstream signalling pathways, and potentially epigenetic changes (D'Sa and Duman 2002; Vialou et al. 2013; Carr and Lucki 2011). The many genes that are reportedly influenced by a course of antidepressant treatment makes it challenging to evaluate the relative importance of individual genes and their products and specific downstream effects. For example, which of the many genes are key to the antidepressant effects and thereby might make reasonable targets for drug development? In this regard, the importance of the gene transcription factor CREB has been highlighted (Blendy 2006; D'Sa and Duman 2002) and there has been much interest in BDNF itself. Thus, the use of gene knock out mice suggests that, at least in preclinical models, antidepressant effects are dependent upon intact BDNF-TrkB signalling (Casarotto et al. 2022).

As noted above, an alternative and quite different account of how SSRIs and other antidepressants induce plasticity changes is not through 5-HT but via a direct action on TrkB itself (Casarotto et al. 2021). If correct, the latter idea suggests that TrkB might make a druggable target for antidepressant development. A number of brain-penetrant TrkB agonists have been identified and shown to produce antidepressant effects in animal models although TrkB blockade seems to have a similar effect (Casarotto et al. 2022). The tolerability and safety of such agents in humans is unknown, and no selective TrkB drugs have as yet progressed into clinical development.

2.4 *Rapid Acting Antidepressants; Ketamine and Psychedelic Drugs*

The majority of molecular and cellular plasticity changes evoked by SSRIs and other antidepressants do not occur acutely but rather require a course of several weeks of treatment. This delay in onset correlates with the timing of the therapeutic effect and raises the possibility that drugs which accelerate these plasticity changes may have a faster onset of action. Interestingly, both the dissociative anaesthetic ketamine and psychedelic drugs such as psilocybin are emerging as rapid onset antidepressants in the clinical setting (see chapters by Joseph Piell and Emma Robinson). Moreover, these drugs are reported to induce changes in markers of plasticity such as spine density in the prefrontal cortex with hours of a single administration (Autry et al. 2011; Kwan et al. 2022; McClure-Begley and Roth 2022; Cameron et al. 2021; Shao et al. 2021). A rapid increase in BDNF expression has also been linked to the antidepressant effects of both ketamine (Kim et al. 2023) and psilocin, the active form of psilocybin (Moliner et al. 2023) (see also below).

Given that both ketamine and psilocin are well characterised as a non-competitive NMDA receptor antagonist and 5-HT_{2A} receptor agonist respectively, it is perhaps surprising that the pharmacological mechanism underlying their antidepressant actions is currently uncertain. In the case of ketamine, recent data suggests that (R)-ketamine and the ketamine metabolite (2R,6R)-HNK may exert antidepressant effects by increasing transmission via AMPA receptors, independent of NMDA receptor blockade (reviewed by Johnston et al. 2023). An opioid receptor agonist action of ketamine has also been linked to the drug's antidepressant action (Williams et al. 2018; Klein et al. 2020).

In the case of psychedelic drugs there is convincing evidence that a 5-HT_{2A} receptor agonist action underpins their psychotropic effects. For instance, in humans the latter effects are prevented by 5-HT_{2A} receptor blockade (Vollenweider et al. 1998), and PET studies demonstrate a correlation between intensity of psychedelic effects and cerebral 5-HT_{2A} receptor occupancy (Madsen et al. 2019). In preclinical studies 5-HT_{2A} receptor agonist-induced head-twitches are thought to be a surrogate marker of the hallucinogenic effects in humans; for example, the potency of 5-HT_{2A}

receptor agonists to induce head-twitches correlates with their hallucinogenic potency, and the head-twitch response is abolished by 5-HT_{2A} receptor knockout or selective antagonists (Halberstadt et al. 2020; Gonzalez-Maeso et al. 2007; Jennings et al. 2008).

Activation of 5-HT_{2A} receptors has also been linked to an antidepressant action in preclinical studies although the link is somewhat indirect, and there are conflicting data. Thus, psychedelic drug-induced alterations in a plasticity marker, specifically changes in dendrite morphology, were found sensitive to ketanserin or the more selective 5-HT_{2A} receptor antagonist MDL 100,907 in some studies (Lewis et al. 2023; Ly et al. 2018) but curiously not others (Moliner et al. 2023; Hesselgrave et al. 2021; Shao et al. 2021). A further complication is evidence that the effect of psychedelic drugs on dendrite morphology is mediated by 5-HT_{2A} receptors but only those receptors with an intracellular location and not membrane bound (Vargas et al. 2023). Finally, a recent report attributed the effect of certain psychedelics on dendrite plasticity to positive allosteric modulation of TrkB and not 5-HT_{2A} receptor agonism (Moliner et al. 2023). Here, the affinity of lysergic acid diethylamide (LSD) and psilocin for TrkB was high (nM) although similar to the affinity of these drugs for the 5-HT_{2A} receptor. Moreover, structure-activity data are needed before a TrkB action can be attributed to all psychedelic drugs and not just LSD and psilocin.

There is, as yet, no clear explanation for these seemingly incompatible findings regarding the role (or not) of 5-HT_{2A} receptors in the antidepressant effects of psychedelic drugs. It is possible that the controversy centres around the use of changes in dendrite morphology of cortical neurons (in culture or in vivo) and might not generalise to other markers of plasticity. For instance, it seems clear that 5-HT_{2A} receptors mediate psychedelic drug-induced increases in expression of plasticity genes such as Arc and Egr-1 in mouse cortex since this effect is abolished by 5-HT_{2A} receptor knock out or blockade (Gonzalez-Maeso et al. 2007; Jennings et al. 2008; Pei et al. 2004). It is also possible that in specific experimental paradigms the degree of 5-HT_{2A} receptor blockade is not always optimal, or that some of the actions of psychedelic drugs are being explored in models with low 5-HT_{2A} receptor expression, thus permitting the detection of effects directly (rather than indirectly) mediated by other mechanisms such as BDNF signalling.

Notwithstanding the above controversies, currently there is a strong interest in developing non-hallucinogenic 5-HT_{2A} receptor agonists as rapid onset antidepressants. Several recent preclinical studies have reported compounds which elicit plasticity changes and antidepressant-like behavioural effects but without inducing head-twitches (Lewis et al. 2023; Cameron et al. 2021; Cunningham et al. 2023). The mechanism underlying the preferential antidepressant actions of such drugs is again not certain, but biased agonism and/or partial agonism at the 5-HT_{2A} receptor, as well as contributions from other 5-HT receptor subtypes, are amongst the possibilities.

2.5 Depression Pathophysiology

The likely contribution of 5-HT and neural plasticity to depression pathophysiology has been extensively reviewed elsewhere, but a few points are worthy of brief mention here. Many strands of evidence for a general deficiency in 5-HT in depression was recently reviewed in an 'umbrella-style' fashion, and the data in favour of the 5-HT hypothesis were held to be inconsistent, failing in meta-analyses, and thereby unreliable (Moncrieff et al. 2022). The latter review was itself deemed inadequate for several reasons including weaknesses in the methods of analysis and exclusion of key data pointing to a 5-HT deficit in depression (Jauhar et al. 2023). Also, some studies emphasised in the review concerned peripheral and/or indirect measures of 5-HT function, such as levels of 5-HT and its metabolites in body fluids or 5-HT gene association studies, whereas studies using more direct measures of central 5-HT function (PET imaging of 5-HT receptors) were down-played.

It is noteworthy that a recent study using a direct measurement of 5-HT release based on a novel PET radioligand displacement approach, which has a dopamine equivalent that has been highly influential in the schizophrenia field, reported a deficiency in 5-HT release in frontal cortex of depressed patients (Erritzoe et al. 2023). It will be interesting to see whether the latter observation holds up when applied to larger and more diverse patient populations. As argued by others (Jauhar et al. 2023), currently it cannot be excluded that 5-HT deficits play a role in depression particularly in vulnerable individuals as well as those suffering from the illness. Of course, the therapeutic effect of SSRIs may be mediated via 5-HT regardless of the role played by 5-HT in depression pathophysiology.

Deficits in markers of neural plasticity have also been observed is association with depression pathophysiology. Thus, many of the same genes that increase in expression in response to antidepressant treatment are reported to be decreased in animal models of depression (e.g. chronic exposure to stress), and similar observations have been made in post-mortem tissue from subjects diagnosed with depression. For example, reduced BDNF-TrkB signalling was observed in the CNS of rodents exposed to chronic stress as well as in the CNS and peripheral tissues of depressed patients (see Castren and Monteggia 2021). Evidence for decreased numbers of synapses in severe depression was also observed both at the high-resolution electron microscopy level (Kang et al. 2012) and using a novel PET marker of synaptic density (Holmes et al. 2019). The latter findings accord with structural imaging studies that consistently report reduced hippocampal volume in depressed subjects, and smaller hippocampal volume is associated with worse depression scores (Schmaal et al. 2016).

Perhaps a shortcoming in the current literature is that evidence of a causal link between these changes in plasticity markers, depression readouts and antidepressant action is somewhat lacking. For instance, genetic knock-down of molecular mediators of plasticity such as BDNF is not typically associated with depressive-like or stress-vulnerable behavioural phenotypes in mouse models (MacQueen et al. 2001; Groves 2007). Similarly, lack of neurogenesis does not consistently elicit depressive-like effects in animal models (Tartt et al. 2022; Malberg et al. 2021). For example, whereas an important early study demonstrated a causal link between increased neurogenesis and an antidepressant effect (Santarelli et al. 2003), this could not be confirmed in different mouse strains (Huang et al. 2008; Holick et al. 2008). Moreover, subsequent work found that many antidepressants, including ketamine, have neurogenesis-independent effects (Malberg et al. 2021). It will be important to show that other key cellular markers of plasticity such as increases neurite outgrowth and dendritic spine formation directly link to depressive-like behavioural changes and antidepressant-like behavioural effects. This said, there remains continuing doubt regarding the validity of many commonly used behavioural readouts of antidepressant effects in animal models (see chapter by Emma Robinson).

2.6 A Neuropsychological Perspective of SSRI Action

The hypothesis that SSRIs induce changes in neural plasticity can be considered in the context of neuropsychological theories of antidepressant action. One popular theory emphasises that altered emotional processing is critical to the actions of drug treatments rather than a direct elevation in mood per se. Specifically, it is thought that antidepressants work by inducing a shift away from a cognitive state in which attention, memory and appraisal of internal and external emotional stimuli are negatively biased (Harmer et al. 2017). An important part of this theory is evidence that 5-HT facilitates this switch in emotional processing. For example, tryptophan depletion caused negative biases in the recognition of facial expressions in recovered depressed patients (Hayward et al. 2005) whereas SSRI treatment had the opposite effect (Harmer et al. 2009). Given that SSRI administration alters emotional bias even after a single dose, it is argued that this provides the basis for the relearning of new positive emotional associations eventually resulting in an improvement in subjective mood (Harmer et al. 2009).

The neural plasticity and neuropsychological hypotheses of antidepressant action are not mutually exclusive. Indeed, increased neural plasticity within key neural circuits of the limbic system might provide the neural substrate through which increases in positive emotional processing permit the relearning of positive emotional associations and eventually improve mood. In this case it is possible that a window of plasticity evoked by a course of SSRI treatment might synergise with other psychological or physiological inputs (Fig. 1). Interestingly, an earlier study found that repeated SSRI treatment, when combined with a visual stimulus, restored developmentally-induced visual deficits in rats (Maya Vetencourt et al. 2008). The latter finding raises the possibility of using SSRIs in the management of conditions associated with abnormal visual development early in life (amblyopia) although initial small-scale clinical trials have been disappointing (Lagas et al. 2019). It is currently unclear if the plasticity-inducing effects of SSRIs might generalise to other medical conditions that could benefit from a restoration of neural plasticity. Thus, the

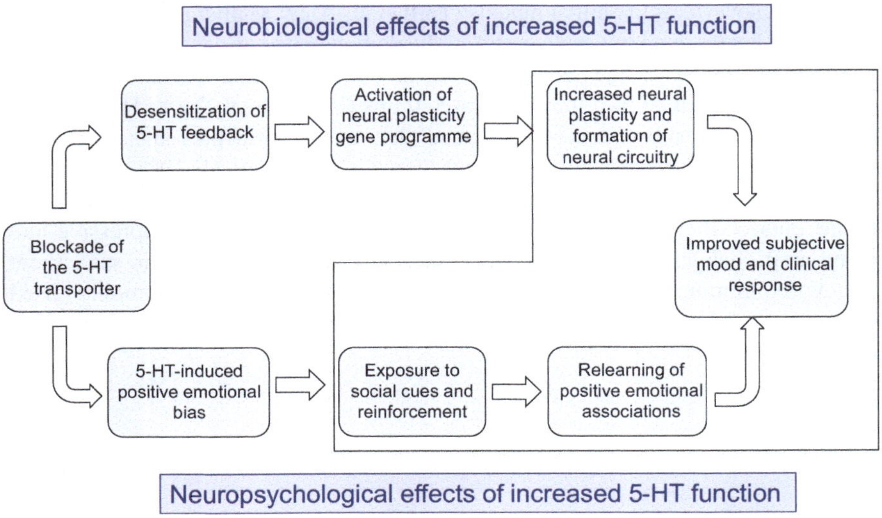

Fig. 1 Theoretical convergence of adaptive molecular, cellular and neuropsychological changes that occur during a course of SSRI treatment. In response to 5-HT autoreceptor desensitization and increased 5-HT availability, a gene expression programme(s) is activated resulting in increased neural plasticity within emotion processing neural networks (shadowed box). At the neuropsychological level, following a switch in emotional bias this increase in plasticity is thought to allow positive emotional stimuli (e.g. social cues) to be relearned and bring about improved mood. Emerging, fast acting antidepressants such as ketamine and psilocybin may trigger changes in gene expression by circumventing 5-HT autoreceptor desensitization. The latter drugs may also act via a different neuropsychological mechanism to SSRIs (see chapter by Emma Robinson). An alternative but not yet widely held view is that SSRIs and fast-acting antidepressants induce changes in neural plasticity directly by enhancing BDNF signalling (see text for details). Figure redrafted from Sharp and Cowen 2011

results of combining SSRIs with motor rehabilitation in stroke patients have had limited success thus far (Legg et al. 2021; Mead et al. 2012). An interesting prediction of the neural plasticity hypothesis of antidepressant action, as noted by others (Branchi 2022), is that it may be a 'double edged sword' in which the induction of plasticity has the capacity to produce beneficial or detrimental effects according to context. For instance, this concept has been used to explain reports that unfavourable living conditions and low socioeconomic status associate with poor response to SSRI treatment (Branchi 2022).

3 SSRI Discontinuation

3.1 SSRI Discontinuation Syndrome

Despite the therapeutic value of SSRIs noted above, when treatment is suddenly stopped patients often experience disabling psychological and somatic effects. These adverse effects are common to all SSRIs and most other antidepressant drugs, and are characterised by unpleasant central and peripheral symptoms including heightened anxiety and agitation, sleep disruption, irritability, anorexia, flu-like symptoms, and sensory disturbances (Table 1) (Haddad 1997, 2007; Horowitz and Taylor 2019). Although there is some overlap between these symptoms and those of clinical depression and anxiety, many symptoms are specific to discontinuation and the syndrome is accepted to be distinct from relapse. People suffer the adverse effects of SSRI discontinuation at all stages in life, from the neonates of SSRI-treated mothers (Levinson-Castiel et al. 2006) through to adolescent and adult psychiatric patients across all age groups, and the syndrome is independent of clinical diagnosis (Horowitz and Taylor 2019; Jha et al. 2018).

SSRI discontinuation symptoms commonly emerge within 2–3 days of treatment cessation but may appear as early as 24 h after the last dose, and usually resolve within 2–3 weeks but can last longer (Haddad 1997; Fava et al. 2015).

Antidepressant discontinuation rose to prominence in the UK through a Parliamentary report and linked systematic review (APPG 2018; Davies and Read 2019). This analysis concluded that, contrary to prior expectations, a majority of patients

Table 1 Clinical symptoms of SSRI discontinuation broken down as CNS and peripheral features. Information taken from Delgado (2006), Haddad and Anderson (2007) and Horowitz and Taylor (2019). Symptoms with asterisk (*) are reported effects of tryptophan depletion in patients treated with an SSRI, which in isolation tend to be considered as depression relapse and not discontinuation (Delgado 2006)

CNS features	Peripheral features
Affective Irritability Anxiety* or agitation Low mood* Tearfulness Dread Suicidal ideation	**General somatic** Flu-like symptoms Lethargy or fatigue Headache Tremor Sweating Anorexia*
Sleep disruption Insomnia* Nightmares Excessive dreaming	**Gastrointestinal** Nausea* Vomiting* Diarrhoea
Sensory symptoms Paraesthesia Numbness Electric shock-like sensations Rushing noises Palinopsia (visual trails)	**Disequilibrium** Dizziness (most common) Light-headedness Vertigo
Cognitive symptoms Confusion Decreased concentration* Amnesia	**Sexual** Genital hypersensitivity Premature ejaculation

undergoing discontinuation experience adverse effects with nearly half reporting effects as severe. Many patients also reported that withdrawal symptomatology lasted many months or longer. However, data from randomised placebo-controlled trials suggest that these reviews may have overestimated the incidence of discontinuation with an incidence of around 20% being thought more likely (Jauhar et al. 2019; Baldwin et al. 2006). Nevertheless, even this lower figure represents a high degree of morbidity given the many millions of individuals likely to be undergoing treatment cessation at any point in time.

3.2 'Discontinuation' vs. 'Withdrawal'

Here, the term 'discontinuation' is used when referring to the syndrome associated with stopping antidepressant treatment, but in the literature the terms 'discontinuation' and 'withdrawal' are used interchangeably. There is an ongoing discussion regarding the most appropriate terminology (e.g. Jauhar et al. 2019; Cosci et al. 2023). From a research perspective the use of 'discontinuation' could be taken to imply that the mechanism of antidepressant withdrawal may in some way be different from that of other psychotropic drugs, thereby potentially missing important comparisons. It has also been argued that 'discontinuation' better indicates the potential for symptoms to be enduring and extended beyond the time of immediate drug 'withdrawal' (Cosci et al. 2023). On the other hand, from the perspective of the patients and those involved in their care and management, 'withdrawal' is often considered a stigmatised term due to its association with other psychotropic drugs such as alcohol, psychostimulants, benzodiazepines and opioid analgesics, which are in widespread recreational use and have a high level of addiction liability. Thus, the term withdrawal can be taken to imply that a drug is addictive and associated with compulsive use, tolerance and cravings, which is not the case for SSRIs and most other conventional antidepressants (Jauhar et al. 2019; Haddad 2005; Lerner and Klein 2019).

3.3 Discontinuation Risk Factors

The severity of the antidepressant discontinuation syndrome varies widely between individuals, with some patients experiencing few or mild symptoms whereas hospitalisation may even be required in extreme cases (Bhanji et al. 2006). There are likely to be many sources of this variation.

In the case of SSRIs some drugs are more prone to generating a discontinuation syndrome, and with greater severity, than others. Specifically, evidence suggests that paroxetine is particularly problematic whereas fluoxetine is much less so (Gastaldon et al. 2022; Fava et al. 2015). This is likely due in part to paroxetine having a relatively short half-life of 21 h in humans (Bourin et al. 2001) compared to 4 days for fluoxetine and even longer for its active metabolite norfluoxetine (Altamura et al. 1994). Moreover, paroxetine has non-linear metabolism kinetics and inhibits its own metabolising enzyme (CYP2D6) such that when plasma paroxetine levels fall its metabolism increases and its washout accelerated (Bourin et al. 2001). In comparison, the earlier generation of antidepressant drugs, tricyclics and monoamine oxidase inhibitors, are also associated with discontinuation syndromes as are the more modern serotonin-noradrenaline reuptake inhibitors venlafaxine and duloxetine. In fact, evidence suggests that venlafaxine has a high rate of discontinuation comparable with that of paroxetine (Gastaldon et al. 2022), a major contributing factor being venlafaxine's short half-life in humans (3–4 h; Schweizer et al. 1997).

Other important discontinuation risk factors include abrupt treatment cessation, which is more likely to cause discontinuation symptoms than gradual tapering (Horowitz and Taylor 2019). SSRI treatment duration is also a likely risk factor in discontinuation. For example, in clinical trial 6–8 weeks of treatment with paroxetine was sufficient to produce symptoms (Baldwin et al. 2006) while only 1 week of treatment was not (Yonkers et al. 2015). There is evidence that younger patients are more likely to experience discontinuation symptoms than older patients (Fava et al. 2015; Gastaldon et al. 2022), and prior experience of discontinuation symptoms may also be a predisposing factor (Jha et al. 2018; Horowitz et al. 2023). Interestingly, a report from the World Health Organisation found that compared to women, men were more likely to experience discontinuation symptoms and symptoms were more likely to be severe (Gastaldon et al. 2022).

Genetic risk factors have not been systematically investigated although a small-scale study found a weak association between the incidence of paroxetine discontinuation and a polymorphism in the 5-HT_{1A} receptor gene (Murata et al. 2010). Other polymorphisms of interest include those of the cytochrome P450 enzymes as well as variants of the 5-HT transporter gene itself which likely contribute to natural variations in expression of about 10-fold within the human population (see Jennings et al. 2012). Ongoing clinical trials are aimed to understand factors that determine a patient's risk of developing discontinuation symptoms (Vinkers et al. 2021; Duffy et al. 2021).

Overall, much evidence suggests that SSRI discontinuation is common and likely carries a higher level of disability than previously expected. Moreover, it is apparent that current methods for managing SSRI discontinuation are sub-optimal and without a clear understanding of the risk factors and mechanisms involved. Indeed, present guidelines for managing SSRI discontinuation are based on dose tapering and not a knowledge of the neural mechanisms underlying discontinuation, which typically informs the clinical management of other syndromes associated with cessation of drug-taking (e.g. opiates, alcohol).

3.4 Modelling SSRI Discontinuation in Animals

Despite evidence that SSRI discontinuation can cause debilitating symptoms in patients, there have been surprisingly few attempts to model the effects of SSRI treatment cessation in animals or establish its underlying mechanisms. Although the symptoms in patients are highly diverse and often subjective in nature, some specific symptoms are accessible for measurement and quantification in animals. Moreover, since discontinuation appears to be independent of a specific illness diagnosis, the complication of combining drug treatment in a model of a particular disease pathophysiology can arguably be circumvented.

In one of the few preclinical studies attempting to model SSRI discontinuation, the abrupt cessation of citalopram administration to rats was reported to increase acoustic startle responsivity compared to animals maintained on treatment (Bosker et al. 2010). This anxiogenic effect was evident within 2 days of discontinuation and lasted for at least 7 days. In support of this earlier finding, it was recently found that mice discontinued from repeated treatment with paroxetine exhibited evidence of increased anxiety-like behaviour (Collins et al. 2022). Specifically, mice showed evidence of increased avoidance behaviour on the elevated plus maze 2 days after discontinuation from 2 weeks treatment from paroxetine or citalopram; this effect was not apparent after treatment with paroxetine for 1 week, suggesting a requirement for prolonged treatment. Moreover, the increase in anxiety was evident at 2 days but not 5 days after discontinuation from paroxetine (Collins et al. 2022). This seemingly short-lasting effect in mice is consistent with findings of studies reporting no change in anxiety- or depressive-like behaviour in rats 1–3 weeks after the last dose of repeated SSRI administration (e.g. Elizalde et al. 2008; Bouet et al. 2012; Strekalova et al. 2013).

Interestingly, in the case of paroxetine the discontinuation-evoked anxiety-like response in mice was observed in male and not female mice (Collins et al. 2022), potentially mapping onto evidence that SSRI discontinuation may be more common in male patients (see above). A sexually dimorphic effect of SSRI discontinuation could relate to sex differences in the 5-HT system which are commonly reported in animals (e.g. Renoir et al. 2011) as well as humans (e.g. Jovanovic et al. 2008). A sex difference in drug metabolism is an alternative explanation, and findings that female mice metabolise an SSRI faster than males due to greater cytochrome P450 activity may be relevant (Hodes et al. 2010).

Taken together, these preclinical findings suggest that SSRI discontinuation in rodents produces a short-lasting increase in anxiety-like behaviour, potentially mirroring the anxiety often experienced by patients discontinuing SSRI treatment. In the case of paroxetine and citalopram, the rapid emergence of this behavioural effect of discontinuation is consistent with the short half-life of these drugs in rodents ($t_{1/2} = 6$ h and 1.5–3 h, respectively: Kreilgaard et al. 2008; Fredricson Overo 1982). The behavioural data are also in line with clinical observations that patients can experience discontinuation symptoms within just 2 days of placebo substitution for paroxetine (Fava et al. 2015) which, as noted above, has a short half-life in humans.

Recent preclinical studies have investigated whether there are longer lasting behavioural effects of cessation of a course of SSRI treatment by continuously monitoring mouse home cage activity over 2 weeks after paroxetine discontinuation using non-invasive passive infrared sensors (Collins et al. 2023b). This approach is emerging as a highly sensitive tool to assess the effects of experimental interventions on mouse activity, and especially for monitoring phenotypic changes in circadian behaviour and the sleep-wake cycle. Notably, these experiments detected evidence that sleep structure subtly changed following paroxetine discontinuation, in that the duration of sleep bouts in the dark phase increased in length and this effect was evident for up to 9 days after paroxetine discontinuation (Collins et al. 2023b).

The latter findings link to various observational studies as well as clinical trials of SSRI discontinuation in patients which collectively report sleep disturbances, including insomnia, interrupted sleep and vivid dreams (Haddad 1997; Jha et al. 2018; Fava et al. 2015). In addition, electroencephalogram studies have shown that SSRI discontinuation in patients can produce a rebound increase in REM sleep (Staner et al. 1995) that might contribute a mechanistic explanation for the sleep disturbances. Alternative explanations include disrupted circadian regulation of the sleep-wake cycle by SSRI discontinuation (Collins et al. 2023b).

4 Mechanisms of SSRI Discontinuation

4.1 Role of 5-HT

As the 5-HT transporter is the primary target of SSRIs, and since 5-HT neurons adapt to continuous SSRI exposure (see above), changes in 5-HT function in response to SSRI discontinuation seem likely as noted in previous reviews (Blier and Tremblay 2006; Harvey and Slabbert 2014; Renoir 2013). Surprisingly, however, the effect of SSRI discontinuation on the 5-HT system has been relatively little investigated to date and (see Table 2).

In one of few such studies (that was not in itself directed at the mechanisms of SSRI discontinuation) it was reported that continuous treatment of rats with fluoxetine was associated with reduced levels of the major 5-HT metabolite 5-HIAA in the hippocampus and other brain regions (Trouvin et al. 1993). Importantly, a striking increase in 5-HIAA was evident over 2 weeks following treatment cessation. Similar changes in 5-HT synthesis were more recently observed in mice administered repeated fluoxetine (Stenfors and Ross 2002), as well as rats administered citalopram (Bosker et al. 2010). Continuous exposure to an SSRI treatment is well known to induce an adaptive fall in 5-HT metabolism and synthesis through indirect activation of 5-HT$_{1A}$ and 5-HT$_{1B}$ autoreceptors (e.g. Barton and Hutson 1999). These earlier findings with fluoxetine are suggestive that this adaptive fall in 5-HT

Table 2 Reported effect of SSRI discontinuation on measures of 5-HT neuronal function in rodent brain. 'Rebound increase' is defined as relative to saline controls combined with evidence of decreased 5-HT neuronal function in animals on continuous SSRI treatment. Timing of effect refers to time post discontinuation

Measurement	Treatment	Effect of discontinuation	Timing of effect	Reference
Regional brain 5-HIAA	Fluoxetine (5–15 mg/kg, twice daily, 21 days, rats)	Rebound increase	3–14 days	Trouvin et al. (1993)
Regional brain 5-HT synthesis	Fluoxetine (7 mg/kg, once daily, 21 days, mice)	Rebound increase	2–8 days	Stenfors and Ross (2002)
Regional brain 5-HT synthesis	Citalopram (32 mg/kg, constant infusion, 14 days, rats)	Rebound increase	2 days	Bosker et al. (2010)
Regional brain 5-HIAA	Paroxetine (10 mg/kg, once daily, 12 days, mice)	Rebound increase	5 days	Collins et al. (2023a)
Depolarisation-evoked 5-HT release in vivo	Paroxetine (10 mg/kg, once daily, 12 days, mice)	Rebound increase	2–5 days	Collins et al. (2023a)
c-Fos expression in midbrain 5-HT neurons	Paroxetine (10 mg/kg, once daily, 12 days, mice)	Rebound increase	2–5 days	Collins et al. (2023a)

function is followed by a rebound increase over in the days following drug discontinuation.

Studies from our lab have recently replicated and extended the above findings (Collins et al. 2023a, Table 2). Specifically, continuous administration of paroxetine to mice decreased 5-HT metabolism in various brain regions, whereas abrupt cessation was associated with a rebound increase. This work was followed up by experiments using in vivo microdialysis. It was observed that continuous paroxetine increased extracellular 5-HT in hippocampus as expected, and this effect rapidly reversed within 2 days of discontinuation. Interestingly, however, a rebound increase in depolarisation (high potassium)-evoked release of hippocampal 5-HT was detected shortly (2–5 days) after paroxetine discontinuation. Thus, it appears that discontinuation resulted in increased release of 5-HT in response to a physiological stimulus. These neurochemical data were supported by immunocytochemical studies which found increased expression of the activity-dependent immediate early gene c-Fos in midbrain 5-HT neurons 2–5 days post discontinuation of paroxetine, also suggesting hyperexcitable 5-HT neurons (Table 2).

Thus, taking the above findings together, a plausible scenario is that upon abrupt cessation of SSRI administration and the reversal of 5-HTT blockade, extracellular 5-HT rapidly falls. These changes would relieve inhibitory 5-HT feedback systems (e.g. somatodendritic and nerve terminal 5-HT autoreceptors) which then could trigger a rebound increase in 5-HT metabolism, release and neuronal excitability (Fig. 2). It is also likely that 5-HT-glutamate co-release at the 5-HT synapse will change (see above). Thus, in the presence of 5-HT transporter blockade, transmission of 5-HT is likely to be promoted over glutamate. On the other hand, abrupt

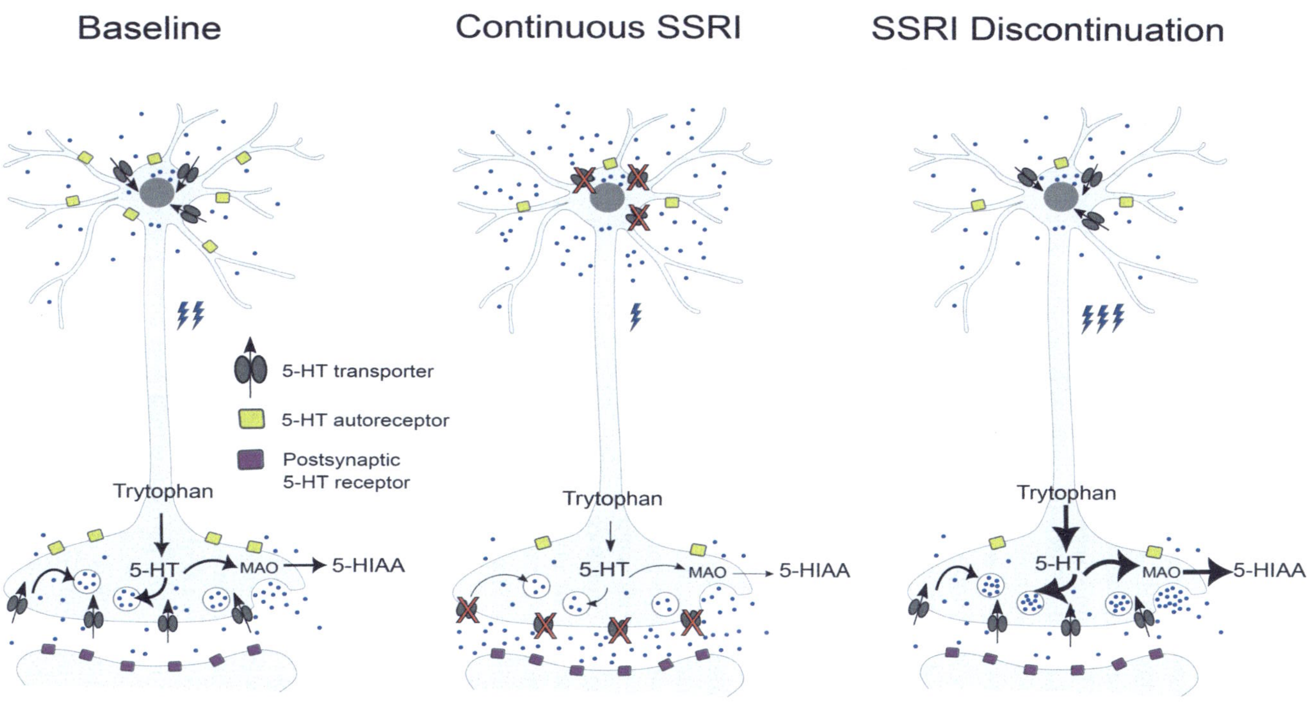

Fig. 2 Theoretical adaptive changes in 5-HT neuron function associated with discontinuation of SSRI treatment. Left, diagram showing key mechanisms of 5-HT neuron control at the level of cell body and nerve terminal, including the 5-HT transporter and 5-HT autoreceptors together with a simplified view of postsynaptic 5-HT receptors (i.e. 14 5-HT receptor subtypes and their signalling pathways not show). Middle, continuous SSRI exposure and 5-HT transporter blockade increases availability of extracellular 5-HT. At the same time this increased 5-HT activates autoreceptors on the somatodendrites (5-HT_{1A}) and nerve terminals (5-HT_{1B}) to reduce 5-HT synthesis and suppress 5-HT cell firing. 5-HT autoreceptors also desensitise (illustrated as decreased receptor number but more likely a loss of function). Right, abrupt cessation of SSRI exposure brings about rapid relief from suppression of 5-HT neurons and an unstable state of 5-HT neuron control (5-HT autoreceptor desensitization included) which triggers a rebound increase in 5-HT synthesis, release and neuronal excitability. Both continued and discontinued SSRI treatment may also shift the balance 5-HT/glutamate co-release (not shown; see text for details)

removal of 5-HT transporter inhibition will likely shift the balance of 5-HT-glutamate co-release in favour of glutamate. Interestingly, SSRI discontinuation has been previously attributed to increased glutamatergic neurotransmission albeit on the basis of evidence of adaptive upregulation of NMDA receptors (Harvey and Slabbert 2014).

Whilst relief from negative 5-HT feedback seems a likely contributor to an increase in excitability of 5-HT neurons after SSRI discontinuation, the involvement of other mechanisms such as adaptive changes in ion channels that control the excitability of 5-HT neurons seems possible. This idea contrasts with an earlier theory that the cause of the SSRI discontinuation syndrome reduced extracellular 5 HT due to rapid SSRI washout and relief of 5-HT transporter blockade (see Blier and Tremblay 2006). The latter theory was largely based on the findings that the symptomatic effects of tryptophan depletion in SSRI-treated patients shared features with those of SSRI discontinuation (see Table 1). However, closer examination of the symptoms of tryptophan depletion suggested that they are more characteristic of depression relapse than those of SSRI discontinuation (Delgado 2006). On the other hand, it is plausible that an acute fall in 5-HT induced by tryptophan depletion brings about relief from 5-HT feedback suppression that rapid removal of the SSRI by itself achieves. If correct this suggests that aspects of the effects of tryptophan depletion in SSRI-treated patients may indeed reflect SSRI discontinuation.

How does this idea of a rebound increase in 5-HT fit with the fact that reinstatement of SSRI treatment is a typical approach to managing the discontinuation syndrome? A possible explanation is that re-introduction of the SSRI increases extracellular 5-HT but at the same time reduces the activity of 5-HT neurons by indirectly activating the inhibitory 5-HT autoreceptors that are a major cause of 5-HT neuron instability. In effect, recommencement of SSRI treatment would restore 5-HT neuron feedback to dampen down the 5-HT neuron hyperexcitability that is a potential driver of discontinuation symptoms.

4.2 Causal Link between 5-HT Changes and Behaviour

There is a strong link between changes in 5-HT transmission and anxiety, suggesting that the discontinuation-induced rebound increase in 5-HT neuron excitability and anxiety-like behaviour may be causally linked. For example, optogenetic activation of midbrain 5-HT neurons was found to generate anxiety-like behaviour on the elevated plus maze (Ohmura et al. 2020). However, against this idea is evidence that following discontinuation from paroxetine treatment, anxiety-like behaviour was not temporally related to the increase in 5-HT function; the former being short-lasting compared to the latter (Collins et al. 2023a; Collins et al. 2022). This discrepancy could simply be related to differential sensitivity of the behaviour and neurochemical methods used. It is also possible that the rebound increase in 5 HT neurotransmission after cessation of paroxetine treatment generates other behaviours or physiological changes associated with the discontinuation syndrome. For example, sleep is well known to be regulated by 5-HT neurons, and sleep disturbances were detected in mice discontinued from paroxetine (see above) and are a recognised feature of SSRI discontinuation syndrome.

Moreover, a rebound increase in 5 HT neuron activity could also contribute to other aspects of the discontinuation syndrome, such as the somatic symptoms; increased 5-HT_3 receptor activation in the periphery can induce nausea and vomiting (Navari 2015). Moreover, vertigo and dizziness have also been linked to altered 5-HT transmission in the vestibular system (Soto et al. 2013). Although these symptoms would be more difficult to study in rodents, they present interesting opportunities to explore the role of 5-HT transmission in the broader discontinuation syndrome.

4.3 A Role for Other Neurotransmitter Systems

Although SSRIs target the 5-HT transporter and likely to influence 5-HT function both centrally and peripherally, changes in 5-HT alone might not explain the large diversity of symptoms. Although as a class of drugs, SSRIs have selectivity for the 5-HT transporter, the pharmacological profile of individual SSRIs is subtly different. Thus, specific SSRIs have affinity for targets of other neurotransmitter systems, which may thereby contribute to discontinuation effects. Specifically, similar to the tricyclic antidepressants but unusual amongst SSRIs, paroxetine has modest affinity for cholinergic muscarinic receptors (Owens et al. 1997). This raises the possibility that cholinergic rebound contributes to some of the central and peripheral effects of paroxetine discontinuation as noted in previously (Blier and Tremblay 2006). Paroxetine also has modest affinity for the noradrenaline transporter, and increases extracellular noradrenaline in the rodent hippocampus when administered repeatedly (Hajos-Korcsok et al. 2000). It has been speculated previously that noradrenaline may be involved in the discontinuation effects of antidepressants either through off-target effects such as with paroxetine or interactions between 5-HT and noradrenergic neurons (Blier and Tremblay 2006). This idea is, however, countered by reports that the selective noradrenaline reuptake inhibitors reboxetine and atomoxetine are less prone to discontinuation effects (Mucci 1997; Wernicke et al. 2004).

Aside from noradrenaline, 5-HT neurons interact with numerous other transmitter systems including dopamine, glutamate and GABA, and previous reviews speculate that such transmitters play a role in SSRI discontinuation effects (Blier and Tremblay 2006; Harvey and Slabbert 2014; Renoir 2013). These ideas require systematic investigation; for example, there is much current research interest in interactions between 5-HT and pyramidal neurons (glutamate) and interneurons (GABA) in cerebral cortex (reviewed in Sharp and Barnes 2020). Changes in excitatory-inhibitory balance in cerebral cortex likely play a critical role in cognition and sensory information processing and may be disrupted during discontinuation to generate unwanted symptoms.

As noted above, repeated SSRI administration generates complex adaptive changes downstream of the 5-HT synapse, and these could also contribute to the discontinuation syndrome. However, a role for plasticity in discontinuation symptoms is less easy to imagine since the changes take weeks of SSRI treatment to build up and are unlikely to rapidly reverse with abrupt cessation of treatment. Also, it is unclear how changes in plasticity in mood-related regions like the hippocampus could contribute to the large range of symptoms associated with discontinuation.

4.4 Mechanisms Underpinning Other Drug Withdrawal States

Numerous psychotropic drug classes are associated with drug withdrawal symptoms, with disabling syndromes being described for opiates, alcohol, benzodiazepines and psychostimulants amongst others. There are interesting parallels between these drug withdrawal states and the SSRI discontinuation syndrome in that initial withdrawal symptoms are often immediate in onset and relatively short-lasting but may persist for several months (Cosci et al. 2023). Also, as with SSRIs the withdrawal symptoms from these drugs are typically more frequent and severe with drugs with a short elimination half-life. Although the symptoms are distinct for each drug class, some of the symptoms of the SSRI discontinuation syndrome resemble those of other drug withdrawal states including affective symptoms such as anxiety and agitation, as well as nausea, dizziness, and headaches. These similarities suggest some common underlying neural substrates.

This said, since these CNS drugs have different pharmacological targets, the neurochemical mechanisms underlying the different withdrawal states are likely to be distinct. It is noteworthy, however, that withdrawal reactions to drugs such as opiates, alcohol and benzodiazepines are associated with inhibition of neural activity during drug exposure and then rebound increases in neuronal activity on withdrawal. Put in a simple way, the cause of the withdrawal syndrome is seen as an adaptive neurochemical response designed to oppose the drug's acute effects (Littleton 1998; Cowen and Nutt 1982).

In the case of opioids, evidence suggests that a rebound in noradrenergic activity is triggered by cessation of morphine administration to mediate withdrawal symptoms; indeed, inhibition of noradrenergic activity using the α2-adrenoceptor agonist clonidine is an approved treatment of opiate withdrawal (reviewed in Kosten and George, 2002). As another example, both alcohol and benzodiazepines augment the inhibitory actions of GABA on glutamate neurons, and it is thought that withdrawal effects are mediated by hyperexcitability of glutamate and thereby a shift in inhibitory:excitatory balance (Littleton 1998; Gass and Olive 2008; Koob et al. 2009; Warlick et al. 2023). This research has provided a rationale for managing symptoms associated with alcohol and benzodiazepine withdrawal using gabapentanoids (Anton et al. 2020; Warlick et al. 2023); the latter drugs have a complex pharmacology, which includes interaction with α2δ calcium channel subunits, but are thought to act by modulating inhibitory:excitatory balance.

Thus, there are strong parallels between the emerging evidence of a rebound in 5-HT function during SSRI discontinuation and mechanisms associated with other drug withdrawal states. This suggests perhaps an unanticipated theoretical framework in which an understanding of the discontinuation effects of SSRIs and antidepressants can be better interpreted through consideration with other drug withdrawal states.

5 Conclusion

SSRIs have been a mainline treatment option for depression and anxiety for almost 40 years. A goal now is to develop medicines with a faster onset of action and greater efficacy than SSRIs, and which lack their adverse effects not least a disabling SSRI discontinuation syndrome. There is much evidence from preclinical and clinical studies that 5-HT transporter blockade is key to the therapeutic effects of SSRIs. This view has been challenged by recent evidence that the drugs might act directly via BDNF, although this result requires replication and further validation. Either way, changes in neural plasticity and interactions with neuropsychological processes provide a compelling account of the action of SSRIs to inform further antidepressant drug development. In contrast, the mechanisms underpinning SSRI discontinuation are little explored. Nevertheless, emerging evidence of a rebound increase in 5-HT neuron excitability offers a mechanistic explanation of SSRI discontinuation with interesting parallels to that of withdrawal states of other psychotropic drugs.

References

Altamura AC, Moro AR, Percudani M (1994) Clinical pharmacokinetics of fluoxetine. Clin Pharmacokinet 26(3):201–214

Ansorge MS, Zhou M, Lira A et al (2004) Early-life blockade of the 5-HT transporter alters emotional behavior in adult mice. Science 306(5697):879–881

Anton RF, Latham P, Voronin K et al (2020) Efficacy of gabapentin for the treatment of alcohol use disorder in patients with alcohol withdrawal symptoms: a randomized clinical trial. JAMA Intern Med 180(5):728–736

APPG (2018) Antidepressant withdrawal: a survey of patients' experience. All Party Parliamentary Group for Prescribed Drug Dependence. http://prescribeddrug.org/wp-content/uploads/2018/10/APPG-PDD-Survey-of-antidepressant-withdrawal-experiences.pdf

Artigas F, Romero L, de Montigny C et al (1996) Acceleration of the effect of selected antidepressant drugs in major depression by 5-HT1A antagonists. Trends Neurosci 19(9):378–383

Autry AE, Adachi M, Nosyreva E et al (2011) NMDA receptor blockade at rest triggers rapid behavioural antidepressant responses. Nature 475(7354):91–95

Baldwin DS, Cooper JA, Huusom AK et al (2006) A double-blind, randomized, parallel-group, flexible-dose study to evaluate the tolerability, efficacy and effects of treatment discontinuation

with escitalopram and paroxetine in patients with major depressive disorder. Int Clin Psychopharmacol 21(3):159–169
Barton CL, Hutson PH (1999) Inhibition of hippocampal 5-HT synthesis by fluoxetine and paroxetine: evidence for the involvement of both 5-HT1A and 5-HT1B/D autoreceptors. Synapse 31(1):13–19
Bhanji NH, Chouinard G, Kolivakis T et al (2006) Persistent tardive rebound panic disorder, rebound anxiety and insomnia following paroxetine withdrawal: a review of rebound-withdrawal phenomena. Can J Clin Pharmacol 13(1):e69–e74
Blendy JA (2006) The role of CREB in depression and antidepressant treatment. Biol Psychiatry 59(12):1144–1150
Blier P, de Montigny C (1994) Current advances and trends in the treatment of depression. Trends Pharmacol Sci 15(7):220–226
Blier P, Tremblay P (2006) Physiologic mechanisms underlying the antidepressant discontinuation syndrome. J Clin Psychiatry 67(Suppl 4):8–13
Bosker FJ, Tanke MA, Jongsma ME et al (2010) Biochemical and behavioral effects of long-term citalopram administration and discontinuation in rats: role of serotonin synthesis. Neurochem Int 57(8):948–957
Bouet V, Klomp A, Freret T et al (2012) Age-dependent effects of chronic fluoxetine treatment on the serotonergic system one week following treatment. Psychopharmacology (Berl) 221(2): 329–339
Bourin M, Chue P, Guillon Y (2001) Paroxetine: a review. CNS Drug Rev 7(1):25–47
Branchi I (2022) Plasticity in mental health: a network theory. Neurosci Biobehav Rev 138:104691
Cameron LP, Tombari RJ, Lu J et al (2021) A non-hallucinogenic psychedelic analogue with therapeutic potential. Nature 589(7842):474–479
Carr GV, Lucki I (2011) The role of serotonin receptor subtypes in treating depression: a review of animal studies. Psychopharmacology (Berl) 213(2–3):265–287
Cartwright C, Gibson K, Read J et al (2016) Long-term antidepressant use: patient perspectives of benefits and adverse effects. Patient Prefer Adherence 10:1401–1407
Casarotto PC, Girych M, Fred SM et al (2021) Antidepressant drugs act by directly binding to TRKB neurotrophin receptors. Cell 184(5):1299–1313 e1219
Casarotto P, Umemori J, Castren E (2022) BDNF receptor TrkB as the mediator of the antidepressant drug action. Front Mol Neurosci 15:1032224
Castren E (2005) Is mood chemistry? Nat Rev Neurosci 6(3):241–246
Castren E, Monteggia LM (2021) Brain-derived neurotrophic factor signaling in depression and antidepressant action. Biol Psychiatry 90(2):128–136
Cipriani A, Furukawa TA, Salanti G et al (2018) Comparative efficacy and acceptability of 21 antidepressant drugs for the acute treatment of adults with major depressive disorder: a systematic review and network meta-analysis. Lancet 391(10128):1357–1366
Collins HM, Pinacho R, Ozdemir D et al (2022) Effect of selective serotonin reuptake inhibitor discontinuation on anxiety-like behaviours in mice. J Psychopharmacol 36(7):794–805
Collins HM, Ozdemir D, Gullino LS et al (2023a) Rebound activation of 5-HT neurons following SSRI discontinuation. bioRxiv 2023.09.23.558888. https://doi.org/10.1101/2023.09.23.558888
Collins HM, Pinacho R, Tam SKE et al (2023b) Continuous home cage monitoring of activity and sleep in mice during repeated paroxetine treatment and discontinuation. Psychopharmacology 240 (11), 2403–2418.
Cosci F, Chouinard VA, Chouinard G (2023) Discontinuation of antidepressant medications: a significant healthcare problem insufficiently addressed by the NICE guidelines. Psychother Psychosom 92(3):148–151
Cowen PJ, Nutt DJ (1982) Abstinence symptoms after withdrawal of tranquillising drugs: is there a common neurochemical mechanism? Lancet 2(8294):360–362
Cunningham MJ, Bock HA, Serrano IC et al (2023) Pharmacological mechanism of the non-hallucinogenic 5-HT(2A) agonist ariadne and analogs. ACS Chem Nerosci 14(1):119–135

Davies J, Read J (2019) A systematic review into the incidence, severity and duration of antidepressant withdrawal effects: are guidelines evidence-based? Addict Behav 97:111–121
Delgado PL (2006) Monoamine depletion studies: implications for antidepressant discontinuation syndrome. J Clin Psychiatry 67(Suppl 4):22–26
Delgado PL, Charney DS, Price LH et al (1990) Serotonin function and the mechanism of antidepressant action. Reversal of antidepressant-induced remission by rapid depletion of plasma tryptophan. Arch Gen Psychiatry 47(5):411–418
D'Sa C, Duman RS (2002) Antidepressants and neuroplasticity. Bipolar Disord 4(3):183–194
Duffy L, Clarke CS, Lewis G et al (2021) Antidepressant medication to prevent depression relapse in primary care: the ANTLER RCT. Health Technol Assess 25(69):1–62
Elizalde N, Gil-Bea FJ, Ramirez MJ et al (2008) Long-lasting behavioral effects and recognition memory deficit induced by chronic mild stress in mice: effect of antidepressant treatment. Psychopharmacology (Berl) 199(1):1–14
Elsayed M, Banasr M, Duric V et al (2012) Antidepressant effects of fibroblast growth factor-2 in behavioral and cellular models of depression. Biol Psychiatry 72(4):258–265
Erritzoe D, Godlewska BR, Rizzo G et al (2023) Brain serotonin release is reduced in patients with depression: a [(11)C]Cimbi-36 positron emission tomography study with a d-amphetamine challenge. Biol Psychiatry 93(12):1089–1098
Fava GA, Gatti A, Belaise C et al (2015) Withdrawal symptoms after selective serotonin reuptake inhibitor discontinuation: a systematic review. Psychother Psychosom 84(2):72–81
Fischer AG, Jocham G, Ullsperger M (2014) Dual serotonergic signals: a key to understanding paradoxical effects? Trends Cogn Sci. Epub ahead of print2014/12/24. https://doi.org/10.1016/j.tics.2014.11.004
Fredricson Overo K (1982) Kinetics of citalopram in test animals; drug exposure in safety studies. Prog Neuropsychopharmacol Biol Psychiatry 6(3):297–309
Gartside SE, Umbers V, Hajos M et al (1995) Interaction between a selective 5-HT1A receptor antagonist and an SSRI in vivo: effects on 5-HT cell firing and extracellular 5-HT. Br J Pharmacol 115(6):1064–1070
Gass JT, Olive MF (2008) Glutamatergic substrates of drug addiction and alcoholism. Biochem Pharmacol 75(1):218–265
Gastaldon C, Schoretsanitis G, Arzenton E et al (2022) Withdrawal syndrome following discontinuation of 28 antidepressants: pharmacovigilance analysis of 31,688 reports from the WHO spontaneous reporting database. Drug Saf 45(12):1539–1549
Gelenberg AJ, Chesen CL (2000) How fast are antidepressants? J Clin Psychiatry 61(10):712–721
Gonzalez-Maeso J, Weisstaub NV, Zhou M et al (2007) Hallucinogens recruit specific cortical 5-HT(2A) receptor-mediated signaling pathways to affect behavior. Neuron 53(3):439–452
Gras C, Herzog E, Bellenchi GC et al (2002) A third vesicular glutamate transporter expressed by cholinergic and serotoninergic neurons. J Neurosci 22(13):5442–5451
Groves JO (2007) Is it time to reassess the BDNF hypothesis of depression? Mol Psychiatry 12(12):1079–1088
Haddad P (1997) Newer antidepressants and the discontinuation syndrome. J Clin Psychiatry 58 (Suppl 7):17–21. discussion 22
Haddad PM (2005) Do antidepressants cause dependence? Epidemiol Psichiatr Soc 14(2):58–62
Haddad PM, Anderson IM (2007) Recognising and managing antidepressant discontinuation symptoms. Adv Psychiatr Treat 13:447–457
Hajos-Korcsok E, McTavish SF, Sharp T (2000) Effect of a selective 5-hydroxytryptamine reuptake inhibitor on brain extracellular noradrenaline: microdialysis studies using paroxetine. Eur J Pharmacol 407(1–2):101–107
Halberstadt AL, Chatha M, Klein AK et al (2020) Correlation between the potency of hallucinogens in the mouse head-twitch response assay and their behavioral and subjective effects in other species. Neuropharmacology 167:107933

Harmer CJ, Goodwin GM, Cowen PJ (2009) Why do antidepressants take so long to work? A cognitive neuropsychological model of antidepressant drug action. Br J Psychiatry 195(2): 102–108

Harmer CJ, Duman RS, Cowen PJ (2017) How do antidepressants work? New perspectives for refining future treatment approaches. Lancet Psychiatry 4(5):409–418

Harvey BH, Slabbert FN (2014) New insights on the antidepressant discontinuation syndrome. Hum Psychopharmacol 29(6):503–516

Hayward G, Goodwin GM, Cowen PJ et al (2005) Low-dose tryptophan depletion in recovered depressed patients induces changes in cognitive processing without depressive symptoms. Biol Psychiatry 57(5):517–524

Hesselgrave N, Troppoli TA, Wulff AB et al (2021) Harnessing psilocybin: antidepressant-like behavioral and synaptic actions of psilocybin are independent of 5-HT2R activation in mice. Proc Natl Acad Sci U S A 118(17)

Hodes GE, Hill-Smith TE, Suckow RF et al (2010) Sex-specific effects of chronic fluoxetine treatment on neuroplasticity and pharmacokinetics in mice. J Pharmacol Exp Ther 332(1): 266–273

Holick KA, Lee DC, Hen R et al (2008) Behavioral effects of chronic fluoxetine in BALB/cJ mice do not require adult hippocampal neurogenesis or the serotonin 1A receptor. Neuropsychopharmacology 33(2):406–417

Holmes SE, Scheinost D, Finnema SJ, Naganawa M, Davis MT, DellaGioia N, Nabulsi N, Matuskey D, Angarita GA, Pietrzak RH, Duman RS, Sanacora G, Krystal JH, Carson RE, Esterlis I (2019) Lower synaptic density is associated with depression severity and network alterations. Nature Commun. 10(1):1529

Horowitz MA, Taylor D (2019) Tapering of SSRI treatment to mitigate withdrawal symptoms. Lancet Psychiatry 6(6):538–546

Horowitz MA, Framer A, Hengartner MP et al (2023) Estimating risk of antidepressant withdrawal from a review of published data. CNS Drugs 37(2):143–157

Huang GJ, Bannerman D, Flint J (2008) Chronic fluoxetine treatment alters behavior, but not adult hippocampal neurogenesis, BALB/cJ mice. Mol Psychiatry 13(2):119–121

Jauhar S, Hayes J, Goodwin GM et al (2019) Antidepressants, withdrawal, and addiction; where are we now? J Psychopharmacol 33(6):655–659

Jauhar S, Arnone D, Baldwin DS et al (2023) A leaky umbrella has little value: evidence clearly indicates the serotonin system is implicated in depression. Mol Psychiatry. Epub ahead of print 2023/06/16. https://doi.org/10.1038/s41380-023-02095-y

Jennings KA, Sheward WJ, Harmar AJ et al (2008) Evidence that genetic variation in 5-HT transporter expression is linked to changes in 5-HT2A receptor function. Neuropharmacology 54(5):776–783

Jennings KA, Lesch KP, Sharp T et al (2010) Non-linear relationship between 5-HT transporter gene expression and frequency sensitivity of 5-HT signals. J Neurochem 115(4):965–973

Jennings KA, Licht CL, Bruce A et al (2012) Genetic variation in 5-hydroxytryptamine transporter expression causes adaptive changes in 5-HT(4) receptor levels. Int J Neuropsychopharmacol 15(8):1099–1107

Jha MK, Rush AJ, Trivedi MH (2018) When discontinuing SSRI antidepressants is a challenge: management tips. Am J Psychiatry 175(12):1176–1184

Johnston JN, Kadriu B, Kraus C et al (2023) Ketamine in neuropsychiatric disorders: an update. Neuropsychopharmacology. Epub ahead of print 2023/06/21. https://doi.org/10.1038/s41386-023-01632-1

Jovanovic H, Lundberg J, Karlsson P et al (2008) Sex differences in the serotonin 1A receptor and serotonin transporter binding in the human brain measured by PET. Neuroimage 39(3): 1408–1419

Kang HJ, Voleti B, Hajszan T et al (2012) Decreased expression of synapse-related genes and loss of synapses in major depressive disorder. Nat Med 18(9):1413–1417

Kim JW, Suzuki K, Kavalali ET et al (2023) Bridging rapid and sustained antidepressant effects of ketamine. Trends Mol Med 29(5):364–375
Klein ME, Chandra J, Sheriff S et al (2020) Opioid system is necessary but not sufficient for antidepressive actions of ketamine in rodents. Proc Natl Acad Sci U S A 117(5):2656–2662
Koob GF, Kenneth Lloyd G, Mason BJ (2009) Development of pharmacotherapies for drug addiction: a Rosetta stone approach. Nat Rev Drug Discov 8(6):500–515
Kosten TR, George TP (2002) The neurobiology of opioid dependence: implications for treatment. Sci Pract Perspect 1(1):13–20
Kreilgaard M, Smith DG, Brennum LT et al (2008) Prediction of clinical response based on pharmacokinetic/pharmacodynamic models of 5-hydroxytryptamine reuptake inhibitors in mice. Br J Pharmacol 155(2):276–284
Kwan AC, Olson DE, Preller KH et al (2022) The neural basis of psychedelic action. Nat Neurosci 25(11):1407–1419
Lagas AK, Black JM, Russell BR et al (2019) The effect of combined patching and citalopram on visual acuity in adults with amblyopia: a randomized, crossover, placebo-controlled trial. Neural Plast 2019:5857243
Legg LA, Rudberg AS, Hua X et al (2021) Selective serotonin reuptake inhibitors (SSRIs) for stroke recovery. Cochrane Database Syst Rev 11(11):CD009286
Lerner A, Klein M (2019) Dependence, withdrawal and rebound of CNS drugs: an update and regulatory considerations for new drugs development. Brain Commun 1(1):fcz025
Levinson-Castiel R, Merlob P, Linder N et al (2006) Neonatal abstinence syndrome after in utero exposure to selective serotonin reuptake inhibitors in term infants. Arch Pediatr Adolesc Med 160(2):173–176
Lewis V, Bonniwell EM, Lanham JK et al (2023) A non-hallucinogenic LSD analog with therapeutic potential for mood disorders. Cell Rep 42(3):112203
Licznerski P, Duman RS (2013) Remodeling of axo-spinous synapses in the pathophysiology and treatment of depression. Neuroscience 251:33–50
Lima J, Sharp T, Bannerman DM et al (2019) Enhanced discriminative aversive learning and amygdala responsivity in 5-HT transporter mutant mice. Transl Psychiatry 9(1):139
Lindefors N, Brodin E, Metsis M (1995) Spatiotemporal selective effects on brain-derived neurotrophic factor and trkB messenger RNA in rat hippocampus by electroconvulsive shock. Neuroscience 65(3):661–670
Littleton J (1998) Neurochemical mechanisms underlying alcohol withdrawal. Alcohol Health Res World 22(1):13–24
Ly C, Greb AC, Cameron LP et al (2018) Psychedelics promote structural and functional neural plasticity. Cell Rep 23(11):3170–3182
MacQueen GM, Ramakrishnan K, Croll SD et al (2001) Performance of heterozygous brain-derived neurotrophic factor knockout mice on behavioral analogues of anxiety, nociception, and depression. Behav Neurosci 115(5):1145–1153
Madsen MK, Fisher PM, Burmester D et al (2019) Psychedelic effects of psilocybin correlate with serotonin 2A receptor occupancy and plasma psilocin levels. Neuropsychopharmacology 44(7): 1328–1334
Malberg JE, Hen R, Madsen TM (2021) Adult neurogenesis and antidepressant treatment: the surprise finding by Ron Duman and the field 20 years later. Biol Psychiatry 90(2):96–101
Maya Vetencourt JF, Sale A, Viegi A et al (2008) The antidepressant fluoxetine restores plasticity in the adult visual cortex. Science 320(5874):385–388
McClure-Begley TD, Roth BL (2022) The promises and perils of psychedelic pharmacology for psychiatry. Nat Rev Drug Discov 21(6):463–473
Mead GE, Hsieh CF, Lee R et al (2012) Selective serotonin reuptake inhibitors (SSRIs) for stroke recovery. Cochrane Database Syst Rev 11(11):CD009286
Meyer JH, Wilson AA, Sagrati S et al (2004) Serotonin transporter occupancy of five selective serotonin reuptake inhibitors at different doses: an [11C]DASB positron emission tomography study. Am J Psychiatry 161(5):826–835

Moliner R, Girych M, Brunello CA et al (2023) Psychedelics promote plasticity by directly binding to BDNF receptor TrkB. Nat Neurosci 26(6):1032–1041
Moncrieff J, Cooper RE, Stockmann T et al (2022) The serotonin theory of depression: a systematic umbrella review of the evidence. Mol Psychiatry. Epub ahead of print 2022/07/20. https://doi.org/10.1038/s41380-022-01661-0
Mucci M (1997) Reboxetine: a review of antidepressant tolerability. J Psychopharmacol 11(4 Suppl):S33–S37
Murata Y, Kobayashi D, Imuta N et al (2010) Effects of the serotonin 1A, 2A, 2C, 3A, and 3B and serotonin transporter gene polymorphisms on the occurrence of paroxetine discontinuation syndrome. J Clin Psychopharmacol 30(1):11–17
Nackenoff AG, Moussa-Tooks AB, McMeekin AM et al (2016) Essential contributions of serotonin transporter inhibition to the acute and chronic actions of fluoxetine and citalopram in the SERT Met172 mouse. Neuropsychopharmacology 41(7):1733–1741
Navari RM (2015) 5-HT3 receptors as important mediators of nausea and vomiting due to chemotherapy. Biochim Biophys Acta 1848(10 Pt B):2738–2746
Nibuya M, Morinobu S, Duman RS (1995) Regulation of BDNF and trkB mRNA in rat brain by chronic electroconvulsive seizure and antidepressant drug treatments. J Neurosci 15(11): 7539–7547
Ohmura Y, Tsutsui-Kimura I, Sasamori H et al (2020) Different roles of distinct serotonergic pathways in anxiety-like behavior, antidepressant-like, and anti-impulsive effects. Neuropharmacology 167:107703
Owens MJ, Morgan WN, Plott SJ et al (1997) Neurotransmitter receptor and transporter binding profile of antidepressants and their metabolites. J Pharmacol Exp Ther 283(3):1305–1322
Parsey RV, Kent JM, Oquendo MA et al (2006) Acute occupancy of brain serotonin transporter by sertraline as measured by [11C]DASB and positron emission tomography. Biol Psychiatry 59(9):821–828
Pei Q, Tordera R, Sprakes M et al (2004) Glutamate receptor activation is involved in 5-HT2 agonist-induced arc gene expression in the rat cortex. Neuropharmacology 46(3):331–339
Renoir T (2013) Selective serotonin reuptake inhibitor antidepressant treatment discontinuation syndrome: a review of the clinical evidence and the possible mechanisms involved. Front Pharmacol 4:45
Renoir T, Zajac MS, Du X et al (2011) Sexually dimorphic serotonergic dysfunction in a mouse model of Huntington's disease and depression. PloS One 6(7):e22133
Riad M, Zimmer L, Rbah L et al (2004) Acute treatment with the antidepressant fluoxetine internalizes 5-HT1A autoreceptors and reduces the in vivo binding of the PET radioligand [18F]MPPF in the nucleus raphe dorsalis of rat. J Neurosci 24(23):5420–5426
Ruhe HG, Mason NS, Schene AH (2007) Mood is indirectly related to serotonin, norepinephrine and dopamine levels in humans: a meta-analysis of monoamine depletion studies. Mol Psychiatry 12(4):331–359
Santarelli L, Saxe M, Gross C et al (2003) Requirement of hippocampal neurogenesis for the behavioral effects of antidepressants. Science 301(5634):805–809
Schmaal L, Veltman DJ, van Erp TG et al (2016) Subcortical brain alterations in major depressive disorder: findings from the ENIGMA major depressive disorder working group. Mol Psychiatry 21(6):806–812
Schweizer E, Thielen RJ, Frazer A (1997) Venlafaxine:a novel antidepressant compound. Expert Opin Investig Drugs 6(1):65–78
Sengupta A, Bocchio M, Bannerman DM et al (2017) Control of amygdala circuits by 5-HT neurons via 5-HT and glutamate cotransmission. J Neurosci 37(7):1785–1796
Shao LX, Liao C, Gregg I et al (2021) Psilocybin induces rapid and persistent growth of dendritic spines in frontal cortex in vivo. Neuron 109(16):2535–2544 e2534
Sharp T (2013) Molecular and cellular mechanisms of antidepressant action. Curr Top Behav Neurosci 14:309–325

Sharp T, Barnes NM (2020) Central 5-HT receptors and their function; present and future. Neuropharmacology 177:108155
Sharp T, Cowen PJ (2011) 5-HT and depression: is the glass half-full? Curr Opin Pharmacol 11(1): 45–51
Soto E, Vega R, Sesena E (2013) Neuropharmacological basis of vestibular system disorder treatment. J Vestib Res 23(3):119–137
Spies M, Knudsen GM, Lanzenberger R et al (2015) The serotonin transporter in psychiatric disorders: insights from PET imaging. Lancet Psychiatry 2(8):743–755
Staner L, Kerkhofs M, Detroux D et al (1995) Acute, subchronic and withdrawal sleep EEG changes during treatment with paroxetine and amitriptyline: a double-blind randomized trial in major depression. Sleep 18(6):470–477
Stenfors C, Ross SB (2002) Evidence for involvement of 5-hydroxytryptamine(1B) autoreceptors in the enhancement of serotonin turnover in the mouse brain following repeated treatment with fluoxetine. Life Sci 71(24):2867–2880
Strekalova T, Anthony DC, Dolgov O et al (2013) The differential effects of chronic imipramine or citalopram administration on physiological and behavioral outcomes in naive mice. Behav Brain Res 245:101–106
Tartt AN, Mariani MB, Hen R et al (2022) Dysregulation of adult hippocampal neuroplasticity in major depression: pathogenesis and therapeutic implications. Mol Psychiatry 27(6):2689–2699
Trouvin JH, Gardier AM, Chanut E et al (1993) Time course of brain serotonin metabolism after cessation of long-term fluoxetine treatment in the rat. Life Sci 52(18):PL187-192
Umemori J, Winkel F, Didio G et al (2018) iPlasticity: induced juvenile-like plasticity in the adult brain as a mechanism of antidepressants. Psychiatry Clin Neurosci 72(9):633–653
Vargas MV, Dunlap LE, Dong C et al (2023) Psychedelics promote neuroplasticity through the activation of intracellular 5-HT2A receptors. Science 379(6633):700–706
Vialou V, Feng J, Robison AJ et al (2013) Epigenetic mechanisms of depression and antidepressant action. Annu Rev Pharmacol Toxicol 53:59–87
Vigneault E, Poirel O, Riad M et al (2015) Distribution of vesicular glutamate transporters in the human brain. Front Neuroanat 9:23
Vinkers CH, Ruhe HG, Penninx BW (2021) Antidepressant discontinuation: in need of scientific evidence. J Clin Psychopharmacol 41(5):512–515
Vollenweider FX, Vollenweider-Scherpenhuyzen MF, Babler A et al (1998) Psilocybin induces schizophrenia-like psychosis in humans via a serotonin-2 agonist action. Neuroreport 9(17): 3897–3902
Warlick HT, Leon L, Patel R et al (2023) Application of gabapentinoids and novel compounds for the treatment of benzodiazepine dependence: the glutamatergic model. Mol Biol Rep 50(2): 1765–1784
Warner-Schmidt JL, Duman RS (2008) VEGF as a potential target for therapeutic intervention in depression. Curr Opin Pharmacol 8(1):14–19
Wernicke JF, Adler L, Spencer T et al (2004) Changes in symptoms and adverse events after discontinuation of atomoxetine in children and adults with attention deficit/hyperactivity disorder: a prospective, placebo-controlled assessment. J Clin Psychopharmacol 24(1):30–35
Williams NR, Heifets BD, Blasey C et al (2018) Attenuation of antidepressant effects of ketamine by opioid receptor antagonism. Am J Psychiatry 175(12):1205–1215
Yonkers KA, Kornstein SG, Gueorguieva R et al (2015) Symptom-onset dosing of sertraline for the treatment of premenstrual dysphoric disorder: a randomized clinical trial. JAMA Psychiatry 72(10):1037–1044

Pramipexole for the Treatment of Depression: Efficacy and Mechanisms

Don Chamith Halahakoon and Michael Browning

Contents

Abstract Dopaminergic mechanisms are a plausible treatment target for patients with clinical depression but are relatively underexplored in conventional antidepressant medications. There is continuing interest in the potential antidepressant effects of the dopamine receptor agonist, pramipexole, with data from both case series and controlled trials indicating that this agent may produce benefit for patients with difficult-to-treat depression. Pramipexole's therapeutic utility in depression is likely to be expressed through alterations in reward mechanisms which are strongly influenced by dopamine pathways and are known to function abnormally in depressed patients. Our work in healthy participants using brain imaging in conjunction with computational modelling suggests that repeated pramipexole facilitates reward learning by inhibiting value decay. This mechanism needs to be confirmed in larger clinical trials in depressed patients. Such studies will also

D. C. Halahakoon and M. Browning (✉)
Department of Psychiatry, Warneford Hospital, Oxford, UK

Oxford Health NHS Foundation Trust, Oxford, UK
e-mail: michael.browning@psych.ox.ac.uk

Curr Topics Behav Neurosci (2024) 66: 49–66
https://doi.org/10.1007/7854_2023_458
Published Online: 21 November 2023

allow assessment of whether baseline performance in reward learning in depression predicts therapeutic response to pramipexole treatment.

Keywords Antidepressant · Computation · D2/D3 receptors · Depression · Dopamine · Pramipexole · Reward learning

1 Dopaminergic Treatments in Depression

Dopamine pathways have long been implicated in motivation and reward learning, two processes reliably impaired in patients with clinical depression. Despite this, conventional antidepressants generally do not have direct effects on dopamine neurotransmission, the exception being the noradrenaline and dopamine reuptake inhibitor, bupropion. Thus, at clinically effective doses, bupropion occupies about 25% of striatal dopamine transporters, an effect regarded as important in its therapeutic mode of action (Stahl et al. 2004).

Non-selective monoamine oxidase inhibitors (MAOIs) such as tranylcypromine increase synaptic levels of dopamine by inhibition of its metabolism. However, the MAOI, selegiline, used in the treatment of Parkinson's disease, is not thought to have antidepressant activity at doses that selectively block MAO-B and increase dopamine availability (Suchting et al. 2021).

There is a small literature on the possible antidepressant effect of the dopamine receptor agonist, bromocriptine, which was shown to be equivalent in therapeutic effect to tricyclic antidepressants in some early trials (see for example Waehrens and Gerlach 1981). More recently, however, interest in this area has focused on pramipexole, a dopamine D2/3/4 agonist licensed for the treatment of Parkinson's disease and at low dose for the treatment of restless legs.

2 Pharmacology of Pramipexole

Pramipexole, an aminobenzothiazole compound, was originally developed as agonist of presynaptic dopamine receptors in a search for an antipsychotic drug that would lower dopamine synthesis and release. Indeed, pramipexole has a high affinity for dopamine D3 receptors (around 0.5 nM) and also significant affinities for D2 and D4 receptors (around 2.0 and 2.75 nM respectively) (Ferraiolo and Hermans 2023). D3 receptors are believed to play an important role in dopamine autoreceptor activity and, consistent with this, acute modest doses of pramipexole decreased the neural response to a taste reward (McCabe et al. 2013) and impair reward learning more generally (see Halahakoon et al. 2020; Huys and Browning in press for review).

In humans, pramipexole is well absorbed reaching peak plasma levels in about 2 h. Pramipexole has a half-life of about 12 h. It is excreted in the urine, largely unchanged. The main use of pramipexole is in the treatment of Parkinson's disease either as a monotherapy or an adjunct to L-DOPA, although it is also licensed for use in restless leg syndrome. In addition, systematic reviews indicated that pramipexole has a significant antidepressant effect in patients with Parkinson's disease, independent of its ability to improve improvement in motor symptoms (Jiang et al. 2021).

The dose of pramipexole may be expressed as either that of the base drug, or of the salt, in this chapter doses are expressed as salt.

3 Pramipexole, an Emerging Treatment for Major Depression?

3.1 Efficacy in Randomised Trials

Three randomised placebo-controlled trials have assessed the efficacy of pramipexole in major depression (MDD). Cusin et al. (2013) employed a flexible dosing regime to a maximum of 1.5 mg pramipexole daily as an adjunctive treatment in 60 patients with treatment-resistant depression (TRD). Over 8 weeks, there was a 40% response rate in patients assigned to pramipexole compared to 27% in those receiving placebo. Remission rates were 33% in the pramipexole group compared to 23% with placebo. However, neither of these differences were statistically significant.

In a further study, MDD participants with previous non-response to at least one antidepressant were randomised to a 6-week course of escitalopram, pramipexole 2.25 mg or a combination of the two (Franco-Chaves et al. 2013). Neither the pramipexole or combination groups achieved statistically higher remission or response rates than the escitalopram monotherapy group. However, this study was small with 13 participants in each group, and only 4 of the 'combination' group completed the study.

In a third MDD study, 174 participants were assigned to 8 weeks of placebo, fluoxetine or three different doses of pramipexole (0.375 mg, 1 mg or 5 mg per day). By 8 weeks, both the fluoxetine and pramipexole 1 mg groups had experienced significantly greater symptom reduction than the placebo group. Though participants assigned to pramipexole 5 mg reported the largest numerical decreases in symptom scores, low completion rates (due to intolerability) precluded statistical comparison to other study arms (Corrigan et al. 2000).

Two RCTs assessed the effect of pramipexole in bipolar depression (Goldberg et al. 2004; Zarate et al. 2004). In one study, 22 treatment-resistant bipolar depressed individuals taking mood stabiliser medication were assigned to either flexibly dosed pramipexole (with a target range of 1.0–2.5 mg/day) or placebo. Response rates among study completers (70%) taking pramipexole were numerically, though not statistically ($p = 0.06$), higher in the pramipexole than in the placebo group (17%) (Goldberg et al. 2004). In the other study, 21 bipolar II depressed individuals taking a mood stabiliser were assigned to either flexibly dosed pramipexole (with a target range of 1-3 mg/day) or placebo. This yielded a significantly higher response rate in the pramipexole group (60%) than in the placebo group (9%).

3.2 *Efficacy in Case Series*

Observational case series tend to reveal a more clinically significant antidepressant effect of pramipexole than the RCTs (see for example Lattanzi et al. 2002). While such studies risk overstating the efficacy of pramipexole for lack of a moderating counterfactual, it is notable that they generally used higher doses of pramipexole.

The most compelling observational study, a case series by Fawcett et al. (2016), suggested a substantial benefit of pramipexole augmentation in 42 patients with treatment-refractory MDD or bipolar depression. Pramipexole doses ranged from 0.25 to 5 mg with a mean of 2.5 mg/day. Of the 42 patients, 20 remitted, 12 responded but did not remit, two did not respond and eight failed to tolerate pramipexole medication. Participants in this study generally continued to benefit from pramipexole over longer-term follow-up (average of 15 months). These response and remission rates were the more impressive given that the patient group had all failed at least four treatment steps and eight had not benefitted from electroconvulsive therapy.

A systematic review and meta-analysis that included both observational studies and clinical trials concluded that the antidepressant effect of pramipexole is greater than that of a placebo and comparable to that of an SSRI (Tundo et al. 2019). However, further results from large-scale randomised trials are needed before it can be concluded that pramipexole represents a useful advance in the treatment of MDD, given the somewhat mixed results from previous trials. However, the overall picture is of an effective antidepressant with unconventional pharmacodynamics (for an antidepressant) and particular promise in treatment-resistant patients.

To understand the neuropsychological effects of pramipexole treatment that might be relevant to its antidepressant action, it is necessary first to consider the role of dopamine, and its dedicated neural circuitry in reward learning.

4 The Neural Architecture of Reward Learning

4.1 *Neural Circuitry*

Pramipexole is a full agonist at dopamine D2-like, primarily D3, receptors (Gerlach et al. 2003; Piercey 1998), which are found largely in the striatum (Gurevich and Joyce 1999; Hall et al. 1996). Comprised of the nucleus accumbens, olfactory tubercle, caudate and putamen (Haber and Knutson 2010), the striatum is central to anticipatory hedonic experience and motivation. It is also centrally involved in reward learning, which is the process by which an organism uses feedback to modify behaviour over time (Husain and Roiser 2018; Sutton and Barto 2018). The neuro-anatomical confluence between reward learning and anticipatory hedonia/motivation likely reflects a functional interdependence (Höflich et al. 2019). The study of reward learning may therefore shed light on motivation-related depressive symptoms such as reduced interest and activity (Halahakoon et al. 2020; Husain and Roiser 2018).

From the phylogenetically ancient mesencephalon, large groups of dopamine secreting cells, comprising the ventral tegmental area (VTA) and substantia nigra (SN) (Yamamoto and Vernier 2011), project to a seemingly disparate set of brain regions including, centrally, the striatum (Haber and Knutson 2010), which also receives long-range (non-dopaminergic) projections from frontal cortical regions such as the orbitofrontal cortex (OFC), ventromedial prefrontal cortex (vmPFC) and anterior cingulate cortex (ACC). The striatum projects to the globus pallidus (GP) and to the VTA/SN which project, in turn, back to the frontal cortex via the thalamus, completing a feedback loop (Haber and Knutson 2010). This comprises the neural scaffolding of the brain's 'reward circuit'. Around this scaffold, reward-related experiences and behaviour are implemented by a complex tapestry of neural pathways, executing functions as disparate as eye movement, autonomic arousal and working memory. For a comprehensive account of reward-related neuroanatomy, see the review by Haber and Knutson (2010).

4.2 Computational Accounts of Reward Learning

The primary signal that drives reward learning is a dopaminergic report of the difference between expected and received reward known as the 'reward prediction error' (RPE; δ) (Schultz et al. 1997), which can be expressed as:

$$\delta_t = R_{t(a)} - Q_{t(a)}$$

Here, $Q_{t(a)}$ is the expected value of (that is, the average reward the organism expects it will receive from) performing action a at time t. $R_{t(a)}$ is the actual reward the organism receives from performing action a at time t.

One prominent theory of basal ganglia function (Frank and Claus 1996) proposes that this process of reinforcement occurs via two parallel and complementary pathways: an activating, D1-mediated direct or 'go' pathway and an inhibitory, D2-mediated indirect or 'no go' pathway. Phasic dopamine RPE signals are argued to stimulate D1-like receptors, which excite the direct pathway and reinforce the operations that preceded reward. Simultaneously they stimulate D2-like receptors, which inhibit the indirect pathway and again, effectively reinforces the preceding action (by electrochemically inhibiting this behaviourally inhibitory pathway) (Frank and Claus 1996). Tonic dopamine dips do the opposite, facilitating no-go learning (Frank and Claus 1996). While the effect of dopaminergic RPE activity on learning can be modelled by a number of reinforcement learning algorithms of varying biophysical plausibility (Glimcher 2011), the simplest is known as the Rescorla-Wagner model:

$$Q_{t+1(a)} = Q_{t(a)} + \eta(\delta_t)$$

Here, as before, $Q_{t(a)}$ is the expected value of performing action a at time t. The RPE (δ) modifies the value of performing action a over time at a rate determined by the learning rate parameter (η). The larger the learning rate, the faster a given RPE alters the value of performing the action.

The expected value of performing action a at timepoint t, ($Q_{t(a)}$) influences the probability that the organism will carry out action a at timepoint t, i.e. $P_{t(a)}$. Specifically, the probability of performing action a at timepoint t, ($P_{t(a)}$) depends on the expected value carrying out that action vs the summed expected value of carrying out all available actions at timepoint t. This process can be emulated using a 'softmax' equation:

$$P_{t(a)} = \frac{\exp\left(Q_{t(a)}\right)}{\sum_{j=1}^{K} \exp\left(Q_{t(j)}\right)}$$

where the numerator is determined by the expected value of action a and the denominator is determined by the summed expected values of all available actions. Models such as these serve as algorithmic intermediaries that translate neuronal activity into behavioural output (Marr 2010) and so allow us to test proposed links between the two by omparing internal model metrics such as the RPE (δ_t), to participants' brain activity while fitting model outputs, such as the probability of performing a certain action at a certain time ($P_{t(a)}$), to participants behaviour.

5 Reward Learning in Depression

The preceding section outlined a proposed mechanism by which the reward circuit helps an organism to learn (or at least estimate) underlying reward contingencies in order to make advantageous reward-related choices. Recently, elaborated versions of the outlined model have linked, not only brain activity and behaviour but also affect (Blain and Rutledge 2022; Chew et al. 2021; Keren et al. 2022; Rutledge et al. 2014). These studies demonstrate a consistent correspondence between striatal activity, model-derived RPEs and reported subjective affect (Blain and Rutledge 2022; Chew et al. 2021; Keren et al. 2022; Rutledge et al. 2014). If reward learning underlies affect in health, might alterations in reward learning underlie affective disorders such as depression? A growing body of literature suggests that this might be so.

A number of psychological tasks have probed reward learning differences between individuals with major depressive disorder and healthy controls. One commonly used task, the 'probabilistic reward task' (PRT), rewards individuals probabilistically for correctly identifying stimuli as belonging to one of two possible categories. Correctly identifying stimuli as belonging to one category is rewarded more frequently than correctly identifying them as belonging to the other category. Information relating to potential rewards and probabilities is typically not provided explicitly. The measure of interest in this task is the 'response bias' which is derived from signal detection theory and reflects an individual's tendency to identify stimuli as belonging to the more frequently rewarded category, regardless of accuracy (Pizzagalli et al. 2008b). Healthy controls tend to develop a larger response bias (i.e. are more influenced by reward probability) than depressed individuals (Halahakoon et al. 2020; Huys and Browning in press).

6 Pramipexole and Reward Processing

Given its promising (if mixed) clinical effects, one might expect pramipexole to enhance reward responding in both healthy participants and depressed patients. However, previous experimental studies generally indicate that a single dose of pramipexole blunts rather than enhances participants' behavioural responses to reward (Gallant et al. 2016; Murray et al. 2019; Pizzagalli et al. 2008a; Santesso et al. 2009; McCabe et al. 2013).

A substantial portion of the behavioural evidence for reduced reward responsiveness in MDD comes from the PRT where patients with MDD develop less of a reward-based response bias than do HC (Halahakoon et al. 2020). Behaviourally single doses of pramipexole impede the development of a reward-based response bias (Pizzagalli et al. 2008a) in the PRT; this is similar to the observed effect of depression (Halahakoon et al. 2020). In a different stimulus-response task participants given pramipexole learned correct stimulus-response pairs significantly less well than those given placebo (Gallant et al. 2016). In a probabilistic instrumental learning task participants given a dose of pramipexole made fewer correct choices (numerically, but not statistically) than those that were not (Murray et al. 2019). In a gambling task pramipexole blunted ventral striatal (VS) and midbrain (including VTA) response to unexpected monetary gain (Riba et al. 2008). Finally, pramipexole also reduced striatal activity to rewarding stimuli in a simple reward processing task (McCabe et al. 2013). This behavioural and neuroimaging data suggests that the effects of depression and acute pramipexole administration on reward processing run in the same direction; however, clinically pramipexole appears to be antidepressant.

One explanation for the seeming contradiction between the clinical and experimental evidence is that the above experimental studies examined the effect of a single dose of pramipexole while sustained treatment is required to improve symptoms (Fawcett et al. 2016; Lemke et al. 2006). From a pharmacological perspective, acute treatment with D2/3/4 agonists is believed to primarily influence inhibitory presynaptic auto-receptors, leading to reduced dopaminergic transmission, whereas sustained administration leads to auto-receptor down-regulation and enhanced transmission via agonism at post-synaptic D2-like receptors (Chen et al. 2005; Chernoloz et al. 2008; Grace 1991; Willner et al. 1994). This suggests that the clinically relevant effects of pramipexole on reward learning are likely to become apparent only after sustained administration of the drug.

A few studies have examined the effect of a course of pramipexole in clinical populations. In individuals with remitted bipolar disorder (Burdick et al. 2014) and those with Parkinson's disease (Bodi et al. 2009), a sustained course of pramipexole enhanced reward sensitivity. However, in those with MDD, a sustained course of pramipexole appeared to have no effect on reward sensitivity (Whitton et al. 2020), although no placebo control group was used in this study. Interestingly, in the same study a greater clinical antidepressant response to pramipexole was experienced by those that had larger baseline behavioural and VS RPE responses to rewards (Whitton et al. 2020). This is antithetical to the idea that a sustained course of pramipexole exerts an antidepressant effect by remediating blunted reward sensitivity. Indeed, it suggests that relatively intact reward system might be necessary for pramipexole to produce an antidepressant effect in depressed patients.

7 Pramipexole Preserves Value Estimates

In order to probe the potential neuropsychological mechanisms of pramipexole's antidepressant effect we tested the impact of a 14-day course of pramipexole on reward learning in a group of 40 healthy volunteers in a placebo-controlled design (Halahakoon et al. 2023). Pramipexole was titrated to 1 mg/day over 10 days and this dose maintained until testing was completed (i.e. this dosing regime was selected to produce autoreceptor down-regulation), followed by gradual down-titration and discontinuation. The apparent dose of the placebo was altered in the same manner.

Participants performed a modified version of the probabilistic instrumental learning task (PILT; a reward learning task) (Pessiglione et al. 2006) before and then twice between days 12 and 15 of the course of pramipexole (once with fMRI data collection, once behavioural). In the PILT, the participant is presented with a number of 'trials' in which they must choose between two stimuli, which have reciprocal probabilities (0.7 vs 0.3) of a 'win' outcome (winning money) vs a 'no win' outcome (not winning money) in reward condition trials, or a 'loss' outcome (losing money) vs a 'no loss' outcome (not losing money) in loss condition trials. After making their choice, participants receive feedback on the trial outcome ('win', 'no win', 'loss' or 'no-loss').

In this task, a choice is considered 'accurate' if the participant chooses the 'advantageous' stimulus (i.e. the stimulus associated with the higher probability of a 'win' outcome in the reward condition, or 'no-loss' outcome in the loss condition). As the symbols used in the task are arbitrary, participants are expected to start the task with a choice accuracy of 0.5 (i.e. choosing stimuli indiscriminately), and make

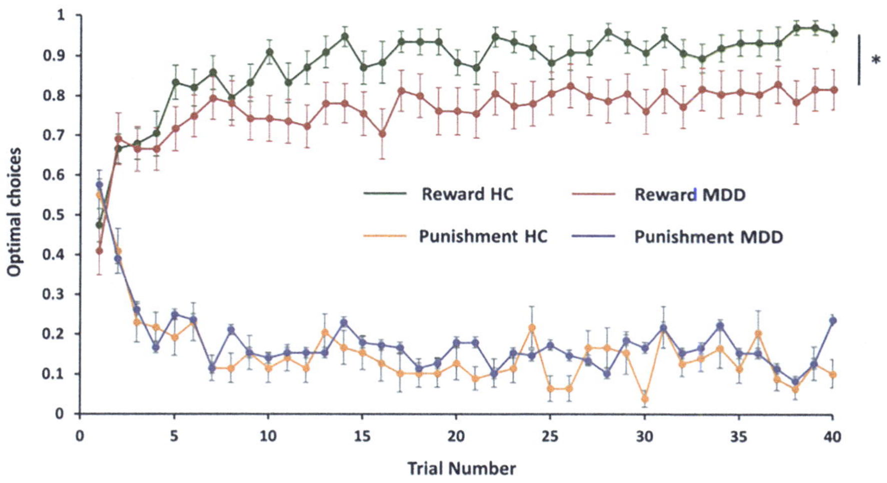

Fig. 1 Reward learning accuracy in depression. This figure depicts the averaged proportion of participants choosing accurately in each reward trial ('accurate' meaning choosing the stimulus associated with the higher probability of winning money), or inaccurately in each punishment trial ('inaccurate' meaning choosing the stimulus associated with the lower probability of winning money). Error bars represent standard error. As can be seen, depressed individuals achieve a lower accuracy asymptote in reward trials (red curve) than healthy controls (green curve), while there is little to distinguish the two groups in punishment trials (purple and yellow curves). Reproduced from Kumar et al. (2018) with permission

more accurate choices over time as they receive increasing amounts of feedback. In practice, participants tend to achieve a certain level of accuracy fairly early in the task and then plateau at that level of accuracy for the rest of the task. As reviewed above, individuals with MDD have been shown to achieve a lower choice accuracy plateau than non-depressed controls (Kumar et al. 2018; Walsh et al. 2018), specifically in the reward condition, as depicted in Fig. 1.

Since depressed individuals have a lower reward accuracy plateau, one might therefore expect 2 weeks of an antidepressant drug like pramipexole to have the opposite effect (i.e. increase individuals' reward accuracy plateau). This is, in fact, what was observed (see Fig. 2) and is in contrast to the majority of previous experimental studies of acute pramipexole which have found that it impairs reward learning (Gallant et al. 2016; Murray et al. 2019; Pizzagalli et al. 2008a; Santesso et al. 2009).

Overall, therefore, the impact of pramipexole on learning is opposite to that associated with depression and is consistent with the antidepressant effects of the drug (Cusin et al. 2013; Fawcett et al. 2016; Tundo et al. 2019).

How might pramipexole exert this effect? In order to answer this question, we turn back to the equations (outlined earlier in the chapter) that describe how organisms update their beliefs about environmental contingencies (the Rescorla-Wagner equation) and how these beliefs influence behaviour (the softmax equation). Using modified versions of these equations, we can formalise the question of how

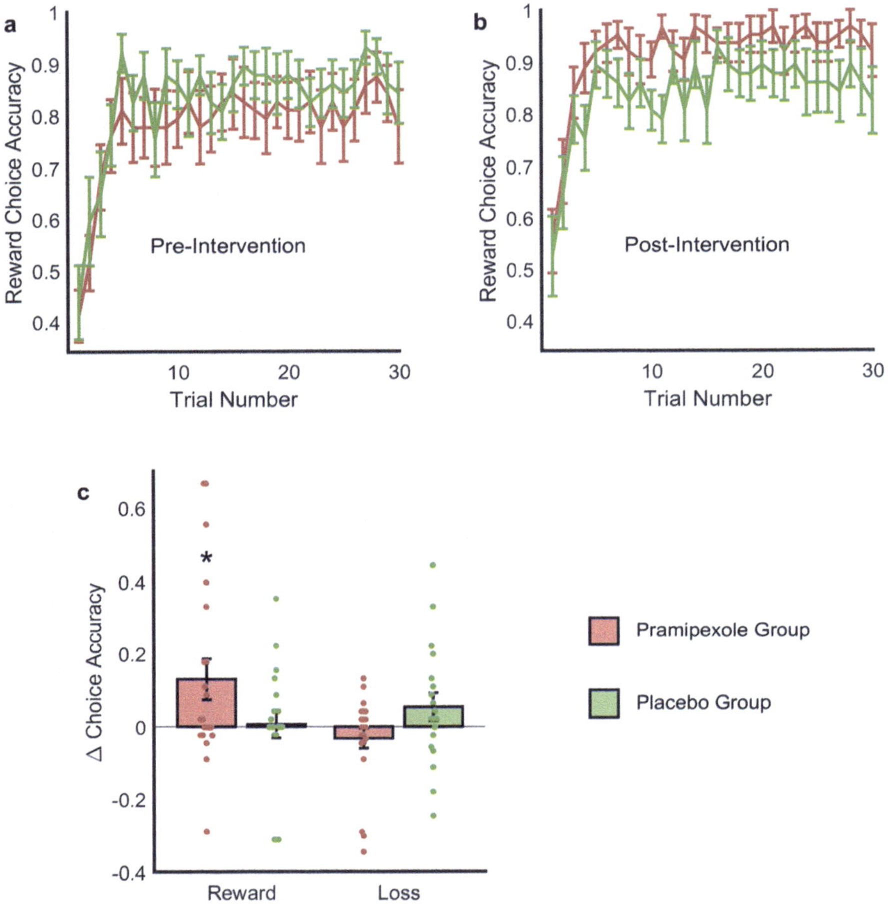

Fig. 2 Reward learning accuracy before and during pramipexole or placebo treatment. (**a**, **b**) Learning curves depicting reward choice accuracy (**a**) before and (**b**) during (i.e. day 12–15) treatment. Green (pink) curves represent the placebo (pramipexole) groups. Error bars represent SEM. The curves represent the mean proportion of advantageous shapes (i.e. the shape associated with a 70% probability of receiving a 'win' outcome) across participants in a given trial. As can be seen the groups do not differ before treatment, whereas during treatment reward learning is increased in the pramipexole group. (**c**) Mean (SEM) of summary accuracy scores showing pre-vs-post intervention change in reward/loss conditions, illustrating a specific effect of pramipexole on rewarded relative to loss trials. Reproduced from (Halahakoon et al. 2023) with permission

pramipexole might affect behaviour in this task. Previous studies have identified three distinct alterations of learning and decision-making that may do so: pramipexole may cause participants to make more deterministic choices (Huys et al. 2013), it may cause participants to treat each reward 'as if' it were of increased value (Huys et al. 2013), or it may reduce the decay of learned value estimates over time (Collins and Frank 2012).

Using the Rescorla-Wagner learning rule introduced earlier in this chapter.

$$Q_{t+1(a)} = Q_{t(a)} + \eta_j(\delta_t)$$

To model the effect of pramipexole, an increase in the perceived value of rewards, may be emulated using a reward sensitivity parameter ρ,

$$\delta_t = \rho_j R_{t(a)} - Q_{t(a)}$$

Here, $Q_{t(s)}$ is again the expectation about the value of shape s on trial t, R_t is the observed outcome (1 for a positive outcome, 0 for a negative outcome), η_j is the learning rate used for condition j (i.e. win or loss condition) and ρ_j is a reward sensitivity parameter for condition j. The reward sensitivity parameter ρ scales the effective size of the rewards – if this parameter is greater than 1, the effective reward increases, if it is less than one then effective reward decreases.

To model the effect of pramipexole reducing the decay of learned value estimates over time, the Rescorla-Wagner model may be further modified using a decay factor ϕ which controls the rate at which the model's expectations drift back towards their initial value. The smaller the decay factor ϕ, the slower the learned values decay over time (Collins and Frank 2012):

$$Q \leftarrow Q + \phi(Q_0 - Q)$$

As described above, to produce choices from the value estimates (Q), a softmax equation can be used. When only two choice options are available, this equation can be simplified to:

$$P_{t(a)} = \frac{1}{1 + \exp^{-\beta\left(Q_{t(a)} - Q_{t(b)}\right)}}.$$

Here an inverse temperature parameter, β, is included. This parameter controls the degree to which the probability of the participant choosing shape a, $P_{t(a)}$, is determined by the difference in Q values and models possible changes in decision determinancy caused by pramixeole. That is, as the value of β increases, the more consistently the model will choose the higher value shape.

Changes to the three parameters ρ, ϕ and β can produce very similar effects on asymptotic choice (as illustrated in Fig. 3).

If each of the three distinct mechanisms can produce the same observed behaviour, how do we arbitrate between them to find the mechanism by which pramipexole increases choice accuracy? To do this we turn back to reward prediction errors (RPEs). In Fig. 3 we saw that increasing reward sensitivity (ρ), increasing inverse temperature (β) or decreasing decay (ϕ) produce equivalent changes in asymptotic choice. However, they produce this effect via distinct changes in RPEs (Fig. 4).

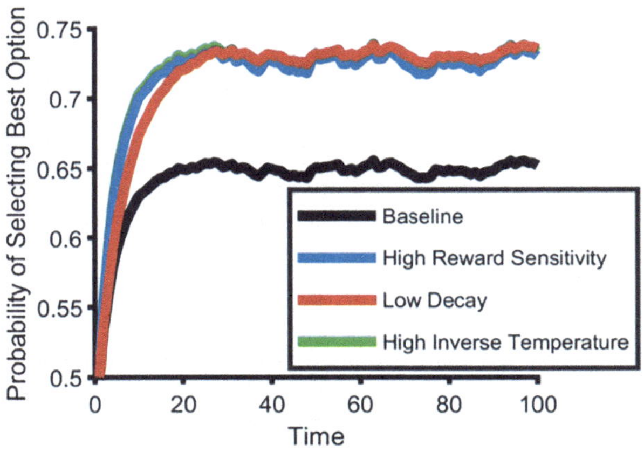

Fig. 3 Model-derived causes of increased reward learning accuracy. Simulated effect of a high reward sensitivity, high inverse temperature or a low decay produce effectively indistinguishable learning curves. The learning curves displayed here were generated by the modified Rescorla-Wagner and Softmax rules described above using a ρ of 0.6, a ϕ 0.12, a β of 10 and an α of 0.1. Increases in either ρ (blue line) or β (green line) and decreases in ϕ (red line) produce equivalent changes in asymptotic choice. Note the green line is not obvious as it is behind the blue line. In other words, three qualitatively distinct processes lead to the same behavioural effect. Reproduced from (Halahakoon et al. 2023) with permission

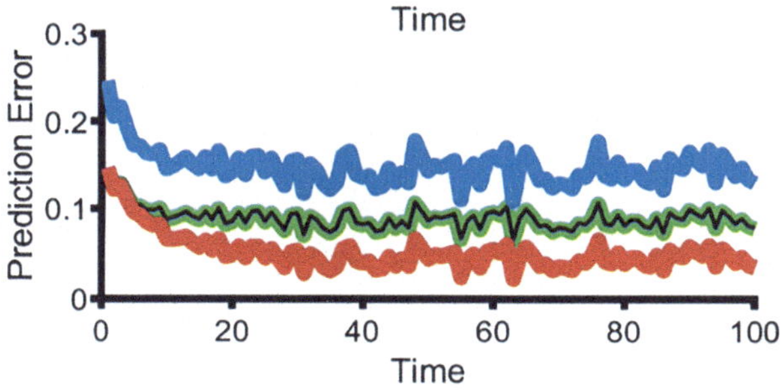

Fig. 4 Model-derived prediction errors. Prediction errors (PE) produced by each of the three models described in Fig. 3 (the colour of the lines used in this figure is the same as those in Fig. 3). While the three models produce effectively indistinguishable behaviour, they produce quite different PEs. Changes in β have no effect on PEs, whereas an increase in ρ leads to increased PEs and a reduction in ϕ leads to decreased PEs. Reproduced from (Halahakoon et al. 2023) with permission

Neuronal activity in a network of brain regions has been found to vary with RPEs (Fouragnan et al. 2018). This activity can be approximated by the blood-oxygen-level-dependent (BOLD) signal, as measured using functional magnetic resonance imaging (fMRI). We contrasted BOLD activity during receipt of rewards (during which participants experience positive RPEs) against BOLD activity during receipt of non-reward (during which participants experience negative RPEs). This contrast

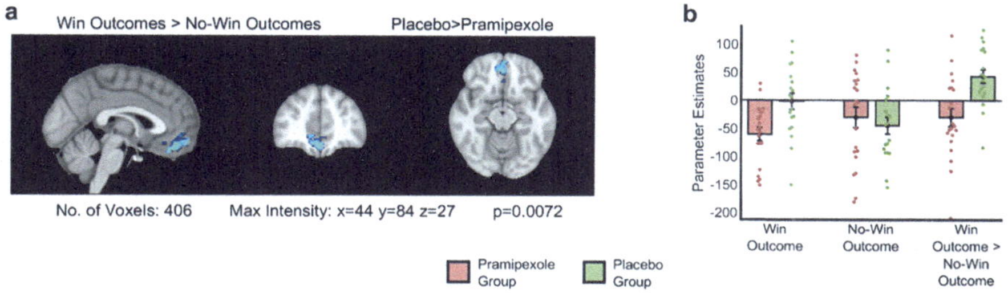

Fig. 5 The effect of pramipexole on RPE-related BOLD signal: (**a**) The (blue) coloured area represents the cluster of significantly decreased activity in the pramipexole vs placebo group in the mPFC associated with win-outcomes > no-win-outcomes. Areas of significantly decreased activity are threshold-free cluster enhancement corrected with a family-wise error cluster significance level of $p \leq 0.05$. (**b**) Parameter estimates extracted from the area of significantly decreased activity in panel a associated with win-outcomes, no-win-outcomes, and win-outcomes > no-win-outcomes. Green bars represent the placebo) group and pink, the pramipexole. Error bars represent SEM. Scatter plots overlaying bar graphs depict corresponding individual parameter estimates. Reproduced from (Halahakoon et al. 2023) with permission

provides a proxy for the magnitude of RPEs. Figure 3 summarises how the predicted effects of pramipexole on this contrast would vary if it influenced reward learning via each of the three proposed mechanisms: If pramipexole increases reward sensitivity (ρ), RPE activity should be increased in the pramipexole group, if it increases the inverse temperature (β), there should be no change in RPE activity, and if it reduces value decay (ϕ), there should be a reduction in RPE activity in the pramipexole group. As illustrated in Fig. 5, relative to placebo, pramipexole reduced (mPFC) activity for this contrast, consistent with the hypothesis that pramipexole lowers the decay (ϕ) of learned value estimates.

These results indicate that pramipexole enhances reward learning by reducing the decay of learned value estimates and suggests a cognitive mechanism by which it may ameliorate the reduced reward learning characteristic of depression (Chen et al. 2015; Halahakoon et al. 2020; Husain and Roiser 2018). The reduced reward learning seen in depression (Halahakoon et al. 2020; Huys and Browning in press) has been associated with reduced BOLD response to rewarding outcomes (Huys and Browning in press) suggesting that it is caused by a reduced reward sensitivity. In other words, pramipexole seems to enhance reward learning, but it does so via a distinct mechanism to that which causes the impairment in depression.

This result may account for why pramipexole appears to be more effective in depressed patients with intact, rather than reduced, reward learning (Whitton et al. 2020). Specifically, pramipexole acts to maintain value estimates, but there will be no benefit to this effect on reward learning, and presumably on symptoms of depression, if these value estimates have been intrinsically lowered, as seems to be the case in some depressed patients with reduced reward learning.

8 Conclusions

This chapter has outlined the evidence for the antidepressant effect of pramipexole. Current data suggests that pramipexole, dosed to around 2 mg daily, can be an effective antidepressant, even in people refractory to more conventional medications. The dopamine receptor action of pramipexole is clearly relevant to its therapeutic effects, but the precise neurobiological mechanism needs further exploration.

An important next step is a larger-scale clinical trial in patients with difficult-to-treat MDD. This will enable clarification of the therapeutic efficacy of pramipexole in this patient group. Such a study could usefully combine ratings of change in symptomatology with measures of reward learning, before and during treatment. This would permit testing of the proposal suggested here that pramipexole improves symptoms of depression by facilitating reward learning. It would also enable elucidation as to whether impaired reward learning prior to pramipexole treatment is a useful predictor of subsequent therapeutic response.

References

Blain B, Rutledge RB (2022) Momentary subjective well-being depends on learning and not reward. Elife 9:e57977. https://doi.org/10.7554/eLife.57977

Bodi N, Keri S, Nagy H, Moustafa A, Myers CE, Daw N, Dibo G, Takats A, Bereczki D, Gluck MA (2009) Reward-learning and the novelty-seeking personality: a between- and within-subjects study of the effects of dopamine agonists on young Parkinson's patients. Brain 132:2385–2395. https://doi.org/10.1093/brain/awp094

Burdick KE, Braga RJ, Gopin CB, Malhotra AK (2014) Dopaminergic influences on emotional decision making in euthymic bipolar patients. Neuropsychopharmacology 39:274–282. https://doi.org/10.1038/npp.2013.177

Chen YCI, Choi JK, Andersen SL, Rosen BR, Jenkins BG (2005) Mapping dopamine D2/D3 receptor function using pharmacological magnetic resonance imaging. Psychopharmacology (Berl) 180:705–715. https://doi.org/10.1007/s00213-004-2034-0

Chen C, Takahashi T, Nakagawa S, Inoue T, Kusumi I (2015) Reinforcement learning in depression: a review of computational research. Neurosci Biobehav Rev 55:247–267. https://doi.org/10.1016/j.neubiorev.2015.05.005

Chernoloz O, El Mansari M, Blier P (2008) Sustained administration of pramipexole modifies the spontaneous firing of dopamine, norepinephrine, and serotonin neurons in the rat brain. Neuropsychopharmacology 34:651–661. https://doi.org/10.1038/npp.2008.114

Chew B, Blain B, Dolan RJ, Rutledge RB (2021) A Neurocomputational model for intrinsic reward. J Neurosci 41:8963–8971. https://doi.org/10.1523/JNEUROSCI.0858-20.2021

Collins AGE, Frank MJ (2012) How much of reinforcement learning is working memory, not reinforcement learning? A behavioral, computational, and neurogenetic analysis. Eur J Neurosci 35:1024–1035. https://doi.org/10.1111/j.1460-9568.2011.07980.x

Corrigan MH, Denahan AQ, Wright CE, Ragual RJ, Evans DL (2000) Comparison of pramipexole, fluoxetine, and placebo in patients with major depression. Depress Anxiety 11:58–65. https://doi.org/10.1002/(sici)1520-6394(2000)11:2<58::aid-da2>3.0.co;2-h

Cusin N, Iovieno N, Iosifescu DV, Nierenberg AA, Fava M, Rush AJ, Perlis RH (2013) A randomized, double-blind, placebo-controlled trial of pramipexole augmentation in treatment-resistant major depressive disorder. J Clin Psychiatry 74:e636–e641. https://doi.org/10.4088/JCP.12M08093

Fawcett J, Rush AJ, Vukelich J, Diaz SH, Dunklee L, Romo P, Yarns BC, Escalona R (2016) Clinical experience with high-dosage pramipexole in patients with treatment-resistant depressive episodes in unipolar and bipolar depression. Am J Psychiatry 173:107–111. https://doi.org/10.1176/appi.ajp.2015.15060788

Ferraiolo M, Hermans E (2023) The complex molecular pharmacology of the dopamine D2 receptor: implications for pramipexole, ropinirole, and rotigotine. Pharmacol Ther 245: 108392. https://doi.org/10.1016/j.pharmthera.2023.108392

Fouragnan E, Retzler C, Philiastides MG (2018) Separate neural representations of prediction error valence and surprise: evidence from an fMRI meta-analysis. Hum Brain Mapp 39:2887–2906. https://doi.org/10.1002/hbm.24047

Franco-Chaves JA, Mateus CF, Luckenbaugh DA, Martinez PE, Mallinger AG, Zarate CA (2013) Combining a dopamine agonist and selective serotonin reuptake inhibitor for the treatment of depression: a double-blind, randomized pilot study. J Affect Disord 149:319–325. https://doi.org/10.1016/j.jad.2013.02.003

Frank MJ, Claus ED (1996) Anatomy of a decision: striato-orbitofrontal interactions in reinforcement learning, decision making and reversal. Beiser & Houk

Gallant H, Vo A, Seergobin KN, MacDonald PA (2016) Pramipexole impairs stimulus-response learning in healthy young adults. Front Neurosci 10. https://doi.org/10.3389/fnins.2016.00374

Gerlach M, Double K, Arzberger T, Leblhuber F, Tatschner T, Riederer P (2003) Dopamine receptor agonists in current clinical use: comparative dopamine receptor binding profiles defined in the human striatum. J Neural Transm 110:1119–1127. https://doi.org/10.1007/s00702-003-0027-5

Glimcher PW (2011) Understanding dopamine and reinforcement learning: the dopamine reward prediction error hypothesis. Proc Natl Acad Sci U S A 108:15647–15654. https://doi.org/10.1073/pnas.1014269108

Goldberg JF, Burdick KE, Endick CJ (2004) Preliminary randomized, double-blind, placebo-controlled trial of pramipexole added to mood stabilizers for treatment-resistant bipolar depression. Am J Psychiatry 161:564–566. https://doi.org/10.1176/appi.ajp.161.3.564

Grace AA (1991) Phasic versus tonic dopamine release and the modulation of dopamine system responsivity: a hypothesis for the etiology of schizophrenia. Neuroscience 41:1–24. https://doi.org/10.1016/0306-4522(91)90196-U

Gurevich EV, Joyce JN (1999) Distribution of dopamine D3 receptor expressing neurons in the human forebrain: comparison with D2 receptor expressing neurons. Neuropsychopharmacology 20:60–80. https://doi.org/10.1016/S0893-133X(98)00066-9

Haber SN, Knutson B (2010) The reward circuit: linking primate anatomy and human imaging. Neuropsychopharmacology. https://doi.org/10.1038/npp.2009.129

Halahakoon DC, Kieslich K, O'Driscoll C, Nair A, Lewis G, Roiser JP (2020) Reward-processing behavior in depressed participants relative to healthy volunteers: a systematic review and meta-analysis. JAMA Psychiatry. https://doi.org/10.1001/jamapsychiatry.2020.2139

Halahakoon DC, Kaltenboeck A, Martens M, Geddes JG, Harmer CJ, Cowen P, Browning M (2023) Pramipexole enhances reward learning by preserving value estimates. Biol Psychiatry. S0006-3223(23)01356-2. https://doi.org/10.1016/j.biopsych.2023.05.023

Hall H, Halldin C, Dijkstra D, Wikström H, Wise LD, Pugsley TA, Sokoloff P, Pauli S, Farde L, Sedvall G (1996) Autoradiographic localisation of D3-dopamine receptors in the human brain using the selective D3-dopamine receptor agonist (+)-[3H]PD 128907. Psychopharmacology (Berl) 128:240–247. https://doi.org/10.1007/s002130050131

Höflich A, Michenthaler P, Kasper S, Lanzenberger R (2019) Circuit mechanisms of reward, anhedonia, and depression. Int J Neuropsychopharmacol 22:105–118. https://doi.org/10.1093/ijnp/pyy081

Husain M, Roiser JP (2018) Neuroscience of apathy and anhedonia: a transdiagnostic approach. Nat Rev Neurosci 19:470–484. https://doi.org/10.1038/s41583-018-0029-9

Huys QJM, Browning M (in press) A computational view on the nature of reward and value in anhedonia, in: anhedonia: preclinical, translational, and clinical integration. Curr Top Behav Neurosci

Huys QJ, Pizzagalli DA, Bogdan R, Dayan P (2013) Mapping anhedonia onto reinforcement learning: a behavioural meta-analysis. Biol Mood Anxiety Disord 3:12. https://doi.org/10.1186/2045-5380-3-12

Jiang D-Q, Jiang L-L, Wang Y, Li M-X (2021) The role of pramipexole in the treatment of patients with depression and Parkinson's disease: a meta-analysis of randomized controlled trials. Asian J Psychiatr 61:102691. https://doi.org/10.1016/j.ajp.2021.102691

Keren H, Zheng C, Jangraw DC, Chang K, Vitale A, Rutledge RB, Pereira F, Nielson DM, Stringaris A (2022) The temporal representation of experience in subjective mood. Elife 10: e62051. https://doi.org/10.7554/eLife.62051

Kumar P, Goer F, Murray L, Dillon DG, Beltzer ML, Cohen AL, Brooks NH, Pizzagalli DA (2018) Impaired reward prediction error encoding and striatal-midbrain connectivity in depression. Neuropsychopharmacology 43:1581–1588. https://doi.org/10.1038/s41386-018-0032-x

Lattanzi L, Dell'Osso L, Cassano P, Pini S, Rucci P, Houck PR, Gemignani A, Battistini G, Bassi A, Abelli M, Cassano GB (2002) Pramipexole in treatment-resistant depression: a 16-week naturalistic study. Bipolar Disord 4:307–314. https://doi.org/10.1034/j.1399-5618.2002.01171.x

Lemke MR, Brecht HM, Koester J, Reichmann H (2006) Effects of the dopamine agonist pramipexole on depression, anhedonia and motor functioning in Parkinson's disease. J Neurol Sci 248:266–270. https://doi.org/10.1016/j.jns.2006.05.024

Marr D (2010) Vision: a computational investigation into the human representation and processing of visual information. The MIT Press. https://doi.org/10.7551/mitpress/9780262514620.001.0001

McCabe C, Harwood J, Brouwer S, Harmer CJ, Cowen PJ (2013) Effects of pramipexole on the processing of rewarding and aversive taste stimuli. Psychopharmacology (Berl) 228:283–290. https://doi.org/10.1007/s00213-013-3033-9

Murray GK, Knolle F, Ersche KD, Craig KJ, Abbott S, Shabbir SS, Fineberg NA, Suckling J, Sahakian BJ, Bullmore ET, Robbins TW (2019) Dopaminergic drug treatment remediates exaggerated cingulate prediction error responses in obsessive-compulsive disorder. Psychopharmacology (Berl) 236:2325–2336. https://doi.org/10.1007/s00213-019-05292-2

Pessiglione M, Seymour B, Flandin G, Dolan RJ, Frith CD (2006) Dopamine-dependent prediction errors underpin reward-seeking behaviour in humans. Nature 442:1042–1045. https://doi.org/10.1038/nature05051

Piercey MF (1998) Pharmacology of pramipexole, a dopamine D3-preferring agonist useful in treating Parkinson's disease. Clin Neuropharmacol 21:141–151

Pizzagalli E, Eden Evins A, Schetter EC, Frank MJ, Pajtas PE, Santesso DL, Culhane M (2008a) Single dose of a dopamine agonist impairs reinforcement learning in humans: behavioral evidence from a laboratory-based measure of reward responsiveness. Psychopharmacology (Berl) 196:221–232. https://doi.org/10.1007/s00213-007-0957-y

Pizzagalli DA, Iosifescu D, Hallett LA, Ratner KG, Fava M (2008b) Reduced hedonic capacity in major depressive disorder: evidence from a probabilistic reward task. J Psychiatr Res 43:76–87. https://doi.org/10.1016/j.jpsychires.2008.03.001

Riba J, Kramer UM, Heldmann M, Richter S, Munte TF (2008) Dopamine agonist increases risk taking but blunts rewardrelated brain activity. PLoS One 3:e2479

Rutledge RB, Skandali N, Dayan P, Dolan RJ (2014) A computational and neural model of momentary subjective well-being. Proc Natl Acad Sci 111:12252–12257. https://doi.org/10.1073/pnas.1407535111

Santesso DL, Evins AE, Frank MJ, Schetter EC, Bogdan R, Pizzagalli DA (2009) Single dose of a dopamine agonist impairs reinforcement learning in humans: evidence from event-related potentials and computational modeling of striatal-cortical function. Hum Brain Mapp 30: 1963. https://doi.org/10.1002/HBM.20642

Schultz W, Dayan P, Montague PR (1997) A neural substrate of prediction and reward. Science 275:1593–1599. https://doi.org/10.1126/science.275.5306.1593

Stahl SM, Pradko JF, Haight BR, Modell JG, Rockett CB, Learned-Coughlin S (2004) A review of the neuropharmacology of bupropion, a dual norepinephrine and dopamine reuptake inhibitor. Prim Care Companion J Clin Psychiatry 6:159–166. https://doi.org/10.4088/pcc.v06n0403

Suchting R, Tirumalajaru V, Gareeb R, Bockmann T, de Dios C, Aickareth J, Pinjari O, Soares JC, Cowen PJ, Selvaraj S (2021) Revisiting monoamine oxidase inhibitors for the treatment of depressive disorders: a systematic review and network meta-analysis. J Affect Disord 282: 1153–1160. https://doi.org/10.1016/j.jad.2021.01.021

Sutton RS, Barto AG (2018) Reinforcement learning: an introdcution, 2nd edn. MIT Press, Cambridge

Tundo A, de Filippis R, Crescenzo FD (2019) Pramipexole in the treatment of unipolar and bipolar depression. A systematic review and meta-analysis. Acta Psychiatr Scand 140:116–125. https://doi.org/10.1111/ACPS.13055

Waehrens J, Gerlach J (1981) Bromocriptine and imipramine in endogenous depression. A double-blind controlled trial in out-patients. J Affect Disord 3:193–202. https://doi.org/10.1016/0165-0327(81)90044-6

Walsh AEL, Browning M, Drevets WC, Furey M, Harmer CJ (2018) Dissociable temporal effects of bupropion on behavioural measures of emotional and reward processing in depression. Philos Trans R Soc B Biol Sci 373. https://doi.org/10.1098/RSTB.2017.0030

Whitton AE, Reinen JM, Slifstein M, Ang Y-S, McGrath PJ, Iosifescu DV, Abi-Dargham A, Pizzagalli DA, Schneier FR (2020) Baseline reward processing and ventrostriatal dopamine function are associated with pramipexole response in depression. Brain 143:701–710. https://doi.org/10.1093/BRAIN/AWAA002

Willner P, Lappas S, Cheeta S, Muscat R (1994) Reversal of stress-induced anhedonia by the dopamine receptor agonist, pramipexole. Psychopharmacology (Berl) 115:454–462. https://doi.org/10.1007/bf02245568

Yamamoto K, Vernier P (2011) The evolution of dopamine systems in chordates. Front Neuroanat 5

Zarate CA, Payne JL, Singh J, Quiroz JA, Luckenbaugh DA, Denicoff KD, Charney DS, Manji HK (2004) Pramipexole for bipolar II depression: a placebo-controlled proof of concept study. Biol Psychiatry 56:54–60. https://doi.org/10.1016/j.biopsych.2004.03.013

Opioid Mechanisms and the Treatment of Depression

Luke A. Jelen, Allan H. Young, and Mitul A. Mehta

Contents

Abstract Opioid receptors are widely expressed in the brain, and the opioid system has a key role in modulating mood, reward processing and stress responsivity. There is mounting evidence that the endogenous opioid system may be dysregulated in depression and that drug treatments targeting mu, delta and kappa opioid receptors may show antidepressant potential. The mechanisms underlying the therapeutic effects of opioid system engagement are complex and likely multi-factorial. This chapter explores various pathways through which the modulation of the opioid system may influence depression. These include impacts on monoaminergic systems, the regulation of stress and the hypothalamic-pituitary-adrenal axis, the

L. A. Jelen (✉) and A. H. Young
Institute of Psychiatry, Psychology and Neuroscience, King's College London, London, UK

South London and Maudsley NHS Foundation Trust, London, UK
e-mail: luke.jelen@kcl.ac.uk

M. A. Mehta
Institute of Psychiatry, Psychology and Neuroscience, King's College London, London, UK

Curr Topics Behav Neurosci (2024) 66: 67–100
https://doi.org/10.1007/7854_2023_448
Published Online: 4 November 2023

immune system and inflammation, brain-derived neurotrophic factors, neurogenesis and neuroplasticity, social pain and social reward, as well as expectancy and placebo effects. A greater understanding of the diverse mechanisms through which opioid system modulation may improve depressive symptoms could ultimately aid in the development of safe and effective alternative treatments for individuals with difficult-to-treat depression.

Keywords Antidepressants · Delta opioid receptor (DOR) · Depression · Kappa opioid receptor (KOR) · Mu opioid receptor (MOR) · Opioid system

1 Introduction

Opium is an extract of the exudate derived from seedpods of the opium poppy, *Papaver somniferum*, and contains high concentrations of the prototypical opiate, morphine. For centuries, opioids have been used medicinally to treat pain and various other physical ailments (Rosenblum et al. 2008). Historically, opioids have also been utilised in the treatment of a range of psychiatric disorders (Weber and Emrich 1988). Since the time of Emil Kraepelin, the so-called opium cure was recommended in the treatment of depression, employing slowly increasing and later decreasing dosages of opium tincture, and was praised as a rapidly effective treatment for refractory melancholic depression (Kraepelin 1901; Tenore 2008). However, the use of opioid drugs for depression has been limited due to their high potential for abuse and dependence. Opioids have been commonly prescribed for pain relief, but their non-medical use has resulted in a significant public health crisis, leading to widespread addiction, overdose and death in many countries, particularly in the United States. Therefore, prescribing opioids for depression has become largely obsolete due to restrictive opioid prescription legislation and the availability of alternative non-addictive antidepressant options. Nevertheless, the connection between depression and the opioid system has been far from being discarded by the field (Dagher et al. 2022). Increasing lines of evidence directly implicate the endogenous opioid system in the regulation of mood and suggest that this system may be dysregulated in depressive disorders, leading to a resurgence of interest in exploring the opioid system as a target for the treatment of depression (Jelen et al. 2022; Pecina et al. 2019). A greater understanding of the potential opioid mechanisms underlying antidepressant effects will be key in the development of novel opioidergic antidepressants that reduce the risk of abuse, while maintaining efficacy, and help expand our knowledge of the neurobiological processes underlying depression.

In this chapter, for context, we first provide a brief overview of the endogenous opioid system, the potential role of this system in depression and a concise summary of the clinical studies investigating the use of opioidergic agents for depression. The remainder of the chapter focusses on emerging neurobiological antidepressant mechanisms of opioid system engagement. We consider opioid receptors and their effects on monoaminergic systems related to mood, the regulation of the stress response and interactions with the hypothalamic-pituitary-adrenal (HPA) axis and the immune system, and the potential roles of brain-derived neurotrophic factor (BDNF), intracellular signalling, neurogenesis and neuroplasticity. We further explore how the opioid-dependent regulation of social reward, social pain and social behaviours may contribute to depressive illness before finally considering the role of the opioid system in expectancy and antidepressant placebo effects.

2 The Endogenous Opioid System

The human opioid system is composed of four G-protein-coupled receptors, namely the three 'classical' opioid receptors, μ-opioid receptor (MOR), κ-opioid receptor (KOR) and δ-opioid receptor (DOR), and the non-opioid nociceptin/orphanin FQ receptor (NOPR), alongside a set of endogenous opioid neuropeptides, including endorphins, endomorphins, enkephalins, dynorphins and nociceptin that show distinct selectivity toward their respective opioid receptors (Wang et al. 2023). Despite a significant sequence similarity with the classical opioid receptors, the NOPR exhibits negligible binding affinity towards opioid peptides or morphine-like compounds. Likewise, the classical opioid receptors demonstrate minimal binding affinity for NOP's natural ligand nociceptin. The endogenous opioid system is widely distributed in the central and peripheral nervous systems, together with the immune system (Benarroch 2012). It plays a central role in pain processing but is also involved in other diverse aspects of physiology, including respiratory, gastrointestinal, endocrine and immune functions (Bodnar 2020). Importantly, the opioid system also modulates reward processing (Le Merrer et al. 2009), mood (Lutz and Kieffer 2013) and responses to stress (Bodnar 2020) and contributes to the regulation of both positive and negative social experiences (Lutz et al. 2020), which together has provided rationale and motivation to investigate this relevant target in the treatment of depressive disorders.

3 The Opioid System and Depression

There are several comprehensive reviews that summarise the pre-clinical and clinical evidence for the involvement of the opioid system in the pathophysiology of major depressive disorder (MDD) (Jelen et al. 2022; Lutz and Kieffer 2013; Pecina et al. 2019; Puryear et al. 2020). Initial pre-clinical evidence from genetic and pharmacological studies has provided preliminary evidence indicating that the opioid system may be linked to a range of behaviours related to depression, with encouraging results from studies exploring the use of opioid-modulating drugs in rodent models of depression (Lutz and Kieffer 2013). Specifically, pre-clinical work has highlighted the distinct functions of opioid receptor sub-types in regulating a range of processes relevant to depression, including reward, motivation and social bonding (Lutz et al. 2020; Lutz and Kieffer 2013). Pre-clinical evidence has suggested that the activation of MORs and DORs has antidepressant-like effects (Jelen et al. 2022). Importantly, MORs are key mediators of both natural and artificial rewards (Le Merrer et al. 2009) and play a significant role in the development of dependence. In contrast to MORs and DORs, the stimulation of KORs has more aversive or pro-depressive effects (Anderson et al. 2014; Bruchas et al. 2007; Carlezon et al. 2006), with pre-clinical studies suggesting the antidepressant potential of KOR antagonist compounds (Grimwood et al. 2011; Jacobson et al. 2020; Rorick-Kehn et al. 2014).

It is important to acknowledge that there are limitations to pre-clinical models, and their translational validity has been questioned (Stanford 2020). Direct measurement in humans is possible with the use of positron emission tomography (PET), which is sensitive to opioid receptor availability and changes in endogenous opioid levels. These studies have confirmed that the opioid system is involved in emotional processing (Hsu et al. 2013; Zubieta et al. 2003), and there is further evidence to suggest that this system may be dysregulated in depression (Hsu et al. 2015; Kennedy et al. 2006; Nummenmaa et al. 2020). In all of these studies, PET scans evidenced reduced opioid receptor expression in patients with MDD. Opioid system activation was attenuated in a study of social rejection in patients with MDD compared to controls (Hsu et al. 2015), and it increased in a state of sustained sadness in a study of women with depression (Kennedy et al. 2006). In a large-scale retrospective cross-sectional study, sub-clinical depressive symptoms were consistently associated with lower MOR availability in brain regions implicated in the regulation of mood, cognitive control and emotional processing, including the amygdala, insula, hippocampus, ventral striatum and cingulate cortices, which may plausibly contribute to the characteristic alterations in mood, dysfunctional processing of negative social interactions and social anhedonia seen in depression (Nummenmaa et al. 2020). Together, pre-clinical and clinical work exploring the opioid dysregulation hypothesis has encouraged efforts to develop opioid modulating drugs as novel therapies for major depressive disorder (MDD).

4 Opioidergic Agents in Depression

There are a growing number of clinical trials investigating the use of opioidergic agents in depression. While a comprehensive discussion is beyond the scope of this review, we have provided a summary of some of the key studies exploring the clinical utility of putative opioid-based antidepressants in Table 1. Clinical studies investigating the combination of MOR/KOR agonists and antagonists (specifically buprenorphine/samidorphan) in patients diagnosed with MDD have yielded preliminary evidence supporting antidepressant efficacy while minimising the likelihood of abuse. Further progress in our understandings of the functions of unique opioid

Table 1 Clinical studies of opioidergic agents in depressive disorders

Study	Compound	Design	Patient group (n)	Dosing	Outcome measure	Findings
Weber and Emrich (1988)	Buprenorphine	Double-blind, placebo controlled	MDD (10)	Sublingual 0.2 mg buprenorphine for 5–8 days	HAM-D	Buprenorphine treatment resulted in a notable decrease in HAM-D scores ($P < 0.02$) compared to the placebo. 50% of the patients responded to buprenorphine
Mongan and Callaway (1990)	Buprenorphine	Some subjects open label, others double-blind	MDD (7) BD (1)	Sublingual 0.15–0.3 mg buprenorphine	Profile of mood scale (POMS)	6/8 demonstrated antidepressant response
Bodkin et al. (1995)	Buprenorphine	Open label	TRD (10)	Buprenorphine 0.15 mg initiated intranasally or sublingually once daily titrated according to tolerance and clinical benefit, with a maximum daily dosage of 1.8 mg for 4 weeks	HAM-D	Out of seven subjects (three dropouts), six achieved significant clinical improvement at the end of the trial, but three dropped out. The mean endpoint HAM-D score was 10.7 (SD = 9.3), a 60.7% decrease from baseline ($p = 0.006$)
Nyhuis et al. (2008)	Buprenorphine	Open label	TRD (6)	Sublingual buprenorphine as monotherapy beginning with 0.4 mg/d increased up to a final dosage of 0.8 to 2.0 mg once daily for 1 week	HAM-D	After 1 week of buprenorphine treatment, the mean HAMD score decreased from 22.8 (range: 17–28) to 6.0 (range: 2–10). Five out of six patients achieved complete remission

(continued)

Table 1 (continued)

Study	Compound	Design	Patient group (n)	Dosing	Outcome measure	Findings
Karp et al. (2014)	Buprenorphine	Open label	TRD (15)	Titrated dose of sublingual buprenorphine ranging from 0.2–1.6 mg/d for 8 weeks	MADRS	Baseline MADRS was 27.0 (SD = 7.3), and by week 8, it dropped significantly to 9.5 (SD = 9.5). 61.54% of participants (95% CI = 32.3 to 84.9) exhibited a response by the end of the 8-week period
Ehrich et al. (2015)	Buprenorphine + Samidorphan *(adjunct to antidepressant)*	Randomised double-blind placebo controlled	TRD (32)	– BUP/SAM 8:1 dose-ratio (2 mg/0.25 mg for 3 days and 4 mg/0.5 mg for 4 days) ($n = 14$) – BUP/SAM 1:1 dose-ratio (4 mg/4 mg for 3 days and 8 mg/8 mg for 4 days) ($n = 14$) – Placebo ($n = 4$)	HAM-D	The group with a 1:1 ratio had the greatest antidepressant effects. Difference in HAM-D scores between the 1:1 treatment group and the placebo was significant ($p = 0.032$)
Yovell et al. (2016)	Buprenorphine (*monotherapy or adjunct to antidepressant*)	Randomised double-blind placebo controlled	Severely suicidal patients (88)	– Sublingual buprenorphine (initial dosage, 0.1 mg once or twice daily; mean final dosage = 0.44 mg/day; $n = 57$) for 4 weeks – Placebo ($n = 31$)	– BSSI – BDI	Compared to the placebo group, the buprenorphine group had a greater reduction in BSSI at week 2 (mean difference = 24.3, 95% CI = 28.5, 20.2; $p = 0.04$), and at the end of week 4 (mean difference = 27.1, 95% CI = 212.0, 22.3;

						$p = 0.004$). Although the difference did not reach significance, the buprenorphine group had lower BDI scores than the placebo group (mean difference = 26.1, 95% CI = 213.2, 1.0; $p = 0.09$)
Fava et al. (2016)	Buprenorphine + samidorphan (*adjunct to antidepressant*)	Randomised double-blind placebo controlled	TRD (142)	– BUP/SAM 2 mg/2 mg; ($n = 24$) – BUP/SAM 8 mg/8 mg ($n = 19$) – Placebo ($n = 98$) – 4 weeks	– HAM-D – MADRS – CGI	Significantly greater improvements in the 2 mg/2 mg dosage group across the three outcome measures compared with placebo (HAM-D: −2.8, 95% CI = −5.1, −0.6; MADRS: −4.9, 95% CI = −8.2, −1.6; CGI-S: −0.5, 95% CI = −0.9, −0.1)
Zajecka et al. (2019)	Buprenorphine + samidorphan (*adjunct to antidepressant*)	Phase 3 multi-centre randomised, placebo controlled	TRD (295)	– BUP/SAM 2 mg/2 mg ($n = 147$) – Placebo ($n = 148$) – 6 weeks	MADRS	At the end of the treatment period, the least squares mean change in MADRS was −4.8 (SE 0.67) for the BUP/SAM 2 mg/2 mg group and −4.6 (SE 0.66) for the placebo group (mean difference −0.3 [SE 0.95], $p = 0.782$). Additionally, there were no differences observed in response or remission rates based on MADRS scores

(continued)

Table 1 (continued)

Study	Compound	Design	Patient group (n)	Dosing	Outcome measure	Findings
Fava et al. (2020b)	Buprenorphine + samidorphan (*adjunct to antidepressant*)	Two phase 3, randomised, double-blind, placebo-controlled studies: *FORWARD-4 (FW4)* *FORWARD-5 (FW5)*	*FW4:* TRD (168) *FW5:* TRD (187)	– BUP/SAM (0.5 mg/0.5 mg) ($n = 56$) – BUP/SAM (2 mg/2 mg) ($n = 56$) – Placebo ($n = 56$) – BUP/SAM (1 mg/1 mg) ($n = 62$) – BUP/SAM (2 mg/2 mg) ($n = 63$) – Placebo ($n = 62$) – 6 weeks	MADRS	*FW5*: adjunctive BUP/SAM 2 mg/2 mg was superior to placebo (average difference change from baseline to week 6 in MADRS versus placebo: −1.9, $p = 0.026$) *FW4*: did not achieve the primary endpoint (change from baseline in MADRS-10 at week 5 versus placebo: −1.8, $p = 0.109$) In a pooled analysis, change from baseline on the MADRS in the BUP/SAM 2 mg/2 mg group was greater compared with placebo at all time points from week 3 and later, including week 6 (−1.8; $p = 0.010$; 95% CI: −3.2 to −0.4)
Fava et al. (2020a)	CERC-501 (JNJ-67953964/aticaprant) (kappa selective opioid receptor antagonist)	Sequential parallel comparison design study	TRD (8)	(a) 10 mg/d of CERC-501 for 6 days (b) 20 mg/d of CERC-501 for 6 days (c) Placebo for 3 days, followed by 10 mg/d of	HAMD-6 MADRS	Due to slow enrolment, the study was terminated early, and the limited sample size makes it difficult to draw conclusions. However, the

				CERC- 501 for 3 days (d) Placebo for 3 days, followed by 20 mg/d of CERC-501 for 3 days (e) Placebo for 6 days		weighted mean differences in HAMD-6 (1.28) and MADRS (2.33) scores were numerically greater for CERC-501 than for the placebo, although these differences were not statistically significant
ClinicalTrials.gov NCT03559192	JNJ-67953964 (aticaprant)	Phase 2a randomised, double-blind, placebo-controlled, parallel-group, multi-centre study	MDD (181)	Placebo or 10 mg JNJ-67953964 capsules for 6 weeks	MADRS	Recruitment completed. Results not yet published
ClinicalTrials.gov NCT05455684 VENTURA-1	JNJ-67953964 (aticaprant) (*adjunct to antidepressant SSRI or SNRI*)	Phase 3 randomised, double-blind, multi-centre, parallel group, placebo controlled	MDD with moderate to severe anhedonia (estimated $n = 538$)	Placebo or 10 mg JNJ-67953964 capsules for 6 weeks	MADRS	Currently recruiting
ClinicalTrials.gov NCT05550532 VENTURA-2	JNJ-67953964 (aticaprant) (*adjunct to antidepressant SSRI or SNRI*)	Phase 3 randomised, double-blind, multi-centre, parallel-group, placebo-controlled study	MDD with moderate to severe anhedonia (estimated $n = 710$)	Placebo or 10 mg JNJ-67953964 capsules for 6 weeks	MADRS	Currently recruiting
Richards et al. (2016)	AZD2327 (a selective DOR agonist)	Double-blind, parallel-group design, placebo-controlled pilot study	Anxious MDD (22)	– AZD2327 3 mg BD ($n = 13$) – Placebo ($n = 9$) – 4 weeks	HAM-D HAM-A	54% responded to the active drug, while 33% responded to the placebo. However, there was no significant main drug effect observed for either the HAM-D ($p = 0.39$) or the HAM-A ($p = 0.15$)

receptor sub-types and the production of selective opioid ligands, including KOR selective antagonists (such as aticaprant) and DOR selective agonists, offers hope for the potential of opioidergic agents as viable treatment options for MDD, without the risk of abuse potential.

There are also potential opioid system interactions between various other treatments that are currently understood to exert antidepressant effects through non-opioidergic actions. While the data are somewhat limited, we have summarised relevant findings in Table 2 and refer the reader to Jelen et al. (2022) for a further narrative discussion.

5 Antidepressant Mechanisms of Opioid System Engagement

5.1 Monoaminergic Systems

Monoaminergic systems are involved in the regulation of behavioural processes, including emotion, motivation and arousal, and are a target of most of the current typical antidepressant treatment strategies (Liu et al. 2018). There are a number of functional interactions between opioid receptor systems and monoaminergic systems that are relevant to mood control, representing a mechanism that likely contributes to the antidepressant effects of opioids (Lutz and Kieffer 2013). Monoaminergic neurons synthesising serotonin (5-HT), dopamine (DA) and noradrenaline (NA) originate in the brainstem, project extensively to the forebrain and hindbrain and are regulated by opioid receptors at various sites (Lutz and Kieffer 2013). The activity of 5-HT neurons is closely controlled by MORs. In vivo micro-dialysis data have demonstrated that the activation of MORs in the dorsal raphe nucleus suppresses the GABAergic (gamma-aminobutyric acid) mediated inhibition of serotonergic neurons, thereby leading to an excess release of 5-HT in the frontal cortex and other brain regions involved in emotional integration, including the thalamus, hippocampus, nucleus accumbens, amygdala and striatum (Fadda et al. 2005; Tao and Auerbach 1995, 2002a, b). Conversely, the activation of KORs by dynorphin or other KOR agonists decreases 5-HT release (Grilli et al. 2009; Tao and Auerbach 2002b). Opioid receptors also play a role in modulating the mesolimbic dopaminergic pathway. The activation of MORs and DORs in the ventral tegmental area disinhibit DA release in the nucleus accumbens via GABAergic interneurons (Devine et al. 1993; Pentney and Gratton 1991), while in contrast, the activation of KOR reduces dopamine release in this region (Carlezon and Krystal 2016; Spanagel et al. 1992). Furthermore, a selective blockade of KORs markedly increases basal dopamine release (Spanagel et al. 1992). Finally, the locus coeruleus-noradrenaline system is a major stress response system that is regulated by the endogenous opioids enkephalin and dynorphin and the corresponding MORs and KORs (Van Bockstaele and Valentino 2013). Noradrenergic neurons are directly

Table 2 Antidepressant treatments with potential opioid system interactions

Drug	Opioid system interactions
NMDA-targeting antidepressants • Ketamine/esketamine • Esmethadone	• In a small cross-over trial in TRD, pre-treatment with naltrexone (a MOR/KOR antagonist) significantly attenuated the antidepressant and anti-suicidal effects of ketamine (Williams et al. 2018, 2019) • In rodents, the activation of opioid receptors is 'necessary but not sufficient' for ketamine to reduce depressive-like behavioural symptoms and to reduce activity in the lateral habenula, a critical anti-reward centre implicated in the pathophysiology of depression (Klein et al. 2020) • In mice, naltrexone pre-treatment attenuated acute and sustained effects of ketamine-induced reductions in immobility in the TST (Zhang et al. 2021) • Ketamine is a racemic mixture, composed of (S)- and (R)-enantiomers that may exert antidepressant effects via distinct mechanisms with potential differences in abuse potential (Jelen et al. 2021). While both activate MORs and KORs, (S)-ketamine has a higher affinity and potency. Antidepressant-like doses of (S)-ketamine induced locomotor activity and psychomotor sensitisation, conditioned place preference in mice and increased metabolic activity and dopamine tone in the medial prefrontal cortex of rats, in an opioid receptor-dependent manner (Bonaventura et al. 2021) • Esmethadone is the d-isomer of the opioid drug methadone, and an early trial suggests rapid and sustained antidepressant effects compared with placebo in TRD (Fava et al. 2022). Studies indicate that its affinity for the MOR is 20 times lower than that of levomethadone, albeit still significantly higher (70 times) than its affinity for NMDA receptors (Codd et al. 1995; Gorman et al. 1997)
Tianeptine	• Tianeptine identified as an efficacious MOR full agonist. Also found to act as a full agonist at DOR, although with much lower potency (Gassaway et al. 2014) • MOR required for the acute and chronic antidepressant-like effects of tianeptine in mice (Samuels et al. 2017)
Tricyclic antidepressants • Amitriptyline • Clomipramine • Nortriptyline • Desipramine • Imipramine	• TCAs, such as amitriptyline, nortriptyline, desipramine, and imipramine, exert direct agonist activity at distinct opioid receptor sub-types. Amitriptyline, nortriptyline, imipramine and desipramine show higher selectivity for DOR. All the TCAs examined had relatively low affinity for the MOR (Onali et al. 2010) • Chronic treatment with amitriptyline in rats increases levels of immunoreactive leu-enkephalin and met-enkephalin immunoreactivity in several brain areas (Hamon et al. 1987) • Amitriptyline increases proenkephalin and prodynorphin mRNA levels in the nucleus accumbens in naive mice (Böhm et al. 2006) • Opioid receptor antagonist naloxone blocks the effect of repeated or long-term TCA antidepressant treatments in models of neuropathic pain (Benbouzid et al. 2008; Marchand et al. 2003; Wattiez et al. 2011)
SSRI: • Fluoxetine	• Effects of fluoxetine in the forced-swim test in mice completely blocked by pre-treatment with naloxone (a non-selective opioid receptor antagonist) (Zomkowski et al. 2005) • Chronic fluoxetine treatment in mice upregulates proenkephalin expression in the dentate gyrus of the hippocampus, associated with treatment responsiveness. DOR knockout mice have attenuated

(continued)

Table 2 (continued)

Drug	Opioid system interactions
	responses to fluoxetine following stress-induced corticosterone treatment compared to wild-type mice, whereas MOR knockout mice had similar responses to fluoxetine exposure as controls. These findings suggest that DORs, but not MORs, contribute to antidepressant-like behaviours produced by fluoxetine, as measured in the forced swim test (Carazo-Arias et al. 2022)
Other antidepressants: • Mirtazapine • Trazodone • Mianserin • Venlafaxine	• Antinociceptive effect of trazodone and in mice partially mediated through the opioid system, particularly MOR and KOR, respectively (Schreiber et al. 2000, 2002) • Antihyperalgesic and antiallodynic effects of mianserin reversed by naloxone administration in rats (Üçel et al. 2015) • In different cell systems, mianserin directly activates KORs, displaying partial agonist activity at brain receptors (Olianas et al. 2012) • Antidepressant-like effect of venlafaxine is abolished in MOR knockout mice (Ide et al. 2010)

inhibited by MORs, whereas KORs regulate locus coeruleus neurons at a presynaptic level, inhibiting both excitatory and dynorphin afferents to the locus coeruleus, taking the locus coeruleus 'off-line', which may be important in its link to depression and another target for opioidergic antidepressants (Kreibich et al. 2008; Van Bockstaele and Valentino 2013). In addition to this modulation of noradrenergic function, MOR/DOR agonists and KOR antagonists may act to enhance serotonergic and dopaminergic functions. This indicates a potential mechanism through which opioid drugs can enhance neurochemistry, similar to that seen with current antidepressants, but through a different regulatory mechanism.

5.2 Regulation of the Stress Response and the Hypothalamic-Pituitary-Adrenal (HPA) Axis

The maintenance of homeostasis during stressful conditions is mediated through a complex, interactive organisation of neuroanatomical pathways across the brain, which act to initiate a range of endocrine, metabolic, immune, physiological and behavioural changes. There is significant evidence to support an important contribution of the endogenous opioid systems in the mediation, modulation and regulation of stress responses (Drolet et al. 2001). The activation of the HPA axis is a hallmark of the stress response in humans, representing a key indicator of psychosocial stress (Iob et al. 2018), and is an important biological system implicated in the aetiology of depression (Pariante and Lightman 2008). Briefly, HPA axis activity is controlled by the secretion of adrenocorticotrophic-hormone-releasing factor (CRF) and vasopressin in the hypothalamus, which stimulate adrenocorticotrophic hormone (ACTH) from the pituitary, finally stimulating the release of glucocorticoids (i.e. cortisol) from the adrenal glands, which in turn leads to the feedback inhibition of the secretion of ACTH and CRF (Pariante 2006). In depression, the hyperactivity of the HPA axis and increased levels of cortisol are the result of an impairment of glucocorticoid-receptor-mediated negative feedback (glucocorticoid resistance) (Pariante 2006; Pariante and Lightman 2008). Endogenous opioids are released in response to stress and interact with the HPA axis at several levels to modulate responsivity to stress (Drolet et al. 2001), with β-endorphin and dynorphin exerting tonic inhibition and the stimulation of HPA activity (acting on MORs and KORs, respectively) (Zhou and Leri 2016). CRF neurons in the hypothalamus receive inhibitory input from β-endorphin neurons, which act to reduce ACTH release, and CRF activity may also be indirectly inhibited via mu opioid inhibition of CRF stimulatory noradrenergic neurons (Valentino and Van Bockstaele 2001, 2015; Van Bockstaele and Valentino 2013). In contrast, an increasing body of evidence suggests that KORs play a significant role in transducing the effects of stress (Van't Veer and Carlezon 2013). Although the mechanisms by which KOR activation produces stress-like effects are not fully understood, studies have highlighted a critical role of KORs in interacting with CRF systems in the brain (Lalanne et al. 2014). Indeed, stress-induced release of dynorphin and the activation of KORs leads to the activation of the HPA axis through an increase in CRF by the paraventricular nucleus in the hypothalamus (Van't Veer and Carlezon 2013).

Animal models and human studies have demonstrated the impact of exogenous opioidergic agents on the HPA axis. In primates, the administration of the MOR agonist fentanyl results in the inhibition of HPA activity with reductions in plasma ACTH and cortisol (Broadbear et al. 2004), whereas the MOR antagonist naltrexone increased levels of ACTH and cortisol (Williams et al. 2003). Similarly, in humans, the administration of morphine, a MOR agonist, and buprenorphine, a MOR partial agonist and KOR antagonist, led to both the suppression of ACTH and cortisol release (Allolio et al. 1987; Bershad et al. 2015), while naltrexone administration led to significant increases in ACTH and cortisol levels (King et al. 2002). Considering the KOR system, the administration of a selective KOR agonist led to dose-dependent increases in ACTH and cortisol in rhesus monkeys (Pascoe et al. 2008) and dose-dependent increases in cortisol release in humans (Ur et al. 1997). Crucially, in the study by Pascoe et al. (2008), the stimulatory effects on the HPA axis were blocked by a selective KOR antagonist. Furthermore, several other behavioural and molecular effects of stress and CRF are blocked by selective KOR antagonists (Land et al. 2008; McLaughlin et al. 2003, 2006). Finally, early pre-clinical work using the chronic vicarious social defeat stress (cVSDS) mouse paradigm (an animal model of depression) demonstrated that repeated administrations of the DOR selective agonist (KNT-127) improved their decreased social interaction behaviours and reversed stress-induced increases in serum corticosterone levels (reduced hyperactivation of the HPA axis) (Yoshioka et al. 2023). Based on these findings, it is plausible that the antidepressant effects of MOR/DOR agonists and KOR antagonists may also arise from their ability to reduce the impact of stress and inhibit the HPA axis, providing broad stress-dampening effects.

5.3 Immune System and Inflammation

There is evidence that immune system activation and the inflammatory response have key pathophysiological roles in several psychiatric disorders, including MDD (Gibney and Drexhage 2013). Although there is considerable heterogeneity in alterations in immune responses, some MDD patients show increases in pro-inflammatory cytokines, including interleukin-6 (IL-6) and tumour necrosis factor alpha (TNF-α) (Dowlati et al. 2010), whereas others may also show a heightened compensatory immune-regulatory system (CRIS) response that exerts negative feedback through increased levels of anti-inflammatory/immune-regulatory cytokines (e.g. IL-10) (Anjum et al. 2020; Müller et al. 2019). Potential mechanisms linking cytokine-mediated immune activation to the pathophysiology of depression include altered monoamine metabolism, HPA axis activation, kynurenine pathway activation, reduced neurogenesis and microglia activation (Rengasamy et al. 2021; Zunszain et al. 2013). The opioid system also has several functional interactions with the immune system that may be relevant in regulating depression pathophysiology.

Opioidergic markers are present in a range of different immune cells, including monocytes, macrophages, granulocytes and T and B cells, that express opioid peptides, including β-endorphin, its precursor proopiomelanocortin (POMC) and all necessary processing enzymes (Mousa et al. 2004; Ninković and Roy 2013). Furthermore, met-enkephalin, dynorphin and endomorphins are also found in leukocytes of inflamed tissues (Rittner et al. 2005). During conditions of inflammation and stress, pro-inflammatory cytokines stimulate CRF release, and in response, leukocytes secrete opioid peptides that exert immune stimulatory and regulatory activities and also produce analgesic effects (Rittner et al. 2005). Overall, β-endorphins appear to have immune-regulatory effects by acting to attenuate the stress response, reducing levels of inflammatory cytokines while increasing levels of anti-inflammatory cytokines, primarily through interaction with MORs (Pilozzi et al. 2021). In addition, MORs have been shown to be up-regulated in animal models of inflammation (Ji et al. 1995; Pol and Puig 2004), and MOR activation has been linked to immunosuppressive effects in the peripheral immune system (Zhang et al. 2020). There is evidence that, directly or indirectly, opioids can reduce the activity of a range of immune cells, including natural killer, T and B cells, neutrophils, mast cells and dendritic cells that can be reversed by treating with MOR antagonists (Cuitavi et al. 2023; Plein and Rittner 2018). In contrast, recent evidence from animal studies suggests that in the central nervous system, microglial MORs act as promoters of neuro-inflammation (Cuitavi et al. 2023); however, further work is needed to understand the relevance of this link in humans. Supporting a link between MORs and immune system function in MDD, Prossin et al. (2011) found that in MDD patients, but not healthy subjects, plasma levels of the pro-inflammatory cytokine IL-18 was positively correlated with baseline MOR availability (non-displaceable binding potential measured with PET) and with MOR activation during a sadness challenge in the subgenual anterior cingulate, ventral basal ganglia and amygdala. Considering the KOR system, activation of KORs by dynorphins mediates a number of immunosuppressive and anti-inflammatory effects. This includes a reduction in antibody production and suppressed cytokine production, although the precise mechanisms responsible for many of these immunomodulatory effects have not been investigated (Rogers 2022). Work in humans has demonstrated that serum dynorphin and KOR (alongside β-endorphin and MOR) are significantly higher in MDD compared to healthy controls and that increases in both dynorphin/KOR and β-endorphin/MOR are strongly associated with increased IL-6 and IL-10 levels (Al-Hakeim et al. 2020). The authors suggest that immune activation in MDD is associated with a simultaneous activation of dynorphin/KOR and β-endorphin/MOR signalling and that these opioid systems may participate in the pathophysiology of depression by exerting immune-regulatory activities attenuating the primary immune response (Al-Hakeim et al. 2020). Finally, the DOR system may also be involved in the modulation of neuro-inflammation. In cVSDS mice, the administration of the selective DOR agonist (KNT-127) led to antidepressant-like effects that was associated with a reduced number of Iba-1- and CD11b-positive cells in the sub-granular zone and the granule cell layer of the hippocampal dentate gyrus, indicating a suppression of cVSDS-induced microglial overactivation (Yoshioka et al. 2023). Although the data indicate significant impacts of the opioid system on immune cell functioning, further work is needed to understand the precise interactions between opioid agents, the immune system and inflammation in depressive disorders and exactly if/how these mechanisms contribute to potential antidepressant effects or whether they could be exploited in the development of novel therapies.

5.4 Brain-Derived Neurotrophic Factor (BDNF)

Brain-derived neurotrophic factor (BDNF), one of the major neurotrophic factors, has emerged as a key regulator of neural circuit development and function and has been implicated in the pathophysiology of depression. BDNF and its associated tyrosine kinase receptor (TrkB) have an essential role in mediating activity-dependent processes, including neuronal differentiation and growth, synapse formation and plasticity (Park and Poo 2013). The 'neurotrophin hypothesis of depression' is primarily based on observations that reduced hippocampal BDNF levels are correlated with stress-induced depressive behaviours (Duman and Monteggia 2006; Martinowich et al. 2007) and that typical antidepressant treatments enhance BDNF expression in hippocampal and prefrontal brain regions in pre-clinical models (Zhang et al. 2010). There is evidence that BDNF activity is regulated by opioid receptor systems and is another relevant mediator when considering the potential antidepressant mechanisms of opioidergic treatments. In pre-clinical models, the central administration of endogenous opioid peptides significantly upregulate BDNF mRNA expression, through activation of DOR and MORs, in regions including the frontal cortex and amygdala (Zhang et al. 2006). Further, the administration of DOR agonists, at doses that produced antidepressant-like effects in rodent models, also led to elevated BDNF mRNA in the frontal cortex (Zhang et al. 2006), hippocampus and amygdala (Torregrossa et al. 2004, 2006). Additional work has demonstrated antidepressant-like effects of the DOR-specific agonist SNC80 in the tail suspension test, forced swimming test and sucrose preference test in chronic-restraint-stressed (CRS) mice, associated with enhanced BDNF-TrkB signalling in the hippocampus and amygdala (Wu et al. 2023). Importantly, both the antidepressant-like effects and promoting effects of BDNF-TrKB by the DOR agonist were blocked by a TrkB inhibitor, highlighting a key role for the BDNF/TrkB signalling pathway in the hippocampus and amygdala in the potential antidepressant effects of DOR agonists (Wu et al. 2023). Finally, pre-clinical work has also demonstrated antidepressant-like effects of the KOR antagonist nor-Binaltorphimine (nor-BNI) in rodents that was associated with increased BDNF mRNA in regions including the hippocampus, amygdala and frontal cortex, suggesting that BDNF signalling also has a mechanistic role in the potential antidepressant effects of KOR antagonists (Zhang et al. 2007). Taken together, these data provide preliminary evidence that opioids have potentially beneficial effects on BDNF signalling in brain regions relevant to depressive disorders (hippocampus, amygdala and frontal cortex); however a significant limitation is that the work described has been restricted to animal models, and whether this mechanism holds relevance in humans remains to be determined.

5.5 Intracellular Signalling, Neurogenesis and Neuroplasticity

Beyond BDNF, there are various intracellular signalling cascades involved in the regulation of neurogenesis and neuroplasticity, both key processes implicated in the pathogenesis of depression. Exposure to stress decreases neurogenesis and neuroplastic markers, including dendritic length, dendritic arborisation and spine density, whereas antidepressant treatments can reverse these effects (Duman 2004; Hanson et al. 2011). There is evidence that opioid receptors modulate aspects of neurogenesis and neuroplasticity and interact with key intracellular signalling pathways implicated in these processes, including extracellular-signal-related kinase/mitogen-activated protein kinase (ERK/MAPK), cyclic adenosine monophosphate (cAMP)-response element binding protein (CREB) and activator protein-1 (AP-1) (Puryear et al. 2020).

There is evidence that endogenous opioids such as β-endorphins act as neurogenesis promoters (Koehl et al. 2008; Persson et al. 2003). Although most pre-clinical studies of exogenous MOR agonists such as morphine demonstrate that these negatively affect or disrupt hippocampal neurogenesis (Eisch et al. 2000; Harburg et al. 2007; Kahn et al. 2005), other work has found MOR-mediated increases in neural stem cell neuronal differentiation and neuronal dendrite growth in the hippocampal dentate gyrus (Zhang et al. 2019). KORs are expressed on neural stem/progenitor cells in the hippocampus, suggesting that KORs are also involved in the modulation of neurogenesis (Tan et al. 2018). KOR agonists, including dynorphin, appear to inhibit neurogenesis (Xu et al. 2021), an effect blocked by a KOR antagonist. Finally, pre-clinical work has demonstrated that DOR agonists are also implicated in neurogenesis. DOR agonists stimulate the proliferation of adult hippocampal progenitors in vitro (Persson et al. 2003) and promote the neural differentiation of multi-potent neural stem cells (Narita et al. 2006). In cVSDS mice, administration of the selective DOR agonist (KNT-127) during the stress administration period suppressed decreases in the hippocampal newborn neuron survival rate and social interaction behaviours (Yoshioka et al. 2023). Interestingly, when administered after stress, KNT-127 attenuates microglial overactivation, eventually modulating the homeostasis of the HPA axis without affecting neurogenesis. Therefore, DOR activation could potentially play a role in neuronal protection against the effects of chronic psychological stress, i.e. the prevention of the attenuation of new neurons, rather than the enhancement of neurogenesis, is important for anti-stress effects (Yoshioka et al. 2023).

Considering intracellular pathways, the ERK/MAPK pathway has a significant role in regulating protein synthesis necessary for synaptic development and plasticity (Wang and Mao 2019). Both typical and novel rapid-acting antidepressants, such as ketamine, have been shown to increase the phosphorylation of ERK1/2 (pERK1/2) (Gourley et al. 2008; Lepack et al. 2016; Qi et al. 2008). Similarly, acute MOR activation leads to the rapid activation of the ERK signalling pathway in vitro (Belcheva et al. 2005; Zheng et al. 2008), and both acute MOR and DOR activation increases ERK1/2 and MAPK activity in vivo (Asensio et al. 2006). The KOR system has also been shown to mediate ERK1/2 phosphorylation in vitro and in vivo (Belcheva et al. 2005; Bruchas et al. 2008). While opioidergic agents have been shown to modulate ERK1/2 in a similar manner to typical and rapid-acting antidepressants, increased ERK/MAPK signalling cannot be assumed to be directly linked to antidepressant effects. The effects of modulating ERK-/MAPK-dependent plasticity as an opioidergic antidepressant mechanism are complex, receptor- and region-specific, and likely context-dependent (Puryear et al. 2020).

The transcription factor CREB is involved in learning, memory and the regulation of synaptic plasticity, which may be disrupted in depression, and is an additional intracellular target for opioid receptor modulation (Carlezon et al. 2005). The activity of CREB is impacted by stress, and its effects on brain regions can vary; these can be adaptive, maladaptive or both simultaneously. Stress exposure reduces CREB expression in the hippocampus (Alfonso et al. 2006), while the chronic administration of various antidepressant treatments (Nibuya et al. 1996), including the MOR agonist tianeptine (Lee et al. 2017), increases its expression in this region. In contrast, stress can activate CREB and can increase dynorphin in the nucleus accumbens (Bruchas et al. 2007; Pliakas et al. 2001), and its activity here may regulate dysphoric states (Knoll and Carlezon 2010). It is suggested that CREB-mediated transcription of dynorphin within the nucleus accumbens decreases dopamine function, which may trigger certain features of depression (Carlezon et al. 2005). Pre-clinical work using a variety of stressors, including foot shock, immobilisation and social defeat, has implicated CREB and KOR activation in the nucleus accumbens in behavioural models of depression (Donahue et al. 2015; Muschamp et al. 2011; Shirayama et al. 2004). Furthermore, KOR antagonists can disrupt CREB and dynorphin signalling in the nucleus accumbens and produce antidepressant-like effects (Newton et al. 2002; Pliakas et al. 2001; Shirayama et al. 2004), supporting the hypothesis that KOR antagonists may be effective in treating depression by reducing elevated CREB within the nucleus accumbens.

Finally, transcription factor AP-1 is a major downstream target of BDNF and ERK/MAPK signalling pathways and regulates gene expression related to neuroplasticity (Chottekalapanda et al. 2020). Chronic treatment with selective serotonin reuptake inhibitors (SSRIs) has been found to induce AP-1 activity, leading to a change in the expression of key neuronal plasticity genes and producing an antidepressant response in rodent models (Chottekalapanda et al. 2020). Acute activation of AP-1 has also been observed with MOR agonists (Bilecki et al. 2004; Shoda et al. 2001), which can be reduced by MAPK inhibitors (Shoda et al. 2001). Therefore, the activation of AP-1 by opioidergic agents may be another speculative mechanism that could contribute to potential antidepressant effects.

In summary, the data discussed in this section suggest that opioid receptor modulation plays a key role in neurogenesis and neuroplastic mechanisms and converges on several signalling pathways that are characteristically altered in MDD.

5.6 Opioid System and the Social Brain: Social Pain and Social Reward

Paul MacLean suggested that *a sense of separation is a condition that makes being a mammal so painful* (MacLean 1993). Indeed, experiences of social rejection, exclusion, disconnection or loss are some of the most 'painful' experiences humans may face and are major mediators of risk for depressive disorders and suicidal behaviours (Meyer-Lindenberg and Tost 2012; Rueger et al. 2016). Emerging evidence suggests that experiences of 'social pain', defined as an unpleasant experience that is associated with actual or potential damage to one's sense of social connection or social value, rely on some of the same neurobiological processes that underlie experiences of physical pain (Eisenberger 2012). Early evidence of an overlap between physical and social pain systems came from the finding that MOR agonists, best known for their analgesic effects, also reduce separation distress behaviours (i.e. isolation calls) in non-human mammals (Panksepp et al. 1978), a finding replicated across different species (Kalin et al. 1988; Warnick et al. 2005). Conversely, the administration of MOR antagonists, including naloxone and naltrexone, increase separation distress behaviours and increase attempts at social connection through grooming (Fabre-Nys et al. 1982), suggesting that MOR blockade enhances the need for social attachment, to increase the experience of social reward. Other work examining the role of the opioid system in social play demonstrated juvenile isolation during the play period causes long term effects on social exploration and interaction in adult rats, associated with increased MOR binding sites in the amygdala and in the bed of nucleus stria terminalis and a general upregulation of KOR binding sites in cortical regions, the hippocampus and substantia nigra (Van den Berg et al. 1999). Additionally, the administration of MOR agonists increases social play, while MOR antagonists and KOR agonists suppress social behaviours in rodents (Vanderschuren et al. 1995), suggesting that the opioidergic effect on social reward is mediated through mu and KOR systems. Finally, the effects of repeated social defeat, often used as a rodent model of depression, have also been demonstrated to be in part mediated by MOR and KOR signalling (Komatsu et al. 2011; McLaughlin et al. 2006).

Findings from the animal data summarised above have driven research efforts dedicated to the understanding of opioid involvement in social processes in humans. Converging evidence from human studies highlights a key role for the opioid system in modulating activity during social exclusion and other social pain-causing situations, in brain regions implicated in the affective component of physical pain (dorsal anterior cingulate cortex (dACC) and anterior insula (AI)). These regions have a particularly high density of MOR receptors (Baumgartner et al. 2006) and are crucial in processing the unpleasantness of physical pain. Studies have shown that analgesics, including acetaminophen (paracetamol), can reduce a self-reported experience of social pain and reduce neural responses to experimentally induced social rejection in the dACC and AI (Dewall et al. 2010), further supporting an overlap between social and physical pain. Hsu et al. (2013) directly tested the hypothesis that social rejection may be regulated by the opioid system by measuring MOR availability during social rejection and acceptance conditions using positron emission tomography (PET) in healthy volunteers. Social rejection was associated with significant reductions in MOR availability (suggesting MOR activation by endogenous opioids) in the ventral striatum, amygdala and thalamus and periaqueductal gray (PAG) (Hsu et al. 2013), a similar activation pattern to that seen during physical pain (Zubieta et al. 2001). Furthermore, MOR activation during rejection in the amygdala, PAG and subgenual anterior cingulate cortex (sgACC) was positively associated with greater trait resiliency (Hsu et al. 2013), suggesting a protective or adaptive role of MOR activation to reduce the experience of social pain. In a later PET study using the same paradigm, MDD patients did not show significant activation during rejection, whereas in healthy controls, MOR activation was found in the nucleus accumbens, amygdala and thalamus (Hsu et al. 2015). During acceptance, only healthy controls showed increased social motivation, which positively correlated with MOR activation in the nucleus accumbens, a region known to be involved in reward, whereas in the MDD group, MOR deactivation was observed in this region. Together, this suggests that abnormal MOR system functioning in depression may hinder emotional recovery from negative social interactions and decrease pleasure derived from positive interactions.

Considering the evidence from animal and human studies implicating the opioid system in the regulation of social behaviours, Bershad et al. (2016) evaluated the effects of buprenorphine (a partial MOR agonist and KOR antagonist) on social processing in healthy adults. The within-subject, double-blind design involved two sessions where participants received 0.2 mg of sublingual buprenorphine or placebo in a randomised order before completing three behavioural tasks: (i) a virtual ball-toss game in which they were first included and then excluded by other players, (ii) an emotional face attention task and (iii) a picture-viewing task, in which they rated images with and without social content. Low-dose buprenorphine reduced perceived social rejection during the ball-toss game, decreased attention to fearful expressions during the attention task and increased positivity ratings of social images during the picture viewing task (Bershad et al. 2016). When extending these findings to individuals with symptoms of depression, buprenorphine again reduced attention to fearful face expressions and increased ratings of positivity for images with social content compared to placebo (Bershad et al. 2018). These results further support the role of the opioid system in mediating responses to both social rejection and reward and provide mechanistic rational for opioidergic antidepressant strategies. Cautious pharmacological manipulation of this system could potentially reduce the severity of social pain, thereby reducing suicidal cognitions and behaviours while also reducing social anhedonia and increasing pleasure derived from positive interactions.

Although this section has focussed on the relationship between the opioid system and social reward/pain processing, the opioid system has a broader role in reward processing more generally that is not limited to social stimuli. For example, in a randomised, double-blind, proof-of mechanism trial evaluating the highly selective KOR antagonist JNJ-6795394 (aticaprant), in patients with anhedonia, a JNJ-67953964-treated group (10 mg/day over 8 weeks) exhibited significantly increased functional magnetic resonance imaging (fMRI) ventral striatal activation during (non-social) reward anticipation in the monetary incentive delay task compared with placebo (Krystal et al. 2020). Furthermore, in the probabilistic reward task (PRT), a non-social reward-learning task, the JNJ-6795394-treated group had a significantly higher mean PRT response bias, alongside significant reductions in clinical anhedonia, as measured by the Snaith-Hamilton Pleasure Scale (SHAPS), compared with placebo. Although results from larger clinical trials are awaited (Table 1), these preliminary findings suggest that KOR agonists may have potential as a general anhedonia therapy, not limited to social anhedonia, with therapeutic implications beyond the treatment of MDD.

5.7 *Expectancy and Placebo Effects*

Our interpretation of reality, including expectancies and appraisals, can directly shape our mood (Ashar et al. 2017; Pecina and Zubieta 2018). Even treatment with drugs that contain no active therapeutic ingredient, such as placebos, can alter one's subjective perception of mood states. There is growing evidence that demonstrates that the endogenous opioid system may be activated by positive individual expectations and maintained by conditioning and reward learning to induce internally mediated changes in physiological effects, i.e. placebo effects, resulting in subjective symptom relief and improvements in the emotional state (Pecina and Zubieta 2015). There have been decades of research implicating the opioid system in placebo analgesia (Levine et al. 1978; Petrovic et al. 2002); however, recently, researchers have been investigating the role of the opioid system in antidepressant placebo effects (Pecina et al. 2015, 2018, 2021).

In a PET study by Pecina et al. (2015) involving medication-free MDD patients ($n = 35$), reductions in depressive symptoms after placebo treatment were associated with placebo-induced mu-opioid neurotransmission in a network of regions implicated in emotion and stress regulation, including the sgACC, nucleus accumbens, thalamus and amygdala. Furthermore, when these patients subsequently received open-label treatment with an SSRI 10 weeks after, placebo-induced endogenous opioid release in the nucleus accumbens was associated with greater antidepressant treatment response (Pecina and Zubieta 2015). In a subsequent experiment, patients with MDD ($n = 20$) were administered an IV placebo described as a 'fast-acting antidepressant' and a trial-by trial sham fMRI neurofeedback manipulation, suggesting mood-relevant neural responses (Pecina et al. 2018). Infusion cues resulted in higher expectancies of mood improvement, and both the infusion cue and the sham neurofeedback induced improvements in reported mood. These expectancy and reinforcement manipulations of the antidepressant placebo effects were associated with increased blood-oxygen-level-dependent (BOLD) responses in the lateral prefrontal cortex (PFC). Additionally, placebo-induced increases in plasma levels of β-endorphin were associated with higher expectancy and mood ratings. Finally, this team used a modified version of this experiment in medication free MDD patients, to investigate the effects of the mu-opioid antagonist naltrexone 50 mg, versus placebo, pre-treatment on this paradigm (Pecina et al. 2021). Naltrexone, compared with placebo, partially abolished the interaction of the expectancy and reinforcement manipulation on mood and blocked reinforcement-induced responses in the right central orbitofrontal cortex (OFC). This brain region has a high expression of MORs, has been shown to encode the value of expected and actual outcomes (Schoenbaum et al. 1998) and plays an essential role in signalling outcome expectancies, crucial for changing established behaviour in the face of unexpected outcomes. These results provide preliminary evidence for the role of MOR OFC modulation in antidepressant placebo effects that result from cue-reward associations (Pecina et al. 2021). Although larger replication studies are needed, speculative implications are that MOR antagonists such as naltrexone could be utilised in clinical trials to reduce placebo effects and that drugs that positively modulate the MOR system could potentially be used to enhance antidepressant expectancy effects, thereby improving clinical response.

6 Conclusions

Depression is a heterogenous psychiatric syndrome, with a multi-factorial and complex aetiology, that encompasses a diverse range of psychopathology arising from distinct pathophysiological mechanisms. Although progress has been made, there are still gaps in the understanding of the complex neurobiology of depression

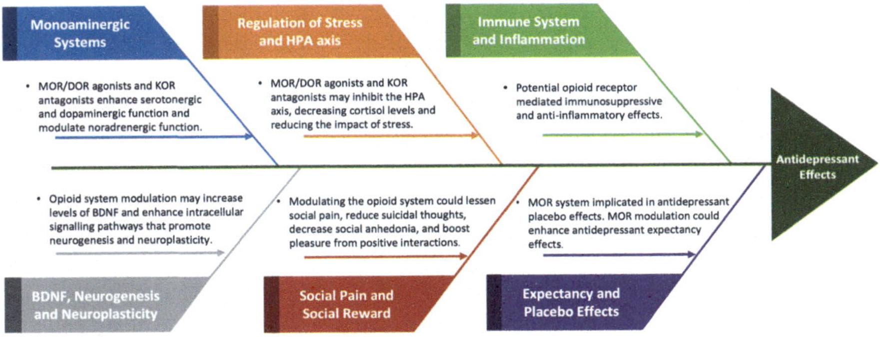

Fig. 1 Potential antidepressant mechanisms of opioid system engagement

and, as a result, in the range of treatment options available. With accumulating evidence that the endogenous opioid system may be dysregulated in depression, the opioid system has re-emerged as a relevant target in the treatment of depression. While the mechanisms underlying the therapeutic effects of opioid modulation in depression are multi-faceted (Fig. 1), there is a growing body of evidence that suggests that targeting opioid receptors may offer potential benefits in the treatment of depression. A better understanding of the complex mechanisms through which opioid system engagement may lead to antidepressant effects will support clinical translation and could ultimately lead to the development of novel, safe and effective treatments for individuals struggling with difficult-to-treat depression. Such treatments could potentially target multiple mechanistic pathways simultaneously, resulting in more comprehensive and effective treatment options. While the opioid system holds promise as a potential target for treating depression, it is important to note that certain opioid drugs, particularly those that target the MOR, are associated with a high risk of abuse and dependence. Therefore, there is a need for caution in the investigation and use of opioid-based treatments for depression, and their use must be closely monitored. Despite the potential risks associated with modulating the MOR system, continued research is necessary to determine the most effective treatment approaches and to develop new and innovative therapies to help those who struggle with depression.

References

Alfonso J, Frick LR, Silberman DM, Palumbo ML, Genaro AM, Frasch AC (2006) Regulation of hippocampal gene expression is conserved in two species subjected to different stressors and antidepressant treatments. Biol Psychiatry 59:244–251. https://doi.org/10.1016/j.biopsych.2005.06.036

Al-Hakeim HK, Al-Fadhel SZ, Al-Dujaili AH, Maes M (2020) In major depression, increased kappa and mu opioid receptor levels are associated with immune activation. Acta Neuropsychiatrica 32:99–108. https://doi.org/10.1017/neu.2019.47

Allolio B, Schulte HM, Deuss U, Kallabis D, Hamel E, Winkelman W (1987) Effect of oral morphine and naloxone on pituitary-adrenal response in man induced by human corticotropin-releasing hormone. Acta Endocrinol 114:509–514. https://doi.org/10.1530/acta.0.1140509

Anderson RI, Morales M, Spear LP, Varlinskaya EI (2014) Pharmacological activation of kappa opioid receptors: aversive effects in adolescent and adult male rats. Psychopharmacology (Berl) 231:1687–1693. https://doi.org/10.1007/s00213-013-3095-8

Anjum S, Qusar MMAS, Shahriar M, Islam SMA, Bhuiyan MA, Islam MR (2020) Altered serum interleukin-7 and interleukin-10 are associated with drug-free major depressive disorder. Ther Adv Psychopharmacol 10:2045125320916655. https://doi.org/10.1177/2045125320916655

Asensio VJ, Miralles A, Garcia-Sevilla JA (2006) Stimulation of mitogen-activated protein kinase kinases (MEK1/2) by mu-, delta- and kappa-opioid receptor agonists in the rat brain: regulation by chronic morphine and opioid withdrawal. Eur J Pharmacol 539:49–56. https://doi.org/10.1016/j.ejphar.2006.04.001

Ashar YK, Chang LJ, Wager TD (2017) Brain mechanisms of the placebo effect: an affective appraisal account. Annu Rev Clin Psychol 13:73–98. https://doi.org/10.1146/annurev-clinpsy-021815-093015

Baumgartner U, Buchholz HG, Bellosevich A, Magerl W, Siessmeier T, Rolke R, Hohnemann S, Piel M, Rosch F, Wester HJ, Henriksen G, Stoeter P, Bartenstein P, Treede RD, Schreckenberger M (2006) High opiate receptor binding potential in the human lateral pain system. Neuroimage 30:692–699. https://doi.org/10.1016/j.neuroimage.2005.10.033

Belcheva MM, Clark AL, Haas PD, Serna JS, Hahn JW, Kiss A, Coscia CJ (2005) Mu and kappa opioid receptors activate ERK/MAPK via different protein kinase C isoforms and secondary messengers in astrocytes. J Biol Chem 280:27662–27669. https://doi.org/10.1074/jbc.M502593200

Benarroch EE (2012) Endogenous opioid systems: current concepts and clinical correlations. Neurology 79:807–814. https://doi.org/10.1212/WNL.0b013e3182662098

Benbouzid M, Choucair-Jaafar N, Yalcin I, Waltisperger E, Muller A, Freund-Mercier MJ, Barrot M (2008) Chronic, but not acute, tricyclic antidepressant treatment alleviates neuropathic allodynia after sciatic nerve cuffing in mice. Eur J Pain 12:1008–1017. https://doi.org/10.1016/j.ejpain.2008.01.010

Bershad AK, Jaffe JH, Childs E, de Wit H (2015) Opioid partial agonist buprenorphine dampens responses to psychosocial stress in humans. Psychoneuroendocrinology 52:281–288. https://doi.org/10.1016/j.psyneuen.2014.12.004

Bershad AK, Seiden JA, de Wit H (2016) Effects of buprenorphine on responses to social stimuli in healthy adults. Psychoneuroendocrinology 63:43–49. https://doi.org/10.1016/j.psyneuen.2015.09.011

Bershad AK, Ruiz NA, de Wit H (2018) Effects of buprenorphine on responses to emotional stimuli in individuals with a range of mood symptomatology. Int J Neuropsychopharmacol 21:120–127. https://doi.org/10.1093/ijnp/pyx077

Bilecki W, Wawrzczak-Bargiela A, Przewlocki R (2004) Activation of AP-1 and CRE-dependent gene expression via mu-opioid receptor. J Neurochem 90:874–882. https://doi.org/10.1111/j.1471-4159.2004.02524.x

Bodkin JA, Zornberg GL, Lukas SE, Cole JO (1995) Buprenorphine treatment of refractory depression. J Clin Psychopharmacol 15:49–57. https://doi.org/10.1097/00004714-199502000-00008

Bodnar RJ (2020) Endogenous opiates and behavior: 2018. Peptides 132:170348. https://doi.org/10.1016/j.peptides.2020.170348

Böhm C, Newrzella D, Herberger S, Schramm N, Eisenhardt G, Schenk V, Sonntag-Buck V, Sorgenfrei O (2006) Effects of antidepressant treatment on gene expression profile in mouse brain: cell type-specific transcription profiling using laser microdissection and microarray analysis. J Neurochem 97(Suppl 1):44–49. https://doi.org/10.1111/j.1471-4159.2006.03750.x

Bonaventura J, Lam S, Carlton M, Boehm MA, Gomez JL, Solis O, Sanchez-Soto M, Morris PJ, Fredriksson I, Thomas CJ, Sibley DR, Shaham Y, Zarate CA Jr, Michaelides M (2021) Pharmacological and behavioral divergence of ketamine enantiomers: implications for abuse liability. Mol Psychiatry. https://doi.org/10.1038/s41380-021-01093-2

Broadbear JH, Winger G, Woods JH (2004) Self-administration of fentanyl, cocaine and ketamine: effects on the pituitary-adrenal axis in rhesus monkeys. Psychopharmacology (Berl) 176:398–406. https://doi.org/10.1007/s00213-004-1891-x

Bruchas MR, Land BB, Aita M, Xu M, Barot SK, Li S, Chavkin C (2007) Stress-induced p38 mitogen-activated protein kinase activation mediates kappa-opioid-dependent dysphoria. J Neurosci 27:11614–11623. https://doi.org/10.1523/JNEUROSCI.3769-07.2007

Bruchas MR, Xu M, Chavkin C (2008) Repeated swim stress induces kappa opioid-mediated activation of extracellular signal-regulated kinase 1/2. Neuroreport 19:1417–1422. https://doi.org/10.1097/WNR.0b013e32830dd655

Carazo-Arias E, Nguyen PT, Kass M, Jee HJ, Nautiyal KM, Magalong V, Coie L, Andreu V, Gergues MM, Khalil H, Akil H, Arcego DM, Meaney M, Anacker C, Samuels BA, Pintar JE, Morozova I, Kalachikov S, Hen R (2022) Contribution of the opioid system to the antidepressant effects of fluoxetine. Biol Psychiatry 92:952–963. https://doi.org/10.1016/j.biopsych.2022.05.030

Carlezon WA Jr, Krystal AD (2016) Kappa-opioid antagonists for psychiatric disorders: from bench to clinical trials. Depress Anxiety 33:895–906. https://doi.org/10.1002/da.22500

Carlezon WA Jr, Duman RS, Nestler EJ (2005) The many faces of CREB. Trends Neurosci 28:436–445. https://doi.org/10.1016/j.tins.2005.06.005

Carlezon WA Jr, Beguin C, DiNieri JA, Baumann MH, Richards MR, Todtenkopf MS, Rothman RB, Ma Z, Lee DY, Cohen BM (2006) Depressive-like effects of the kappa-opioid receptor agonist salvinorin A on behavior and neurochemistry in rats. J Pharmacol Exp Ther 316:440–447. https://doi.org/10.1124/jpet.105.092304

Chottekalapanda RU, Kalik S, Gresack J, Ayala A, Gao M, Wang W, Meller S, Aly A, Schaefer A, Greengard P (2020) AP-1 controls the p11-dependent antidepressant response. Mol Psychiatry 25:1364–1381. https://doi.org/10.1038/s41380-020-0767-8

Codd EE, Shank RP, Schupsky JJ, Raffa RB (1995) Serotonin and norepinephrine uptake inhibiting activity of centrally acting analgesics: structural determinants and role in antinociception. J Pharmacol Exp Ther 274:1263–1270

Cuitavi J, Torres-Pérez JV, Lorente JD, Campos-Jurado Y, Andrés-Herrera P, Polache A, Agustín-Pavón C, Hipólito L (2023) Crosstalk between mu-opioid receptors and neuroinflammation: consequences for drug addiction and pain. Neurosci Biobehav Rev 145:105011. https://doi.org/10.1016/j.neubiorev.2022.105011

Dagher M, Cahill CM, Evans CJ (2022) Opioid systems and depression: the relationship is strengthening. Biol Psychiatry 92:920–922. https://doi.org/10.1016/j.biopsych.2022.09.020

Devine DP, Leone P, Pocock D, Wise RA (1993) Differential involvement of ventral tegmental mu, delta and kappa opioid receptors in modulation of basal mesolimbic dopamine release: in vivo microdialysis studies. J Pharmacol Exp Ther 266:1236–1246

Dewall CN, Macdonald G, Webster GD, Masten CL, Baumeister RF, Powell C, Combs D, Schurtz DR, Stillman TF, Tice DM, Eisenberger NI (2010) Acetaminophen reduces social pain: behavioral and neural evidence. Psychol Sci 21:931–937. https://doi.org/10.1177/0956797610374741

Donahue RJ, Landino SM, Golden SA, Carroll FI, Russo SJ, Carlezon WA Jr (2015) Effects of acute and chronic social defeat stress are differentially mediated by the dynorphin/kappa-opioid receptor system. Behav Pharmacol 26:654–663. https://doi.org/10.1097/FBP.0000000000000155

Dowlati Y, Herrmann N, Swardfager W, Liu H, Sham L, Reim EK, Lanctôt KL (2010) A meta-analysis of cytokines in major depression. Biol Psychiatry 67:446–457. https://doi.org/10.1016/j.biopsych.2009.09.033

Drolet G, Dumont EC, Gosselin I, Kinkead R, Laforest S, Trottier JF (2001) Role of endogenous opioid system in the regulation of the stress response. Prog Neuropsychopharmacol Biol Psychiatry 25:729–741. https://doi.org/10.1016/s0278-5846(01)00161-0

Duman RS (2004) Neural plasticity: consequences of stress and actions of antidepressant treatment. Dialogues Clin Neurosci 6:157–169. https://doi.org/10.31887/DCNS.2004.6.2/rduman

Duman RS, Monteggia LM (2006) A neurotrophic model for stress-related mood disorders. Biol Psychiatry 59:1116–1127. https://doi.org/10.1016/j.biopsych.2006.02.013

Ehrich E, Turncliff R, Du Y, Leigh-Pemberton R, Fernandez E, Jones R, Fava M (2015) Evaluation of opioid modulation in major depressive disorder. Neuropsychopharmacology 40:1448–1455. https://doi.org/10.1038/npp.2014.330

Eisch AJ, Barrot M, Schad CA, Self DW, Nestler EJ (2000) Opiates inhibit neurogenesis in the adult rat hippocampus. Proc Natl Acad Sci U S A 97:7579–7584. https://doi.org/10.1073/pnas.120552597

Eisenberger NI (2012) The pain of social disconnection: examining the shared neural underpinnings of physical and social pain. Nat Rev Neurosci 13:421–434. https://doi.org/10.1038/nrn3231

Fabre-Nys C, Meller RE, Keverne EB (1982) Opiate antagonists stimulate affiliative behaviour in monkeys. Pharmacol Biochem Behav 16:653–659. https://doi.org/10.1016/0091-3057(82)90432-4

Fadda P, Scherma M, Fresu A, Collu M, Fratta W (2005) Dopamine and serotonin release in dorsal striatum and nucleus accumbens is differentially modulated by morphine in DBA/2J and C57BL/6J mice. Synapse 56:29–38. https://doi.org/10.1002/syn.20122

Fava M, Memisoglu A, Thase ME, Bodkin JA, Trivedi MH, de Somer M, Du Y, Leigh-Pemberton-R, DiPetrillo L, Silverman B, Ehrich E (2016) Opioid modulation with buprenorphine/Samidorphan as adjunctive treatment for inadequate response to antidepressants: a randomized double-blind placebo-controlled trial. Am J Psychiatry 173:499–508. https://doi.org/10.1176/appi.ajp.2015.15070921

Fava M, Mazzone E, Freeman M, Flynn M, Judge H, Hoeppner B, Hock RS, Shui A, Macaluso M, Morrison MF, Carpenter LL, Shelton R, Zajecka J, Papakostas GI (2020a) Double-blind, placebo-controlled, proof-of-concept trial of a kappa-selective opioid receptor antagonist augmentation in treatment-resistant depression. Ann Clin Psychiatry 32:18–26. https://doi.org/10.12788/acp.0003

Fava M, Thase ME, Trivedi MH, Ehrich E, Martin WF, Memisoglu A, Nangia N, Stanford AD, Yu M, Pathak S (2020b) Opioid system modulation with buprenorphine/samidorphan combination for major depressive disorder: two randomized controlled studies. Mol Psychiatry 25:1580–1591. https://doi.org/10.1038/s41380-018-0284-1

Fava M, Stahl S, Pani L, De Martin S, Pappagallo M, Guidetti C, Alimonti A, Bettini E, Mangano RM, Wessel T, de Somer M, Caron J, Vitolo OV, DiGuglielmo GR, Gilbert A, Mehta H, Kearney M, Mattarei A, Gentilucci M, Folli F, Traversa S, Inturrisi CE, Manfredi PL (2022) REL-1017 (Esmethadone) as adjunctive treatment in patients with major depressive disorder: a phase 2a randomized double-blind trial. Am J Psychiatry 179:122–131. https://doi.org/10.1176/appi.ajp.2021.21020197

Gassaway MM, Rives ML, Kruegel AC, Javitch JA, Sames D (2014) The atypical antidepressant and neurorestorative agent tianeptine is a mu-opioid receptor agonist. Transl Psychiatry 4:e411. https://doi.org/10.1038/tp.2014.30

Gibney SM, Drexhage HA (2013) Evidence for a dysregulated immune system in the etiology of psychiatric disorders. J Neuroimmune Pharmacol 8:900–920. https://doi.org/10.1007/s11481-013-9462-8

Gorman AL, Elliott KJ, Inturrisi CE (1997) The d- and l-isomers of methadone bind to the non-competitive site on the N-methyl-D-aspartate (NMDA) receptor in rat forebrain and spinal cord. Neurosci Lett 223:5–8. https://doi.org/10.1016/s0304-3940(97)13391-2

Gourley SL, Wu FJ, Kiraly DD, Ploski JE, Kedves AT, Duman RS, Taylor JR (2008) Regionally specific regulation of ERK MAP kinase in a model of antidepressant-sensitive chronic depression. Biol Psychiatry 63:353–359. https://doi.org/10.1016/j.biopsych.2007.07.016

Grilli M, Neri E, Zappettini S, Massa F, Bisio A, Romussi G, Marchi M, Pittaluga A (2009) Salvinorin A exerts opposite presynaptic controls on neurotransmitter exocytosis from mouse brain nerve terminals. Neuropharmacology 57:523–530. https://doi.org/10.1016/j.neuropharm.2009.07.023

Grimwood S, Lu Y, Schmidt AW, Vanase-Frawley MA, Sawant-Basak A, Miller E, McLean S, Freeman J, Wong S, McLaughlin JP, Verhoest PR (2011) Pharmacological characterization of 2-methyl-N-((2′-(pyrrolidin-1-ylsulfonyl)biphenyl-4-yl)methyl)propan-1-amine (PF-04455242), a high-affinity antagonist selective for kappa-opioid receptors. J Pharmacol Exp Ther 339:555–566. https://doi.org/10.1124/jpet.111.185108

Hamon M, Gozlan H, Bourgoin S, Benoliel JJ, Mauborgne A, Taquet H, Cesselin F, Mico JA (1987) Opioid receptors and neuropeptides in the CNS in rats treated chronically with amoxapine or amitriptyline. Neuropharmacology 26:531–539. https://doi.org/10.1016/0028-3908(87)90144-4

Hanson ND, Owens MJ, Nemeroff CB (2011) Depression, antidepressants, and neurogenesis: a critical reappraisal. Neuropsychopharmacology 36:2589–2602. https://doi.org/10.1038/npp.2011.220

Harburg GC, Hall FS, Harrist AV, Sora I, Uhl GR, Eisch AJ (2007) Knockout of the mu opioid receptor enhances the survival of adult-generated hippocampal granule cell neurons. Neuroscience 144:77–87. https://doi.org/10.1016/j.neuroscience.2006.09.018

Hsu DT, Sanford BJ, Meyers KK, Love TM, Hazlett KE, Wang H, Ni L, Walker SJ, Mickey BJ, Korycinski ST, Koeppe RA, Crocker JK, Langenecker SA, Zubieta JK (2013) Response of the mu-opioid system to social rejection and acceptance. Mol Psychiatry 18:1211–1217. https://doi.org/10.1038/mp.2013.96

Hsu DT, Sanford BJ, Meyers KK, Love TM, Hazlett KE, Walker SJ, Mickey BJ, Koeppe RA, Langenecker SA, Zubieta JK (2015) It still hurts: altered endogenous opioid activity in the brain during social rejection and acceptance in major depressive disorder. Mol Psychiatry 20:193–200. https://doi.org/10.1038/mp.2014.185

Ide S, Fujiwara S, Fujiwara M, Sora I, Ikeda K, Minami M, Uhl GR, Ishihara K (2010) Antidepressant-like effect of venlafaxine is abolished in μ-opioid receptor knockout mice. J Pharmacol Sci 114:107–110

Iob E, Kirschbaum C, Steptoe A (2018) Positive and negative social support and HPA-axis hyperactivity: evidence from glucocorticoids in human hair. Psychoneuroendocrinology 96:100–108. https://doi.org/10.1016/j.psyneuen.2018.06.008

Jacobson ML, Wulf HA, Browne CA, Lucki I (2020) The kappa opioid receptor antagonist aticaprant reverses behavioral effects from unpredictable chronic mild stress in male mice. Psychopharmacology (Berl) 237:3715–3728. https://doi.org/10.1007/s00213-020-05649-y

Jelen LA, Young AH, Stone JM (2021) Ketamine: a tale of two enantiomers. J Psychopharmacol 35:109–123. https://doi.org/10.1177/0269881120959644

Jelen LA, Stone JM, Young AH, Mehta MA (2022) The opioid system in depression. Neurosci Biobehav Rev 140:104800. https://doi.org/10.1016/j.neubiorev.2022.104800

Ji RR, Zhang Q, Law PY, Low HH, Elde R, Hokfelt T (1995) Expression of mu-, delta-, and kappa-opioid receptor-like immunoreactivities in rat dorsal root ganglia after carrageenan-induced inflammation. J Neurosci 15:8156–8166. https://doi.org/10.1523/JNEUROSCI.15-12-08156.1995

Kahn L, Alonso G, Normand E, Manzoni OJ (2005) Repeated morphine treatment alters polysialylated neural cell adhesion molecule, glutamate decarboxylase-67 expression and cell proliferation in the adult rat hippocampus. Eur J Neurosci 21:493–500. https://doi.org/10.1111/j.1460-9568.2005.03883.x

Kalin NH, Shelton SE, Barksdale CM (1988) Opiate modulation of separation-induced distress in non-human primates. Brain Res 440:285–292. https://doi.org/10.1016/0006-8993(88)90997-3

Karp JF, Butters MA, Begley AE, Miller MD, Lenze EJ, Blumberger DM, Mulsant BH, Reynolds CF 3rd (2014) Safety, tolerability, and clinical effect of low-dose buprenorphine for treatment-resistant depression in midlife and older adults. J Clin Psychiatry 75:e785–e793. https://doi.org/10.4088/JCP.13m08725

Kennedy SE, Koeppe RA, Young EA, Zubieta JK (2006) Dysregulation of endogenous opioid emotion regulation circuitry in major depression in women. Arch Gen Psychiatry 63:1199–1208. https://doi.org/10.1001/archpsyc.63.11.1199

King AC, Schluger J, Gunduz M, Borg L, Perret G, Ho A, Kreek MJ (2002) Hypothalamic-pituitary-adrenocortical (HPA) axis response and biotransformation of oral naltrexone: preliminary examination of relationship to family history of alcoholism. Neuropsychopharmacology 26:778–788. https://doi.org/10.1016/S0893-133X(01)00416-X

Klein ME, Chandra J, Sheriff S, Malinow R (2020) Opioid system is necessary but not sufficient for antidepressive actions of ketamine in rodents. Proc Natl Acad Sci U S A 117:2656–2662. https://doi.org/10.1073/pnas.1916570117

Knoll AT, Carlezon WA Jr (2010) Dynorphin, stress, and depression. Brain Res 1314:56–73. https://doi.org/10.1016/j.brainres.2009.09.074

Koehl M, Meerlo P, Gonzales D, Rontal A, Turek FW, Abrous DN (2008) Exercise-induced promotion of hippocampal cell proliferation requires beta-endorphin. FASEB J 22:2253–2262. https://doi.org/10.1096/fj.07-099101

Komatsu H, Ohara A, Sasaki K, Abe H, Hattori H, Hall FS, Uhl GR, Sora I (2011) Decreased response to social defeat stress in μ-opioid-receptor knockout mice. Pharmacol Biochem Behav 99:676–682. https://doi.org/10.1016/j.pbb.2011.06.008

Kraepelin E (1901) Psychiatrie. Рипол Классик

Kreibich A, Reyes BAS, Curtis AL, Ecke L, Chavkin C, Van Bockstaele EJ, Valentino RJ (2008) Presynaptic inhibition of diverse afferents to the locus ceruleus by kappa-opiate receptors: a novel mechanism for regulating the central norepinephrine system. J Neurosci 28:6516–6525. https://doi.org/10.1523/JNEUROSCI.0390-08.2008

Krystal AD, Pizzagalli DA, Smoski M, Mathew SJ, Nurnberger J, Lisanby SH, Iosifescu D, Murrough JW, Yang H, Weiner RD, Calabrese JR, Sanacora G, Hermes G, Keefe RSE, Song A, Goodman W, Szabo ST, Whitton AE, Gao K, Potter WZ (2020) A randomized proof-of-mechanism trial applying the 'fast-fail' approach to evaluating κ-opioid antagonism as a treatment for anhedonia. Nat Med 26:760–768. https://doi.org/10.1038/s41591-020-0806-7

Lalanne L, Ayranci G, Kieffer BL, Lutz PE (2014) The kappa opioid receptor: from addiction to depression, and back. Front Psych 5:170. https://doi.org/10.3389/fpsyt.2014.00170

Land BB, Bruchas MR, Lemos JC, Xu M, Melief EJ, Chavkin C (2008) The dysphoric component of stress is encoded by activation of the dynorphin κ-opioid system. J Neurosci 28:407–414. https://doi.org/10.1523/JNEUROSCI.4458-07.2008

Le Merrer J, Becker JA, Befort K, Kieffer BL (2009) Reward processing by the opioid system in the brain. Physiol Rev 89:1379–1412. https://doi.org/10.1152/physrev.00005.2009

Lee H, Im J, Won H, Nam W, Kim YO, Lee SW, Lee S, Cho IH, Kim HK, Kwon JT, Kim HJ (2017) Effects of tianeptine on symptoms of fibromyalgia via BDNF signaling in a fibromyalgia animal model. Korean J Physiol Pharmacol 21:361–370. https://doi.org/10.4196/kjpp.2017.21.4.361

Lepack AE, Bang E, Lee B, Dwyer JM, Duman RS (2016) Fast-acting antidepressants rapidly stimulate ERK signaling and BDNF release in primary neuronal cultures. Neuropharmacology 111:242–252. https://doi.org/10.1016/j.neuropharm.2016.09.011

Levine JD, Gordon NC, Fields HL (1978) The mechanism of placebo analgesia. Lancet 2:654–657. https://doi.org/10.1016/s0140-6736(78)92762-9

Liu Y, Zhao J, Guo W (2018) Emotional roles of mono-aminergic neurotransmitters in major depressive disorder and anxiety disorders. Front Psychol 9:2201. https://doi.org/10.3389/fpsyg.2018.02201

Lutz PE, Kieffer BL (2013) Opioid receptors: distinct roles in mood disorders. Trends Neurosci 36:195–206. https://doi.org/10.1016/j.tins.2012.11.002

Lutz PE, Courtet P, Calati R (2020) The opioid system and the social brain: implications for depression and suicide. J Neurosci Res 98:588–600. https://doi.org/10.1002/jnr.24269

MacLean PD (1993) Cerebral evolution of emotion. In: Lewis M, Haviland J (eds) Handbook of emotions. Guilford Press, New York, pp 67–83

Marchand F, Ardid D, Chapuy E, Alloui A, Jourdan D, Eschalier A (2003) Evidence for an involvement of supraspinal delta and spinal mu opioid receptors in the antihyperalgesic effect of chronically administered clomipramine in mononeuropathic rats. J Pharmacol Exp Ther. https://doi.org/10.1124/jpet.103.052613

Martinowich K, Manji H, Lu B (2007) New insights into BDNF function in depression and anxiety. Nat Neurosci 10:1089–1093. https://doi.org/10.1038/nn1971

McLaughlin JP, Marton-Popovici M, Chavkin C (2003) Kappa opioid receptor antagonism and prodynorphin gene disruption block stress-induced behavioral responses. J Neurosci 23:5674–5683

McLaughlin JP, Li S, Valdez J, Chavkin TA, Chavkin C (2006) Social defeat stress-induced behavioral responses are mediated by the endogenous kappa opioid system. Neuropsychopharmacology 31:1241–1248. https://doi.org/10.1038/sj.npp.1300872

Meyer-Lindenberg A, Tost H (2012) Neural mechanisms of social risk for psychiatric disorders. Nat Neurosci 15:663–668. https://doi.org/10.1038/nn.3083

Mongan L, Callaway E (1990) Buprenorphine responders. Biol Psychiatry 28:1078–1080. https://doi.org/10.1016/0006-3223(90)90619-d

Mousa SA, Shakibaei M, Sitte N, Schäfer M, Stein C (2004) Subcellular pathways of β-endorphin synthesis, processing, and release from immunocytes in inflammatory pain. Endocrinology 145:1331–1341. https://doi.org/10.1210/en.2003-1287

Müller N, Krause D, Barth R, Myint A-M, Weidinger E, Stettinger W, Zill P, Drexhage H, Schwarz MJ (2019) Childhood adversity and current stress are related to pro- and anti-inflammatory cytokines in major depression. J Affect Disord 253:270–276. https://doi.org/10.1016/j.jad.2019.04.088

Muschamp JW, Van't Veer A, Parsegian A, Gallo MS, Chen M, Neve RL, Meloni EG, Carlezon WA Jr (2011) Activation of CREB in the nucleus accumbens shell produces anhedonia and resistance to extinction of fear in rats. J Neurosci 31:3095–3103. https://doi.org/10.1523/JNEUROSCI.5973-10.2011

Narita M, Kuzumaki N, Miyatake M, Sato F, Wachi H, Seyama Y, Suzuki T (2006) Role of δ-opioid receptor function in neurogenesis and neuroprotection. J Neurochem 97:1494–1505. https://doi.org/10.1111/j.1471-4159.2006.03849.x

Newton SS, Thome J, Wallace TL, Shirayama Y, Schlesinger L, Sakai N, Chen J, Neve R, Nestler EJ, Duman RS (2002) Inhibition of cAMP response element-binding protein or dynorphin in the nucleus accumbens produces an antidepressant-like effect. J Neurosci 22:10883–10890. https://doi.org/10.1523/JNEUROSCI.22-24-10883.2002

Nibuya M, Nestler EJ, Duman RS (1996) Chronic antidepressant administration increases the expression of cAMP response element binding protein (CREB) in rat hippocampus. J Neurosci 16:2365–2372

Ninković J, Roy S (2013) Role of the mu opioid receptor in opioid modulation of immune function. Amino Acids 45:9–24. https://doi.org/10.1007/s00726-011-1163-0

Nummenmaa L, Karjalainen T, Isojarvi J, Kantonen T, Tuisku J, Kaasinen V, Joutsa J, Nuutila P, Kalliokoski K, Hirvonen J, Hietala J, Rinne J (2020) Lowered endogenous mu-opioid receptor availability in subclinical depression and anxiety. Neuropsychopharmacology 45:1953–1959. https://doi.org/10.1038/s41386-020-0725-9

Nyhuis PW, Gastpar M, Scherbaum N (2008) Opiate treatment in depression refractory to antidepressants and electroconvulsive therapy. J Clin Psychopharmacol 28:593–595. https://doi.org/10.1097/JCP.0b013e31818638a4

Olianas MC, Dedoni S, Onali P (2012) The atypical antidepressant mianserin exhibits agonist activity at κ-opioid receptors. Br J Pharmacol 167:1329–1341. https://doi.org/10.1111/j.1476-5381.2012.02078.x

Onali P, Dedoni S, Olianas MC (2010) Direct agonist activity of tricyclic antidepressants at distinct opioid receptor subtypes. J Pharmacol Exp Ther 332:255–265. https://doi.org/10.1124/jpet.109.159939

Panksepp J, Herman B, Conner R, Bishop P, Scott JP (1978) The biology of social attachments: opiates alleviate separation distress. Biol Psychiatry 13:607–618

Pariante CM (2006) The glucocorticoid receptor: part of the solution or part of the problem? J Psychopharmacol 20:79–84. https://doi.org/10.1177/1359786806066063

Pariante CM, Lightman SL (2008) The HPA axis in major depression: classical theories and new developments. Trends Neurosci 31:464–468. https://doi.org/10.1016/j.tins.2008.06.006

Park H, Poo MM (2013) Neurotrophin regulation of neural circuit development and function. Nat Rev Neurosci 14:7–23. https://doi.org/10.1038/nrn3379

Pascoe JE, Williams KL, Mukhopadhyay P, Rice KC, Woods JH, Ko MC (2008) Effects of mu, kappa, and delta opioid receptor agonists on the function of hypothalamic-pituitary-adrenal axis in monkeys. Psychoneuroendocrinology 33:478–486. https://doi.org/10.1016/j.psyneuen.2008.01.006

Pecina M, Zubieta JK (2015) Molecular mechanisms of placebo responses in humans. Mol Psychiatry 20:416–423. https://doi.org/10.1038/mp.2014.164

Pecina M, Zubieta J-K (2018) Chapter two – expectancy modulation of opioid neurotransmission. In: Colloca L (ed) International review of neurobiology, neurobiology of the placebo effect part I. Academic Press, pp 17–37. https://doi.org/10.1016/bs.irn.2018.02.003

Pecina M, Bohnert AS, Sikora M, Avery ET, Langenecker SA, Mickey BJ, Zubieta JK (2015) Association between placebo-activated neural systems and antidepressant responses: neurochemistry of placebo effects in major depression. JAMA Psychiatry 72:1087–1094. https://doi.org/10.1001/jamapsychiatry.2015.1335

Pecina M, Heffernan J, Wilson J, Zubieta JK, Dombrovski AY (2018) Prefrontal expectancy and reinforcement-driven antidepressant placebo effects. Transl Psychiatry 8:222. https://doi.org/10.1038/s41398-018-0263-y

Pecina M, Karp JF, Mathew S, Todtenkopf MS, Ehrich EW, Zubieta JK (2019) Endogenous opioid system dysregulation in depression: implications for new therapeutic approaches. Mol Psychiatry 24:576–587. https://doi.org/10.1038/s41380-018-0117-2

Pecina M, Chen J, Lyew T, Karp JF, Dombrovski AY (2021) Mu opioid antagonist naltrexone partially abolishes the antidepressant placebo effect and reduces orbitofrontal cortex encoding of reinforcement. Biol Psychiatry Cogn Neurosci Neuroimaging. https://doi.org/10.1016/j.bpsc.2021.02.009

Pentney RJ, Gratton A (1991) Effects of local delta and mu opioid receptor activation on basal and stimulated dopamine release in striatum and nucleus accumbens of rat: an in vivo electrochemical study. Neuroscience 45:95–102. https://doi.org/10.1016/0306-4522(91)90106-x

Persson AI, Thorlin T, Bull C, Eriksson PS (2003) Opioid-induced proliferation through the MAPK pathway in cultures of adult hippocampal progenitors. Mol Cell Neurosci 23:360–372. https://doi.org/10.1016/s1044-7431(03)00061-7

Petrovic P, Kalso E, Petersson KM, Ingvar M (2002) Placebo and opioid analgesia – imaging a shared neuronal network. Science 295:1737–1740. https://doi.org/10.1126/science.1067176

Pilozzi A, Carro C, Huang X (2021) Roles of β-endorphin in stress, behavior, neuroinflammation, and brain energy metabolism. Int J Mol Sci 22:338. https://doi.org/10.3390/ijms22010338

Plein LM, Rittner HL (2018) Opioids and the immune system – friend or foe. Br J Pharmacol 175: 2717–2725. https://doi.org/10.1111/bph.13750

Pliakas AM, Carlson RR, Neve RL, Konradi C, Nestler EJ, Carlezon WA Jr (2001) Altered responsiveness to cocaine and increased immobility in the forced swim test associated with elevated cAMP response element-binding protein expression in nucleus accumbens. J Neurosci 21:7397–7403

Pol O, Puig M (2004) Expression of opioid receptors during peripheral inflammation. Curr Top Med Chem 4:51–61. https://doi.org/10.2174/1568026043451519

Prossin AR, Koch AE, Campbell PL, McInnis MG, Zalcman SS, Zubieta J-K (2011) Association of plasma interleukin-18 levels with emotion regulation and μ-opioid neurotransmitter function in major depression and healthy volunteers. Biol Psychiatry 69:808–812. https://doi.org/10.1016/j.biopsych.2010.10.014

Puryear CB, Brooks J, Tan L, Smith K, Li Y, Cunningham J, Todtenkopf MS, Dean RL, Sanchez C (2020) Opioid receptor modulation of neural circuits in depression: what can be learned from preclinical data? Neurosci Biobehav Rev 108:658–678. https://doi.org/10.1016/j.neubiorev.2019.12.007

Qi X, Lin W, Li J, Li H, Wang W, Wang D, Sun M (2008) Fluoxetine increases the activity of the ERK-CREB signal system and alleviates the depressive-like behavior in rats exposed to chronic forced swim stress. Neurobiol Dis 31:278–285. https://doi.org/10.1016/j.nbd.2008.05.003

Rengasamy M, Marsland A, Spada M, Hsiung K, Kovats T, Price RB (2021) A chicken and egg scenario in psychoneuroimmunology: bidirectional mechanisms linking cytokines and depression. J Affect Disord Rep 6:100177. https://doi.org/10.1016/j.jadr.2021.100177

Richards EM, Mathews DC, Luckenbaugh DA, Ionescu DF, Machado-Vieira R, Niciu MJ, Duncan WC, Nolan NM, Franco-Chaves JA, Hudzik T, Maciag C, Li S, Cross A, Smith MA, Zarate CA Jr (2016) A randomized, placebo-controlled pilot trial of the delta opioid receptor agonist AZD2327 in anxious depression. Psychopharmacology (Berl) 233:1119–1130. https://doi.org/10.1007/s00213-015-4195-4

Rittner HL, Machelska H, Stein C (2005) Leukocytes in the regulation of pain and analgesia. J Leukoc Biol 78:1215–1222. https://doi.org/10.1189/jlb.0405223

Rogers TJ (2022) Kappa opioid receptor expression and function in cells of the immune system. Handb Exp Pharmacol 271:419–433. https://doi.org/10.1007/164_2021_441

Rorick-Kehn LM, Witkin JM, Statnick MA, Eberle EL, McKinzie JH, Kahl SD, Forster BM, Wong CJ, Li X, Crile RS, Shaw DB, Sahr AE, Adams BL, Quimby SJ, Diaz N, Jimenez A, Pedregal C, Mitch CH, Knopp KL, Anderson WH, Cramer JW, McKinzie DL (2014) LY2456302 is a novel, potent, orally-bioavailable small molecule kappa-selective antagonist with activity in animal models predictive of efficacy in mood and addictive disorders. Neuropharmacology 77:131–144. https://doi.org/10.1016/j.neuropharm.2013.09.021

Rosenblum A, Marsch LA, Joseph H, Portenoy RK (2008) Opioids and the treatment of chronic pain: controversies, current status, and future directions. Exp Clin Psychopharmacol 16:405–416. https://doi.org/10.1037/a0013628

Rueger SY, Malecki CK, Pyun Y, Aycock C, Coyle S (2016) A meta-analytic review of the association between perceived social support and depression in childhood and adolescence. Psychol Bull 142:1017–1067. https://doi.org/10.1037/bul0000058

Samuels BA, Nautiyal KM, Kruegel AC, Levinstein MR, Magalong VM, Gassaway MM, Grinnell SG, Han J, Ansonoff MA, Pintar JE, Javitch JA, Sames D, Hen R (2017) The behavioral effects of the antidepressant tianeptine require the mu-opioid receptor. Neuropsychopharmacology 42:2052–2063. https://doi.org/10.1038/npp.2017.60

Schoenbaum G, Chiba AA, Gallagher M (1998) Orbitofrontal cortex and basolateral amygdala encode expected outcomes during learning. Nat Neurosci 1:155–159. https://doi.org/10.1038/407

Schreiber S, Backer MM, Herman I, Shamir D, Boniel T, Pick CG (2000) The antinociceptive effect of trazodone in mice is mediated through both μ-opioid and serotonergic mechanisms. Behav Brain Res 114:51–56. https://doi.org/10.1016/S0166-4328(00)00185-6

Schreiber S, Rigai T, Katz Y, Pick CG (2002) The antinociceptive effect of mirtazapine in mice is mediated through serotonergic, noradrenergic and opioid mechanisms. Brain Res Bull 58:601–605. https://doi.org/10.1016/s0361-9230(02)00825-0

Shirayama Y, Ishida H, Iwata M, Hazama GI, Kawahara R, Duman RS (2004) Stress increases dynorphin immunoreactivity in limbic brain regions and dynorphin antagonism produces antidepressant-like effects. J Neurochem 90:1258–1268. https://doi.org/10.1111/j.1471-4159.2004.02589.x

Shoda T, Fukuda K, Uga H, Mima H, Morikawa H (2001) Activation of mu-opioid receptor induces expression of c-fos and junB via mitogen-activated protein kinase cascade. Anesthesiology 95: 983–989. https://doi.org/10.1097/00000542-200110000-00030

Spanagel R, Herz A, Shippenberg TS (1992) Opposing tonically active endogenous opioid systems modulate the mesolimbic dopaminergic pathway. Proc Natl Acad Sci U S A 89:2046–2050. https://doi.org/10.1073/pnas.89.6.2046

Stanford SC (2020) Some reasons why preclinical studies of psychiatric disorders fail to translate: what can be rescued from the misunderstanding and misuse of animal "Models"? Altern Lab Anim 48:106–115. https://doi.org/10.1177/0261192920939876

Tan KZ, Cunningham AM, Joshi A, Oei JL, Ward MC (2018) Expression of kappa opioid receptors in developing rat brain – implications for perinatal buprenorphine exposure. Reprod Toxicol 78: 81–89. https://doi.org/10.1016/j.reprotox.2018.04.006

Tao R, Auerbach SB (1995) Involvement of the dorsal raphe but not median raphe nucleus in morphine-induced increases in serotonin release in the rat forebrain. Neuroscience 68:553–561. https://doi.org/10.1016/0306-4522(95)00154-b

Tao R, Auerbach SB (2002a) GABAergic and glutamatergic afferents in the dorsal raphe nucleus mediate morphine-induced increases in serotonin efflux in the rat central nervous system. J Pharmacol Exp Ther 303:704–710. https://doi.org/10.1124/jpet.102.038133

Tao R, Auerbach SB (2002b) Opioid receptor subtypes differentially modulate serotonin efflux in the rat central nervous system. J Pharmacol Exp Ther 303:549–556. https://doi.org/10.1124/jpet.102.037861

Tenore PL (2008) Psychotherapeutic benefits of opioid agonist therapy. J Addict Dis 27:49–65. https://doi.org/10.1080/10550880802122646

Torregrossa MM, Isgor C, Folk JE, Rice KC, Watson SJ, Woods JH (2004) The delta-opioid receptor agonist (+)BW373U86 regulates BDNF mRNA expression in rats. Neuropsychopharmacology 29:649–659. https://doi.org/10.1038/sj.npp.1300345

Torregrossa MM, Jutkiewicz EM, Mosberg HI, Balboni G, Watson SJ, Woods JH (2006) Peptidic delta opioid receptor agonists produce antidepressant-like effects in the forced swim test and regulate BDNF mRNA expression in rats. Brain Res 1069:172–181. https://doi.org/10.1016/j.brainres.2005.11.005

Üçel Uİ, Can ÖD, Demir Özkay Ü, Öztürk Y (2015) Antihyperalgesic and antiallodynic effects of mianserin on diabetic neuropathic pain: a study on mechanism of action. Eur J Pharmacol 756: 92–106. https://doi.org/10.1016/j.ejphar.2015.02.048

Ur E, Wright DM, Bouloux PM, Grossman A (1997) The effects of spiradoline (U-62066E), a kappa-opioid receptor agonist, on neuroendocrine function in man. Br J Pharmacol 120:781–784. https://doi.org/10.1038/sj.bjp.0700971

Valentino RJ, Van Bockstaele E (2001) Opposing regulation of the locus coeruleus by corticotropin-releasing factor and opioids. Potential for reciprocal interactions between stress and opioid sensitivity. Psychopharmacology (Berl) 158:331–342. https://doi.org/10.1007/s002130000673

Valentino RJ, Van Bockstaele E (2015) Endogenous opioids: the downside of opposing stress. Neurobiol Stress 1:23–32. https://doi.org/10.1016/j.ynstr.2014.09.006

Van Bockstaele EJ, Valentino RJ (2013) Neuropeptide regulation of the locus coeruleus and opiate-induced plasticity of stress responses. Adv Pharmacol 68:405–420. https://doi.org/10.1016/B978-0-12-411512-5.00019-1

Van den Berg CL, Van Ree JM, Spruijt BM, Kitchen I (1999) Effects of juvenile isolation and morphine treatment on social interactions and opioid receptors in adult rats: behavioural and autoradiographic studies. Eur J Neurosci 11:3023–3032. https://doi.org/10.1046/j.1460-9568.1999.00717.x

Van't Veer A, Carlezon WA (2013) Role of kappa-opioid receptors in stress and anxiety-related behavior. Psychopharmacology (Berl) 229:435–452. https://doi.org/10.1007/s00213-013-3195-5

Vanderschuren LJ, Niesink RJ, Spruijt BM, Van Ree JM (1995) Mu- and kappa-opioid receptor-mediated opioid effects on social play in juvenile rats. Eur J Pharmacol 276:257–266. https://doi.org/10.1016/0014-2999(95)00040-r

Wang JQ, Mao L (2019) The ERK pathway: molecular mechanisms and treatment of depression. Mol Neurobiol 56:6197–6205. https://doi.org/10.1007/s12035-019-1524-3

Wang Y, Zhuang Y, DiBerto JF, Zhou XE, Schmitz GP, Yuan Q, Jain MK, Liu W, Melcher K, Jiang Y, Roth BL, Xu HE (2023) Structures of the entire human opioid receptor family. Cell. https://doi.org/10.1016/j.cell.2022.12.026

Warnick JE, McCurdy CR, Sufka KJ (2005) Opioid receptor function in social attachment in young domestic fowl. Behav Brain Res 160:277–285. https://doi.org/10.1016/j.bbr.2004.12.009

Wattiez A-S, Libert F, Privat A-M, Loiodice S, Fialip J, Eschalier A, Courteix C (2011) Evidence for a differential opioidergic involvement in the analgesic effect of antidepressants: prediction for efficacy in animal models of neuropathic pain? Br J Pharmacol 163:792–803. https://doi.org/10.1111/j.1476-5381.2011.01297.x

Weber MM, Emrich HM (1988) Current and historical concepts of opiate treatment in psychiatric disorders. Int Clin Psychopharmacol 3:255–266. https://doi.org/10.1097/00004850-198807000-00007

Williams KL, Ko MC, Rice KC, Woods JH (2003) Effect of opioid receptor antagonists on hypothalamic-pituitary-adrenal activity in rhesus monkeys. Psychoneuroendocrinology 28: 513–528. https://doi.org/10.1016/s0306-4530(02)00037-9

Williams NR, Heifets BD, Blasey C, Sudheimer K, Pannu J, Pankow H, Hawkins J, Birnbaum J, Lyons DM, Rodriguez CI, Schatzberg AF (2018) Attenuation of antidepressant effects of ketamine by opioid receptor antagonism. Am J Psychiatry 175:1205–1215. https://doi.org/10.1176/appi.ajp.2018.18020138

Williams NR, Heifets BD, Bentzley BS, Blasey C, Sudheimer KD, Hawkins J, Lyons DM, Schatzberg AF (2019) Attenuation of antidepressant and antisuicidal effects of ketamine by opioid receptor antagonism. Mol Psychiatry 24:1779–1786. https://doi.org/10.1038/s41380-019-0503-4

Wu S, Ning K, Wang Y, Zhang L, Liu J (2023) Up-regulation of BDNF/TrkB signaling by δ opioid receptor agonist SNC80 modulates depressive-like behaviors in chronic restraint-stressed mice. Eur J Pharmacol 942:175532. https://doi.org/10.1016/j.ejphar.2023.175532

Xu C, Fan W, Zhang Y, Loh HH, Law P-Y (2021) Kappa opioid receptor controls neural stem cell differentiation via a miR-7a/Pax6 dependent pathway. Stem Cells 39:600–616. https://doi.org/10.1002/stem.3334

Yoshioka T, Yamada D, Segi-Nishida E, Nagase H, Saitoh A (2023) KNT-127, a selective delta opioid receptor agonist, shows beneficial effects in the hippocampal dentate gyrus of a chronic vicarious social defeat stress mouse model. Neuropharmacology 232:109511. https://doi.org/10.1016/j.neuropharm.2023.109511

Yovell Y, Bar G, Mashiah M, Baruch Y, Briskman I, Asherov J, Lotan A, Rigbi A, Panksepp J (2016) Ultra-Low-dose buprenorphine as a time-limited treatment for severe suicidal ideation: a randomized controlled trial. Am J Psychiatry 173:491–498. https://doi.org/10.1176/appi.ajp.2015.15040535

Zajecka JM, Stanford AD, Memisoglu A, Martin WF, Pathak S (2019) Buprenorphine/samidorphan combination for the adjunctive treatment of major depressive disorder: results of a phase III clinical trial (FORWARD-3). Neuropsychiatr Dis Treat 15:795–808. https://doi.org/10.2147/NDT.S199245

Zhang H, Torregrossa MM, Jutkiewicz EM, Shi YG, Rice KC, Woods JH, Watson SJ, Ko MC (2006) Endogenous opioids upregulate brain-derived neurotrophic factor mRNA through delta- and micro-opioid receptors independent of antidepressant-like effects. Eur J Neurosci 23:984–994. https://doi.org/10.1111/j.1460-9568.2006.04621.x

Zhang H, Shi YG, Woods JH, Watson SJ, Ko MC (2007) Central kappa-opioid receptor-mediated antidepressant-like effects of nor-Binaltorphimine: behavioral and BDNF mRNA expression studies. Eur J Pharmacol 570:89–96. https://doi.org/10.1016/j.ejphar.2007.05.045

Zhang Y, Gu F, Chen J, Dong W (2010) Chronic antidepressant administration alleviates frontal and hippocampal BDNF deficits in CUMS rat. Brain Res 1366:141–148. https://doi.org/10.1016/j.brainres.2010.09.095

Zhang H, Jia M, Wang X-W, Ye C, Li Y, Wang N, Elefant F, Ma H, Cui C (2019) Dentate gyrus μ-opioid receptor-mediated neurogenic processes are associated with alterations in morphine self-administration. Sci Rep 9:1471. https://doi.org/10.1038/s41598-018-37083-8

Zhang P, Yang M, Chen C, Liu L, Wei X, Zeng S (2020) Toll-like receptor 4 (TLR4)/opioid receptor pathway crosstalk and impact on opioid analgesia, immune function, and gastrointestinal motility. Front Immunol 11:1455. https://doi.org/10.3389/fimmu.2020.01455

Zhang F, Hillhouse TM, Anderson PM, Koppenhaver PO, Kegen TN, Manicka SG, Lane JT, Pottanat E, Van Fossen M, Rice R, Porter JH (2021) Opioid receptor system contributes to the acute and sustained antidepressant-like effects, but not the hyperactivity motor effects of ketamine in mice. Pharmacol Biochem Behav:173228. https://doi.org/10.1016/j.pbb.2021.173228

Zheng H, Loh HH, Law PY (2008) Beta-arrestin-dependent mu-opioid receptor-activated extracellular signal-regulated kinases (ERKs) translocate to nucleus in contrast to G protein-dependent ERK activation. Mol Pharmacol 73:178–190. https://doi.org/10.1124/mol.107.039842

Zhou Y, Leri F (2016) Neuroscience of opiates for addiction medicine: from stress-responsive systems to behavior. Prog Brain Res 223:237–251. https://doi.org/10.1016/bs.pbr.2015.09.001

Zomkowski AD, Santos AR, Rodrigues AL (2005) Evidence for the involvement of the opioid system in the agmatine antidepressant-like effect in the forced swimming test. Neurosci Lett 381:279–283. https://doi.org/10.1016/j.neulet.2005.02.026

Zubieta JK, Smith YR, Bueller JA, Xu Y, Kilbourn MR, Jewett DM, Meyer CR, Koeppe RA, Stohler CS (2001) Regional mu opioid receptor regulation of sensory and affective dimensions of pain. Science 293:311–315. https://doi.org/10.1126/science.1060952

Zubieta JK, Ketter TA, Bueller JA, Xu Y, Kilbourn MR, Young EA, Koeppe RA (2003) Regulation of human affective responses by anterior cingulate and limbic mu-opioid neurotransmission. Arch Gen Psychiatry 60:1145–1153. https://doi.org/10.1001/archpsyc.60.11.1145

Zunszain PA, Hepgul N, Pariante CM (2013) Inflammation and depression. Curr Top Behav Neurosci 14:135–151. https://doi.org/10.1007/7854_2012_211

Neuropsychological Effects of Antidepressants: Translational Studies

Katie Kamenish and Emma S. J. Robinson

Contents

Abstract Pharmacological treatments that improve mood were first identified serendipitously, but more than half a century later, how these drugs induce their antidepressant effects remains largely unknown. With the help of animal models, a detailed understanding of their pharmacological targets and acute and chronic effects on brain chemistry and neuronal function has been achieved, but it remains to be elucidated how these effects translate to clinical efficacy. Whilst the field has been dominated by the monoamine and neurotrophic hypotheses, the idea that the maladaptive cognitive process plays a critical role in the development and perpetuation of mood disorders has been discussed since the 1950s. Recently, studies using objective methods to quantify changes in emotional processing found acute effects with conventional antidepressants in both healthy volunteers and patients. These positive effects on emotional processing and cognition occur without a change in the

K. Kamenish and E. S. J. Robinson (✉)
School of Physiology, Pharmacology and Neuroscience, University of Bristol, University Walk, Bristol, UK
e-mail: pmesjr@bris.ac.uk

Curr Topics Behav Neurosci (2024) 66: 101–130
https://doi.org/10.1007/7854_2023_446
Published Online: 14 November 2023

subjective ratings of mood. Building from these studies, behavioural methods for animals that quantify similar cognitive affective processes have been developed. Integrating these behavioural approaches with pharmacology and targeted brain manipulations, a picture is beginning to emerge of the underlying mechanisms that may link the pharmacology of antidepressants, these neuropsychological constructs and clinical efficacy. In this chapter, we discuss findings from animal studies, experimental medicine and patients investigating the neuropsychological effects of antidepressant drugs. We discuss the possible neural circuits that contribute to these effects and discuss whether a neuropsychological model of antidepressant effects could explain the temporal differences in clinical benefits observed with conventional delayed-onset antidepressants versus rapid-acting antidepressants.

Keywords Affective biases · Amygdala · Delayed-onset antidepressant · Prefrontal cortex · Rapid-acting antidepressant

1 Introduction

Understanding exactly how antidepressant drugs achieve their clinical benefits remains unknown despite more than half a century of research and billions of dollars of investment. There is also a debate regarding how effective antidepressant drugs are in patients with different groups, making arguments for and against their clinical efficacy (Ioannidis 2008; Pigott et al. 2010; Gartlehner et al. 2017; Ormel et al. 2022). While the consensus suggests that antidepressant drugs have benefits over placebo, particularly in longer-term outcomes and the prevention of relapse (Geddes et al. 2003; Montgomery et al. 2007; Baldessarini et al. 2015; Cipriani et al. 2018), there remains a significant proportion of patients who do not respond, and why this occurs is not known. The major problem for all psychiatry research is that the drug treatments being used today were discovered serendipitously and long before methods existed to study where in the brain they were working and which neurotransmitter systems and neural circuits were involved. As scientific methodology advanced, the biological targets for these drugs were identified. Fundamental research also revealed the role of certain neurotransmitter systems in modulating emotional behaviour, and based on this, a focus on monoamines and particularly serotonin emerged. From this work, second-generation antidepressants were developed but still without any clear understanding of how a drug that causes a global change in serotonin and/or noradrenaline in the brain would lead to a reduction in symptoms of low mood, rumination of negative thoughts as well as all the somatic/physiological symptoms many patients experience. It is also unclear why most antidepressants are effective in both anxiety disorders and major depressive disorders (MDD) or why their clinical benefits take time to develop, i.e. they have a delayed onset of action. To further add to the complexity of the picture, in 2000, Berman and colleagues reported on the rapid and sustained antidepressant effects of a single low dose of the N-methyl-D-aspartate (NMDA) antagonist ketamine in treatment-resistant patients with MDD. Prior to the findings with ketamine, the established dogma in the field was that conventional first- and second-generation antidepressants had a delayed onset of action, and most of the preceding research had focused on understanding the mechanisms that underlie these delays. Most recently, the field has seen a resurgence in interest in psychedelics, with evidence for rapid and sustained antidepressant effects following a single treatment with psilocybin recently reported in a phase 2b clinical trial (Goodwin et al. 2022). A final point to consider is that for many patients with mild to moderate depression, psychological and behavioural therapies can be as effective as pharmacological treatments. In this article, we focus on neuropsychological mechanisms and findings from animal models, as well as experimental medicine studies in healthy volunteers and patients with MDD.

2 The Role of Animal Models in Research into Antidepressant Mechanisms

Since the discovery of drugs that could treat the emotional symptoms associated with mood disorders, the challenge has been to try to link these to the biochemical effects of the drugs. As is common with other disciplines in medicine, the use of animal models is critical to this endeavour. Although there have been huge advances in human brain imaging, the resolution of these methods as well as the ethical and practical complexities of studying the living human brain means methods in animals are essential. Animal models of depression were first developed to provide a pharmacological screen to develop new drugs based on the in vivo behavioural effects of drugs that were known to be effective in patients – *predictive validity*. Probably the most widely used methods are the forced swim test (FST) and tail suspension test (TST) (Steru et al. 1985; Castagné et al. 2010). When first published in the 1970s, the FST was reported to differentiate between the antidepressants of that era (tricyclic antidepressants and monoamine oxidase inhibitors) and other psychoactive drugs. However, as research shifted from identifying drugs with similar pharmacological and behavioural effects, and hence with potential as anti-depressants, to trying to understand underlying mechanisms and to identify novel drug targets, the way these behavioural tests were used also evolved. As the limitations of these models started to become apparent, other methods, such as the sucrose preference test (a measure aligned to anhedonia) and the novelty-suppressed feeding test (a measure of anxiety-related behaviour), were incorporated into pre-clinical research. A detailed discussion relating to the validity and use of different animal models of depression and behavioural readouts is beyond the scope of this chapter, but reviews on the topic have been published elsewhere (Willner 1984; Robbins 1998; Belzung and Lemoine 2011). Much of what we think we know about the molecular mechanisms that underlie the efficacy of antidepressant drugs has arisen from studies using these behavioural tasks. Both the monoamine hypothesis (Richelson 1996; Hirschfeld 2000) and the neurotrophic hypothesis (Boku et al. 2018; Levy et al. 2018) are strongly supported by findings in these pre-clinical models. For example, the ability of antidepressants to induce acute changes in monoamine transmitter levels as well as chronic treatments leading to adaptive changes in both pre- and post-synaptic monoamine receptors is strongly supported by studies testing antidepressants in these behavioural tests (Bodnoff et al. 1988; Detke et al. 1997; Rénéric and Lucki 1998; Cryan et al. 2005). Studies using exposure to chronic stress in rodents suggest a causal role in the neurodegenerative effects of stress in the brain as well as a reversal of these deficits by antidepressants (Willner 1997; Rygula et al. 2008; Hare et al. 2017; Duman 2022). However, because the behavioural readouts these studies rely on do not readily align with any of the measures used in patients, it is not possible to know if these same mechanisms are involved in human mood disorders and the clinical effects seen with antidepressants. It is also impossible to directly relate the behavioural changes observed in an animal model with the subjective self-reported symptoms which define emotional impairments in clinical populations, and so the clinical relevance of these findings is difficult to substantiate. Despite a huge body of literature arising from animal studies using these tests, findings have largely failed to translate to clinical benefits, raising concerns about exactly what these models are measuring and highlighting the need for alternative pre-clinical approaches (Nestler and Hyman 2010; Berton et al. 2012; O'Leary and Cryan 2013; Harro 2019). Most studies into the effects of antidepressants on human subjects use subjective self-report measures based on questionnaires or structured clinical interviews; however, these can never be directly translated to animal studies. Although some attempts to align these self-reported symptoms with behaviours in animals have been made (Slattery and Cryan 2017), this inevitably involves a degree of anthropomorphism, and the lack of either a reliable biomarker for MDD (Holmes 2003; Jentsch et al. 2015; Slaney et al. 2018) or a well-defined clinical phenotype means that neither a *construct* nor *face* validity is readily achieved (Willner 1991; Anisman and Matheson 2005; Hao et al. 2019).

3 Neuropsychological Impairments in Mood Disorders

In the same era as the emergence of pharmacological treatments for MDD, a cognitive theory of depression was also proposed and laid the foundations for the development of cognitive behavioural therapies (Beck 1979; Wright and Beck 1983). Aaron Beck (1970) hypothesised that depressed patients became engaged in a negative cognitive triad involving distorted thoughts regarding the self, relationships with others and the future. It is thought that adverse life events give rise to the formation of these dysfunctional thought patterns, and the repeated activation of these cognitive distortions in response to stressors is what drives the development and perpetuation of depressive symptoms in vulnerable individuals (Beck 1967; Dozois and Beck 2008). A similar concept underlies the hopelessness theory

(Abramson et al. 1989; Liu et al. 2015), which suggests that those with depression have developed a stable and global shift in expectation that negative events will occur and cannot be changed, resulting in a cognitive processing bias that drives symptoms such as a loss of interest in rewarding activities, feelings of hopelessness and a pessimistic view of future events, which are also linked to self-worth (Abramson et al. 1989). Like the more biologically focused hypotheses of MDD, advances in methodologies have seen the evolution of these cognitive theories of depression (Pessoa and Adolphs 2010; Elliott et al. 2011; Roiser and Sahakian 2017). With the development of brain imaging techniques, particularly functional magnetic resonance imaging (fMRI) and computer-based behavioural tasks, it is now possible to use the objective method to quantify the impact of emotions on cognition and brain function. Using these approaches, evidence of cognitive impairments in mood disorders has emerged that reveals how emotional state directly affects the way information is processed (Leppänen 2006; Price and Drevets 2011; Beck and Clark 2015). These biases in cognition can be observed and quantified across many different cognitive domains, including evidence that MDD and anxiety disorders lead to biases in attention, learning and, memory and decision-making (Leppänen et al. 2004; Broomfield et al. 2007; Disner et al. 2011). Traditionally, information processing linked to the emotional state has been described as 'hot' cognition, whilst 'cold' cognition relates to the processing of information in the absence of an emotionally salient response or cue (Roiser and Sahakian 2013). However, in patients with mood disorders, these internal negative representations generate an emotional state that may impact all aspects of cognition, leading to distortions within domains of both hot and cold cognition and a pervasive negative bias in information processing, i.e. an *affective bias*. Given a person's mood arises from their past experiences and memories and how these influence current and future expectations, these negative biases represent a potentially important neuropsychological mechanism in mood disorders and a target for antidepressant treatment.

There are several studies reporting an increased likelihood of depressed patients exhibiting negative affective biases across multiple domains (Beck and Clark 2015; LeMoult and Gotlib 2019). Depressed individuals are more likely to recall negative information concerning themselves, interpret ambiguous stimuli as negative and demonstrate an attentional bias towards negatively valenced emotional information (Kuiper et al. 1982; Mathews and Macleod 2005; Lee et al. 2016; Orchard et al. 2016; Disner et al. 2017). In a task of emotional recognition, patients were slower to recognise happy faces and demonstrated heightened sensitivity to faces displaying negative emotions such as fear and sadness. Negative affective biases are also associated with a higher risk of relapse and can persist even in recovered patients. In addition, increased negative affective processing has been observed in healthy volunteers deemed to be at high risk of developing depression. These neuropsychological impairments may therefore represent a *cognitive biomarker* that can support the development of a translational approach to studying emotional disorders across species.

Neuropsychological mechanisms may also be useful in understanding the dysfunctional reward processing thought to play a key role in the pathophysiology of depression. Anhedonia, defined as a reduced capacity to experience pleasure from previously rewarding activities, is one of the two core symptoms required by the Diagnostic and Statistical Manual of Mental Disorders, Fifth Edition (DSM-5), for a depression diagnosis. Reward-related behaviours in depressed patients have been studied using a probabilistic reversal learning task (PRLT, Dickstein et al. 2010; Mukherjee et al. 2020) and a probabilistic reward task (PRT, Pizzagalli et al. 2008a; Huys et al. 2013). These tasks present participants with stimuli that predict a higher or lower probability of receiving a positive or rewarding outcome and record learning, investigate responses to positive and negative feedback, or the development of a response bias. People with depression, both medicated and unmedicated, often fail to develop an expected bias towards the more frequently rewarded stimulus, and this deficit is also seen in healthy individuals who demonstrate high levels of anhedonia (Murphy et al. 2003; Pizzagalli et al. 2005; Rizvi et al. 2016; Slaney et al. 2022). Reward learning has also been studied in humans using tasks that measure the motivation to receive a reward. The effort expenditure for reward task (EEfRT) measures this construct by allowing participants to choose between exerting more or less effort in order to gain rewards (Treadway et al. 2012). Depressed patients typically exhibit a reduced willingness to expend effort for reward compared to healthy participants, and patients were also less able to effectively use information from previous trials to improve their chances of receiving a reward. Interestingly, one study reported that depressed patients failed to exert as much effort to receive a reward as healthy controls, despite both groups giving similar ratings regarding the reward's perceived value. The same study also found that even if depressed patients did not expend more effort for a high-value reward, they rated their perceived efforts much higher compared to their ratings for lower-value rewards.

These developments highlight the breadth of cognitive domains impacted by an affective state and the potential to utilise these behavioural deficits to provide an objective assessment of the symptoms of depression and the response to antidepressant treatment. Because these tasks are not based on subjective self-report measures, they also provide an opportunity to develop translational methods that can be used in animal models to provide more clinically relevant behavioural readouts.

4 Neuropsychological Effects of Antidepressants: Healthy Volunteer and Patient Studies

Emotional processing tasks: the cognitive neuropsychological hypothesis of antidepressant action (Harmer et al. 2009a, b; Warren et al. 2015; Godlewska and Harmer 2021) was developed to try to reconcile the delay in clinical benefits observed with conventional antidepressants, which occur despite evidence of their acute molecular, cellular and neurochemical effects. The delayed onset of action of antidepressants has been the focus of research for decades, with most studies suggesting that a sustained clinical benefit is not evident until after 4–6 weeks of administration (Frazer and Benmansour 2002; Mitchell 2006). Although some studies suggest that benefits may arise after a week (Taylor 2007), patients do not report any awareness of improvements, although the rapid emergence of side effects further supports that the drugs are achieving clinically relevant concentrations and have impacts on their primary targets. There is also the observation that cognitive therapies are similarly effective in patients with mild to moderate depression, but how can these very different treatment approaches achieve similar outcomes? In the cognitive neuropsychological hypothesis, these differences are resolved by the proposal that clinical benefits are mediated through effects on negative affective biases. These can be achieved either through a pharmacologically induced change in neurochemistry that facilitates more positive emotional processing or by targeting cognitive mechanisms through behavioural therapies. These changes in affective biases then impact patients' subsequent experiences, attenuating the negative biases, enabling more positive emotional processing and associated learning and memory, which build to then impact the subjective mood of the patient (Harmer et al. 2009a, b; Tranter et al. 2009; Godlewska 2019). Initial studies to test this hypothesis were carried out in healthy volunteers, where a single dose of either a serotonin-specific reuptake inhibitor (SSRI) or a noradrenaline-specific reuptake inhibitor (NaRI) was shown to induce a positive shift in emotional processing (Harmer et al. 2003a, b). These studies were extended with evidence that 1 week of treatment in healthy volunteers induced similar positive shifts in emotional processing, as well as a study in patients with MDD where a single dose of reboxetine was found to induce a similarly positive bias (Harmer et al. 2003b, 2004). Since the first characterisation of these objective changes in emotional behaviour was associated with antidepressant treatment, a range of antidepressants as well as possible pro-depressant drugs have been investigated with evidence consistent with the predictions of this hypothesis. Drugs with established clinical efficacy in depression positively bias at least one emotional processing domain tested in these tasks, e.g. the interpretation of emotional faces or emotional memory with some treatments affecting multiple domains (Pringle et al. 2011a). Importantly, these tasks also show sensitivity to pro-depressant treatments with an acute dose of the CB1 cannabinoid antagonist/inverse agonist rimonabant, found to induce negative affective biases (Horder et al. 2009). Rimonabant was developed as a treatment for obesity but was withdrawn from the market following evidence of pro-depressant side effects, including induction of suicidal ideation and behaviour (Christensen et al. 2007; Buggy et al. 2011).

As well as pharmacological studies that support this cognitive neuropsychological model, combining emotional processing tasks with brain imaging has also provided evidence linking the effects of antidepressant treatment with the remediation of activity changes in brain regions linked to emotional behaviour and MDD. Functional brain imaging studies suggest a fairly consistent picture of depression with increased amygdala activity in response to negative emotional stimuli (Hamilton and Gotlib 2008; Wessa and Lois 2015). Patients with MDD also exhibit hyperactivity in regions of the prefrontal cortex, which is remediated following successful antidepressant treatment and was the basis for deep brain stimulation (Fales et al. 2009; Lemogne et al. 2009, 2010). Imaging studies using fMRI found that acute clinical doses of conventional antidepressants resulted in a decreased amygdala response to negative facial expression in depressed patients (Norbury et al. 2009; Murphy et al. 2009; Williams et al. 2015). This work has been extended with numerous publications reporting similar findings and suggesting that antidepressant treatments alter activity in networks associated with emotional processing and that these changes occur without subjective effects on mood (for a review, see Godlewska 2019). Key regions identified are the amygdala and the anterior cingulate cortex (ACC), as well as a wider network, including the medial prefrontal cortex (mPFC) and the associated regions (for a meta-analysis, see Ma 2015).

An emotional processing task test battery encompassing objective measures, including facial emotion recognition, emotional categorisation and emotional memory, has provided important evidence to support the acute and short-term impacts of pharmacological treatments with conventional antidepressants, which are predictive of their clinical efficacy. There is some evidence that drugs that lack clinical benefits also fail to show effects in these tasks, with the failed antidepressant aprepitant (NK1 antagonist) showing limited effects in these emotional processing tasks, which aligns with its mixed picture in clinical trials (McCabe et al. 2009; Chandra et al. 2010; Pringle et al. 2011b). This may be useful going forward as an experimental medicine approach to predict the efficacy of novel treatments as well as a method to assess potential psychiatric risks, e.g. the P1vital® Oxford Emotional Test Battery (ETB) (Thomas et al. 2016). Using an objective behavioural task with early sensitivity to antidepressant treatment may also be a valuable tool in the clinic to identify responders versus non-responders to conventional antidepressant treatments (Browning et al. 2019).

The discovery of rapid-acting antidepressants (RAADs) poses an interesting challenge for the cognitive neuropsychological hypothesis. The NMDA antagonist ketamine has been shown to induce rapid and sustained antidepressant effects in patients (Zarate et al. 2006; Murrough et al. 2013; Corriger and Pickering 2019), with similar findings reported for the muscarinic antagonist scopolamine (Hasselmann 2014; Drevets et al. 2020) and the serotonergic psychedelic psilocybin (Yu et al. 2022; Prouzeau et al. 2022; Rosenblat et al. 2023). In contrast to conventional antidepressants, patients achieve an almost immediate reduction in depressive symptoms following an acute administration of these pharmacologically diverse treatments. What they all share, however, is the ability to generate an altered state of consciousness, which may have an impact on relevant emotional circuits. The effects of these treatments on emotional processing are still limited; however, imaging studies suggest that ketamine may normalise neural activity changes in MDD during an emotional processing task (Reed et al. 2018). Similarly, psilocybin treatment has been associated with changes in activity in emotional circuits (Kraehenmann et al. 2015; Barrett et al. 2020). Interestingly, studies with the NMDA antagonist memantine failed to see effects on emotional processing (Pringle et al. 2012; Sani et al. 2012), which was consistent with its failure in a clinical trial despite pre-clinical evidence from conventional animal models of depression (Quan et al. 2011; Iadarola et al. 2015). Pre-clinical studies suggest that ketamine directly modulates negative affective biases but in a way that is distinct from that seen with conventional antidepressants (see a discussion of the animal model data below).

Reward processing tasks: most psychopharmacology studies in humans have focused on emotional processing tasks, but affective biases are known to have an impact across a wide range of cognitive domains, and one important area for MDD is reward-related cognition. Treatment with the SSRI citalopram resulted in an increased cortical response to pleasant visual stimuli whilst also decreasing response to unpleasant stimuli (Harmer et al. 2004). However, another study of healthy participants found that citalopram decreased the neural processing of both rewarding and aversive stimuli (McCabe et al. 2010). In depressed patients, treatment with the SSRI paroxetine reduced response to negative feedback in a probabilistic reversal learning task but failed to have any effect on the reduced response to positive feedback thought to be associated with depression (Herzallah et al. 2013). Interestingly, treatment with the tricyclic antidepressant imipramine over 6 weeks was able to both increase positive appraisals of reward and decrease negative appraisals of stress (Wichers et al. 2009). This study also found an association between patients who reported a greater increase in positive reward appraisals with greater improvement in clinical symptoms. In the EEfRT, treatment with vortioxetine, another SSRI, was associated with improved task performance (Cao et al. 2019). These results suggest that, in line with the work surrounding affective biases, antidepressants may exert their effects through neuropsychological mechanisms operating outside of the patient's conscious awareness, and the delayed efficacy associated with conventional antidepressants may be due to the time it takes for the patient to accumulate experiences unaffected by impaired reward processing (Harmer et al. 2009a, b; Admon and Pizzagalli 2015).

The work outlined here provides cogent evidence of negative affective biases and reduced reward sensitivity as potential neuropsychological mechanisms contributing to the development and perpetuation of MDD. The ability of antidepressant treatments to remediate these negative biases when quantified at both a behavioural and neural activity level adds support to the hypothesis that a cognitive neuropsychological mechanism could explain the relationship between the acute, molecular and neurochemical effects of these treatments and their impacts on mood. These methods also provide a foundation for the development of more clinically relevant rodent tasks.

5 Neuropsychological Effects of Antidepressants: Animal Studies

Developing animal models of affective processing requires shifting from the emotional cues in the visual or language domains used in human studies to stimuli more relevant for use in non-human species. In relation to affective biases, two key tasks have been developed: the judgement bias task (JBT) and the affective bias task (ABT) (Harding et al. 2004; Mendl et al. 2009; Stuart et al. 2013; Hales et al. 2014). There has also been a shift from conventional models of anhedonia, based on reward sensitivity and threshold in the sucrose preference test, to tasks that focus on more cognitive aspects of reward processing (Lewis et al. 2019). The sucrose preference test has been shown to be sensitive to chronic stress in animals with deficits reversed by antidepressants; however, a human version of the task, the sweet taste test, failed to find deficits in MDD (Scinska et al. 2004; Dichter et al. 2010), although some changes may correlate with anhedonia in healthy populations (Slaney et al. 2018). Reward-related behaviour is complex involving sensitivity and perception of reward value, motivation to obtain reward and learning about reward value. Studies in human participants suggest that the effort for a reward (Treadway et al. 2012) and the cognitive aspects of a reward are more vulnerable to deficits in MDD, which can be detected in tasks such as PRLT and PRT (Murphy et al. 2003; Pizzagalli et al. 2008a; Kunisato et al. 2012; Mukherjee et al. 2020). These tasks also provide opportunities to develop translational methods, and rodent equivalents have been developed (Bari et al. 2010; Der-Avakian et al. 2013; Rygula et al. 2018; Wilkinson et al. 2020). Reward-learning tasks based on a similar method to the affective bias test have also been developed that quantify animals' development of a reward-induced positive affective bias, which, although using a different methodology, may involve similar mechanisms to PRT.

Judgement bias task (JBT): The JBT (also sometimes referred to as an ambiguous cue interpretation task) was initially designed to measure affective states in animals as a means of informing animal welfare research and to specifically measure biases in decision-making and the interpretation of ambiguous information (Harding et al. 2004; Mendl et al. 2010). Although the task has been adapted to incorporate different types of cues and formats, the principle of the protocol is that animals are trained to form associations between two neutral cues and outcomes that are either positive or negative/less positive. Once trained to discriminate these cues, animals are then presented with an ambiguous cue, with the interpretation of the cue indicated as a positive or negative judgement bias. It has been shown that animals in negative affective states are less likely to interpret ambiguous cues as positive, offering the first evidence of cognitive affective biases in animals (Harding et al. 2004). Building from these early studies, versions of the JBT have been tested in several species of animals with similar results and an overall interpretation that animals in a negative affective state are more likely to make pessimistic choices when presented with an ambiguous cue (Lagisz et al. 2020). The task has also been reverse translated for use in human studies, demonstrated by a study of healthy participants reporting a correlation between anxiety and pessimistic interpretation (Paul et al. 2005; Anderson et al. 2012).

Despite good evidence that the JBT is sensitive to changes in animals' core affective state associated with environmental factors, evidence of pharmacological effects has generated a more mixed picture, particularly when considering translational validity (Neville et al. 2020). Acute treatment with a combination of reboxetine and corticosterone to mimic stress and an anxiogenic drug resulted in a negative judgement bias (Enkel et al. 2010; Anderson et al. 2013). Acute treatment with drugs that increase dopamine induced more optimistic choices (Rygula et al. 2014a, b); however, acute treatment with conventional antidepressants has failed to produce consistent effects in this task, with only one study reporting a positive judgement bias following treatment with citalopram (Rygula et al. 2014a, b). Chronic antidepressant treatment induced a positive bias (Anderson et al. 2013). These findings contrast with human emotional interpretation tasks, which find acute effects and suggest that the JBT may involve different underlying mechanisms and perhaps the time course of effects aligns more with the subjective experience of patients. This is supported by studies with the rapid-acting antidepressant ketamine, which induced a positive shift in decision-making biases following acute administration, with similar effects observed with the NMDA receptor NR2B antagonist CP101606 and the muscarinic antagonist scopolamine, which have both also been reported to induce rapid antidepressant effects (Hales et al. 2020). Although further pharmacological studies, and studies using depressed patients are needed, current evidence suggests that the neurobiological mechanisms underlying interpretation biases in human emotional processing tasks are different from those involved in decision-making biases observed in the JBT. However, the ability of this task to distinguish the effects of delayed-onset and rapid-acting antidepressants indicates that this task may be helpful for testing the efficacy of novel treatments (Wilkinson et al. 2020).

Affective bias test (ABT): The ABT was designed to quantify affective biases related to associative learning and memory. The theoretical framework for this task is based on clinical reports that depressed patients tend to attribute less value to rewarding experiences compared to healthy controls (Gonda et al. 2015; Geugies et al. 2019). These observations suggest that the affective state during a rewarding experience may bias how the memory of the experience is encoded and subsequently retrieved, which is observed as an affective bias (Robinson 2018). The protocol for this task first involves training animals to associate specific digging substrates with a food reward. Independent learning experiences are acquired under either treatment/affective or vehicle/control conditions using substrate-reward pairing sessions. The value of the reward is the same for both substrate-reward pairings, so any arising bias can be attributed to the affective state at the time of learning. Once pairing sessions are complete, a choice test is used to quantify any arising bias where animals are presented with both previously rewarded substrates at the same time, and the number of choices made for each substrate is recorded over a series of independent trials.

Acute treatment with a variety of conventional antidepressants has been shown to induce a positive bias towards the treatment-paired substrate (Stuart et al. 2013). In contrast, treatments/manipulations, including psychosocial stress, early life adversity, treatment with pro-depressant compounds, immune challenges or the administration of corticosterone, induce a negative bias in the ABT (Stuart et al. 2013, 2017; Hinchcliffe et al. 2017). Animals exposed to social enrichment exhibit a positive bias, whilst those that experience acute restraint stress exhibit a negative bias. Together, these data suggest that the ABT is sensitive to both pharmacological and psychosocial manipulations of the affective state (Stuart et al. 2013). These studies

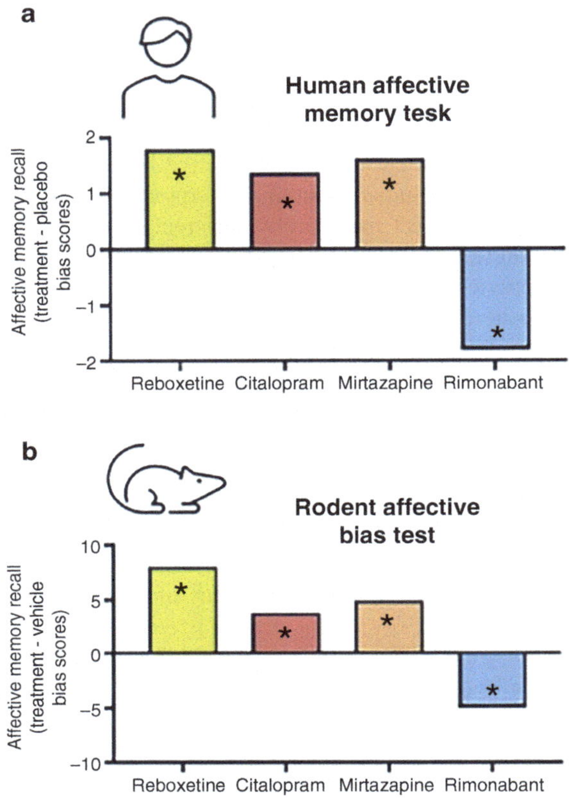

Fig. 1 Similar effects of antidepressant or pro-depressant effects of acute pharmacological treatments in an animal model of affective biases and healthy volunteers. Affective bias tasks have been developed for both human and non-human animal subjects with studies for affective biases associated with learning and memory the more extensively characterised and perhaps with the greatest translational validity. As illustrated here, with the use of acute pharmacological treatments, antidepressant drugs have been shown to induce positive affective biases in both humans (panel **a**) and rodents (panel **b**). Studies with pro-depressant treatments are more limited, but similar negative affective biases have been observed following treatment with the cannabinoid antagonist rimonabant. Data representative of published findings in healthy volunteers or untreated rats

also observed similar effects to those observed in healthy volunteers performing an emotional processing task or memory biases and receiving similar acute treatments (Chan et al. 2007; Williams et al. 2007; Papadatou-Pastou et al. 2012; Robinson and Roiser 2016). (Fig. 1).

The ABT can be used to examine biases associated with learning (when treatment is given prior to substrate-reward pairing sessions) as well as test if treatments can modify biased memories (when administered after learning but prior to the choice test). For example, treatment with the SNRI venlafaxine induces a positive bias during learning but fails to attenuate negatively biased memories. The opposite is true following treatment with rapid-acting antidepressants such as ketamine, scopolamine or psilocybin (Stuart et al. 2015, abstracts for others). These data offer further evidence that the delayed efficacy of conventional antidepressants results from acute changes in both affective and reward-related processing, and an accumulation of positively biased experiences is required for subjective mood improvement to occur. The ability of rapid-acting antidepressants to modulate negative biases in the ABT may explain why patients experience such immediate symptom relief (Stuart et al. 2015). It is important to note that drugs of abuse such as cocaine and amphetamine fail to produce significant effects in the ABT, reaffirming the relationship between the mechanisms underlying the formation of affective biases and those underlying antidepressant drug action. As with the JBT, the ABT is useful for differentiating the effects of delayed versus rapid-acting antidepressants. However, the ABT may be a better tool for understanding the neurobiological mechanisms of affective processing in humans as well as for determining factors that may enhance the clinical efficacy of novel treatments.

6 Animal Models of Reward Learning

Reward learning assay (RLA): the reward learning assay (RLA) shares a similar protocol to the ABT and was used in the initial validation experiments conducted whilst developing the ABT protocol. The RLA involves training animals to discriminate between digging substrates paired with either a high- or a low-value reward (two versus one food reward pellet). Under control conditions, animals demonstrate a preference towards the substrate associated with the high-value, two-pellet reward (Stuart et al. 2013). Animals remain in the same affective state throughout testing, and studies suggest that the RLA may provide a measure sensitive to changes in the core affective state and impairments in reward learning. Exposure to manipulations such as early life adversity as well as the chronic administration of pharmacological and immunological manipulations results in a reduced expression of a reward-induced positive bias (Stuart et al. 2015, 2019; Robinson 2018). A key study by Stuart et al. (2015) found that chronic treatment with IFN-α impaired the formation of a reward bias, and a similar effect was observed following chronic retinoic acid treatment (associated with depressive symptoms in humans). This study tested the same animals using the standard preference test (SPT) and found that neither IFN-α nor retinoic acid produced a difference in sucrose preference compared to control animals, reaffirming a distinction between the consummatory anhedonia measured by the SPT and the reduced sensitivity to reward observed in the RLA (Robinson 2018). Similar results were also observed in a chronic pain model, where an impairment in reward learning in the RLA was observed independent of any change in the SPT (Phelps et al. 2021). The findings of these studies indicate that impairments in reward processing may involve neurobiological mechanisms which alter the learning and retrieval of reward-associated experiences, resulting in decreased motivation to engage in rewarding behaviours. Such mechanisms may align with the anticipatory reward deficits reported by human studies.

Probabilistic reward task (PRT): Depressed patients have been shown to demonstrate a failure to integrate reward-related information and modulate their behaviour accordingly in studies using the probabilistic reward task (PRT) (Pizzagalli et al. 2008a; Huys et al. 2013; Rupprechter et al. 2018). The reverse translated version of the PRT involves training animals to discriminate between two equally rewarded tones. In subsequent testing, two ambiguous tones are presented, with the correct identification of one tone resulting in a much higher frequency of reward compared to the other tone. Animals under control conditions exhibit an increased response bias towards the more frequently rewarded tone (indicated by % of accurate trials), and this bias gradually increases as trials continue (Der-Avakian et al. 2011; Hoffman 2016; Kangas et al. 2020). Treatment with pramipexole, a drug known to blunt reward responsiveness in humans (Pizzagalli et al. 2008b), impaired the formation of a response bias, and treatment with amphetamine potentiated response bias in this task (Der-Avakian et al. 2013). Although antidepressants have yet to be assessed using this protocol, these data show that this version of PRT can effectively quantify behavioural response to reward in animals and is sensitive to the effects of pharmacological manipulations. There are also similarities between this task and the RLA; however, direct comparisons have not been carried out, and there is currently no overlap in pharmacological characterisation.

Probabilistic reversal learning task (PRLT): reversal learning tasks are used to measure sensitivity to reward feedback and can also be used as an assessment of cognitive flexibility. Abnormal responses to performance feedback are thought to be associated with depression as depressed patients often demonstrate a hyposensitivity to rewarding information (Murphy et al. 2003; Rizvi et al. 2016; Mukherjee et al. 2020). Studies using reversal learning tasks have shown that people with depression also exhibit a heightened sensitivity to negative feedback, and this is particularly true when presented with misleading feedback compared to accurate negative feedback (Murphy et al. 2003). In a version of the PRLT developed for rodents, animals are trained to discriminate between a frequently rewarded 'rich' stimulus and a less rewarded 'lean' stimulus. Once animals have made a number of consecutive choices for the 'rich' stimulus, the reward contingencies reversed so that the previously 'lean' stimulus was now associated with the higher frequency of reward and vice versa (Bari et al. 2010). It has been shown that rats reared in social isolation require more trials to reach the criterion for reversal and are slower to adjust their behaviour when reward contingencies are switched (Amitai et al. 2014). Treatment with conventional antidepressants has produced minimal effects in this task, with the exception of studies reporting an increased sensitivity to positive feedback and a greater number of reversals completed following treatment with citalopram (Bari et al. 2010). These results mimic findings from human studies in which the only conventional antidepressants reported to increase reward sensitivity are those selective for 5-HT, suggesting that reversal learning tasks may be sensitive to the manipulations of specific neurotransmitter systems. Ketamine has been shown to decrease sensitivity to negative feedback, but results are inconsistent (Rychlik et al.

2017; Wilkinson et al. 2020). Whilst the format of the PRLT aligns well with the human task, the extended training period required for animals to acquire the task may impact the translational validity of the task and lead to differences in the effects of antidepressants. Studies in depression models are limited, and therefore it is not yet possible to draw any clear conclusions. Comparing the data obtained in the rodent model versus humans also suggests that there may be differences in how animals are performing the task (Skandali et al. 2018).

Effort for reward task: individuals with depression show a decreased willingness to expend effort to obtain a reward in comparison with healthy controls (Treadway et al. 2012; Yang et al. 2014; Hershenberg et al. 2016). One of the more frequently used methods is the progressive ratio breakpoint (PRB) task, where animals are trained to press a lever to obtain a reward, and the number of lever presses required to obtain the reward increases as the session progresses. The 'breakpoint' occurs once the value of the reward fails to outweigh the effort required for lever pressing and the animals stop responding. Although human studies using a PRB procedure report reduced breakpoints in depressed patients (Hershenberg et al. 2016; Klawohn et al. 2022), results from animal studies are highly variable (Barr and Phillips 1998; Olausson et al. 2013; Dieterich et al. 2021). Findings in rodent models of MDD and in the response to pharmacological manipulations, including antidepressants, are not consistent (Nikiforuk and Popik 2009; Yohn et al. 2018). As a result, the focus has shifted to effort-related choice tasks, which may be more appropriate in terms of their translational validity. Effort-based decision-making has been assessed in rodents by giving animals a choice between a reward requiring low effort and a higher value reward that requires more effort to obtain (Salamone et al. 1994; Randall et al. 2012). In contrast to most translational behavioural tasks where animal studies aim to model human behaviour, the EEfRT for humans described earlier was based on these animal tasks (Treadway et al. 2009). Manipulations such as restraint stress and the administration of drugs with pro-depressant effects, e.g. the vesicular monoamine uptake blocker tetrabenazine, reduce the preference for high effort choices in rodents (Nunes et al. 2013; Randall et al. 2014; Jankovic and Beach 1997; Kenney and Jankovic 2006). Conventional antidepressants reverse the effects of tetrabenazine, but studies in disease models are limited, and there has been a tendency towards a focus on dopaminergic mechanisms, which whilst of general interest are less directly relevant to the current antidepressant treatments (Randall et al. 2014; Contreras-Mora et al. 2018; Carratalá-Ros et al. 2023).

7 Discussion and Conclusions

7.1 *Could Different Neuropsychological Effects of Conventional Versus Rapid Acting Antidepressants Explain the Temporal Differences in Their Onset of Clinical Effects?*

One of the most obvious clinical differences between conventional antidepressants and the drugs termed RAADs is the differences they have in the patient's subjective experience of their mood. Studies suggest that conventional antidepressants may take 6–8 weeks before they achieve clinically relevant improvements in mood. It has long been known that mood can be improved more quickly than this with non-pharmacological manipulations such as sleep deprivation; however, the rapid and sustained antidepressant effects of a single dose of ketamine suggest that this can also be achieved pharmacologically and with sustained benefits of at least 7 days (Berman et al. 2000). Since the first clinical study with ketamine was published, a large number of pre-clinical studies in rodent models have sought to explain the underlying mechanism. The main focus of these studies has been the neurotrophic effects, fuelled by publications showing ketamine-induced synaptogenesis in the prefrontal cortex. In this neurotrophic model of antidepressant efficacy, delayed -onset antidepressants function to enhance plasticity in the brain and reverse neuronal loss, but these changes take time to emerge and hence time for clinically meaningful changes to be experienced. In contrast, RAADs have a rapid effect on synapse formation, which is hypothesised to explain their rapid effects on mood. One of the difficulties in translating these findings to the clinic is the limitations of the models available to test the relationship between these underlying molecular mechanisms and human psychiatric symptoms. Although there is no doubt that these effects arise in the animal models and directly interfering with these pathways can prevent behavioural effects in tests such as the FST, whether this same biology underlies the effects in humans is not clear. Most of these studies use high doses of antidepressants which generate plasma levels and hence receptor occupancy which will not be achieved in patients. Furthermore, both conventional antidepressants and RAADs induce similar acute decreases in immobility time in the FST, an effect that is not seen clinically.

In the neuropsychological model of antidepressant efficacy, differences in the way the drug treatments interact with affective biases are hypothesised to explain these temporal differences in clinical experience. First discussed by Harmer (Harmer et al. 2009a, b; Godlewska and Harmer 2021) and developed by Robinson (2018), integrating evidence from animal studies, different neuropsychological effects involving the modulation of affective biases is proposed to underlie the temporal difference in the effects on mood. Conventional antidepressants such as venlafaxine can positively bias new experiences but lack effects on biased memories (Fig. 2). These positive biases can build over time to overcome past, negatively biased experiences, but these take time to generate and hence the delayed onset on subjective mood. In contrast, the RAAD ketamine has been shown to attenuate negative

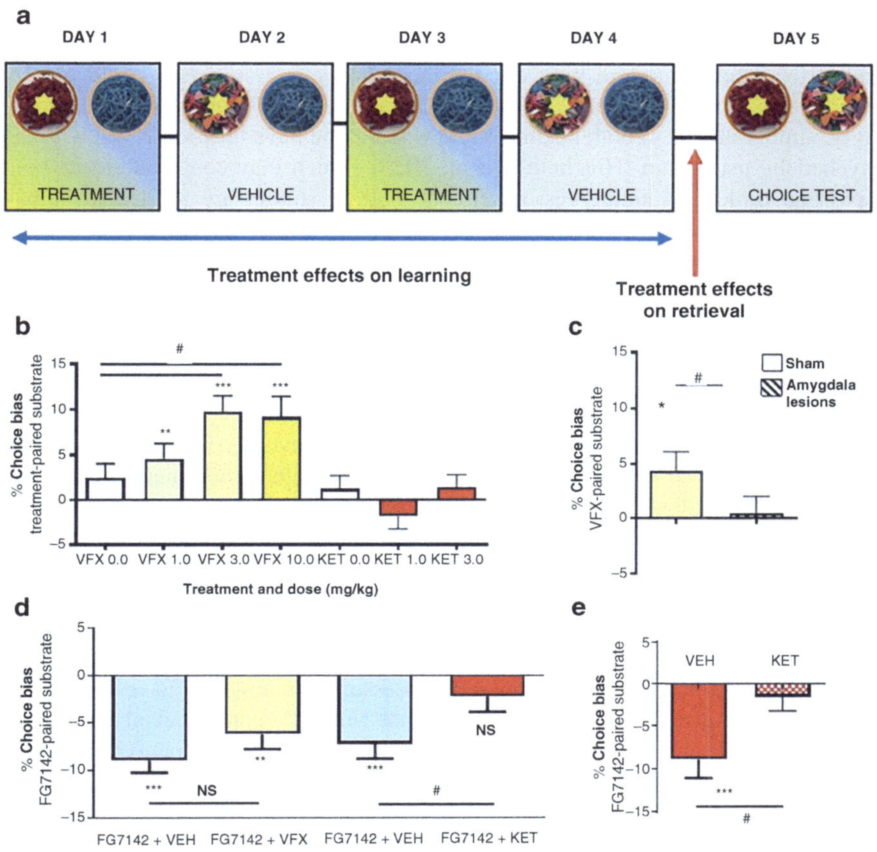

Fig. 2 The rodent affective bias test provides a translation model of affective biases associated with learning and memory and reveals differential effects of conventional delayed-onset versus rapid-acting antidepressants. The affective bias test has been the most widely characterised task for studying the neuropsychological effects of antidepressants. The task is based on animals learning two independent cue-reward associations. One of the cue-reward associations is learnt following an affective manipulation of drug treatment, with the other learnt under control conditions. A choice test is then used to quantify if animals show a preference for either experience. All factors are counterbalanced, and the absolute reward value is kept constant. An outline of a typical experimental protocol is illustrated in panel **a**. The affective bias task has been found to generate reliable positive and negative affective biases consistent with their predicted effects on the affective state. The assay has also provided evidence that conventional delayed-onset antidepressants modulate affective biases in a different way than rapid-acting antidepressants (RAAD), potentially explaining the temporal differences in their onset of clinical benefits. When testing the effects on bias associated with new memories, conventional but not RAADs demonstrate efficacy (panels **b** and **c**), which is linked to the amygdala. In contrast, RAADs but not conventional antidepressants attenuate negative biases associated with past experiences (panels **d** and **e**), an effect linked to the medial prefrontal cortex (Stuart et al. 2013, 2015)

biases associated with past experiences (Fig. 2). Mood is strongly influenced by past experiences that impact the current affective state and future expectations; thus, the ability to directly interfere with circuits mediating affective biases could rapidly reverse the impact of negatively biased experiences that occur in MDD. Interestingly, similar effects to both ketamine and venlafaxine have been observed with the psychedelic psilocybin (Hinchcliffe et al. 2023), which may contribute to its both rapid and prolonged antidepressant effects. Further studies are needed in human subjects to establish if similar effects with RAADs are observed. One important implication of the neuropsychological effects observed with both conventional antidepressants and RAADs is the contribution of experience-dependent factors. In this neuropsychological model, conventional antidepressants require patients to re-engage with normal behaviours to enable the generation of positively biased experiences. RAAD effects may require the re-activation of past experiences and hence benefit from combining with psychological interventions. It is also feasible that the neurotrophic changes discussed above may be involved in the modulation of circuits that regulate affective biases and hence contribute to these neuropsychological effects, particularly the sustained effects of RAADs (Harmer et al. 2017).

7.2 Possible Neural Circuits Mediating Affective Biases

The neural circuits linked to emotional processing and rewards have been quite extensively characterised in human functional MRI (Costafreda et al. 2008; Ma 2015; Pizzagalli and Roberts 2022; Zhang et al. 2022). In participants performing specific tasks of emotional recognition and memory, key regions identified include the amygdala, anterior insula, dorsolateral prefrontal cortex (dlPFC) and pregenual anterior cingulate cortex (pgACC) (Hartling et al. 2021). There have also been extensive studies into the networks that underlie emotions (Underwood et al. 2021); however, it remains to be established how the circuits of emotional regulation relate to affective biases and the neuropsychological effects of antidepressants. The prevailing hypothesis is that dysregulation within these emotional circuits may generate biases in the way emotional information is processed in the brain and hence how it impacts cognitive function, e.g. attention, decision-making, learning and memory. Studies in patients with MDDs suggest that regions of the prefrontal and anterior cingulate cortex may be hyperactive (Dunlop and Mayberg 2022) with evidence of increased activity also seen in the amygdala (Fu et al. 2004). In studies focusing on the neuropsychological effects of antidepressants, the remediation of these activity changes corresponds to a more positive shift in affective biases supporting this idea (Murphy et al. 2009; Norbury et al. 2009). There are strong reciprocal connections between the cortical regions and the amygdala, which further supports the importance of these regions, but they also connect to the wider limbic network, so they may also be influenced by, and may also influence, other brain regions, e.g. regions associated with stored memories. Whilst it is unlikely that any single brain region modulates affective biases, there may be an 'affective bias' circuit that encompasses cortical and amygdala pathways, which become altered in negative affective states and, through their wider connections, contribute to the development and perpetuation of MDD through pervasive negative affective processing. Through these wider connections and influence on learning and memory, these affective biases may also impact how memories are learnt, stored and/or retrieved. As mood is strongly linked to past experiences and hence memory, these biases and the associated circuit represent an important therapeutic target. Current antidepressants seem to be able to remediate hyperactivity in the amygdala and also the prefrontal cortex, and sub-genual cingulate deep brain stimulation has proved to be a successful antidepressant treatment in some patients (Mayberg et al. 2005). However, one of the challenges with fMRI studies and experiments involving patients is understanding whether these effects are causally associated or correlative, and this is an important distinction when considering the potential for novel drug targets. This is where animal models are particularly useful as they can integrate behavioural readouts with direct manipulations of specific neural or neurochemical targets. The number of studies to explore the neural mechanisms underlying affective biases in rodent models is currently very limited, with the majority focused on systemic pharmacology or studies in animals in a putative depression or anxiety-like state. Combining the ABT and targeted brain manipulations, the amygdala was found to be an important locus for the effects of the conventional antidepressant venlafaxine (Stuart et al. 2015). Here, lesions targeting the central nucleus of the amygdala attenuated the ability of acute venlafaxine to generate a positive affective bias. In a separate study, targeted infusions of ketamine into the medial prefrontal cortex induced similar effects as systemic administration in terms of both the acute and sustained effects (Hinchcliffe et al. 2023). Given the reciprocal connections between mPFC and the amygdala, their role in emotional processing and mood disorders and the similarity between the fMRI imaging data and rodent studies, these data provide an early indication of the key brain regions modulating affective biases. As outlined in Fig. 3, however, there is the potential for a wider 'affective bias' circuit involving these structures and connected regions.

7.3 Summary and Future Directions

For more than half a century, scientists have sought to explain the effects of antidepressants. In the last 20 years, studies in both human and non-human animal subjects suggest that affective biases may be an important neuropsychological mechanism, which could explain how their biochemical effects translate into improvements in the emotional experiences of patients. One of the advantages of this model is that it brings together both biological and experience-dependent factors, integrated into one hypothesis. It could help explain the variability in the response rate to conventional antidepressants as the model highlights the need for patients taking conventional delayed-onset antidepressants to re-engage with their environment and social interactions. These are hypothesised to play a critical role in

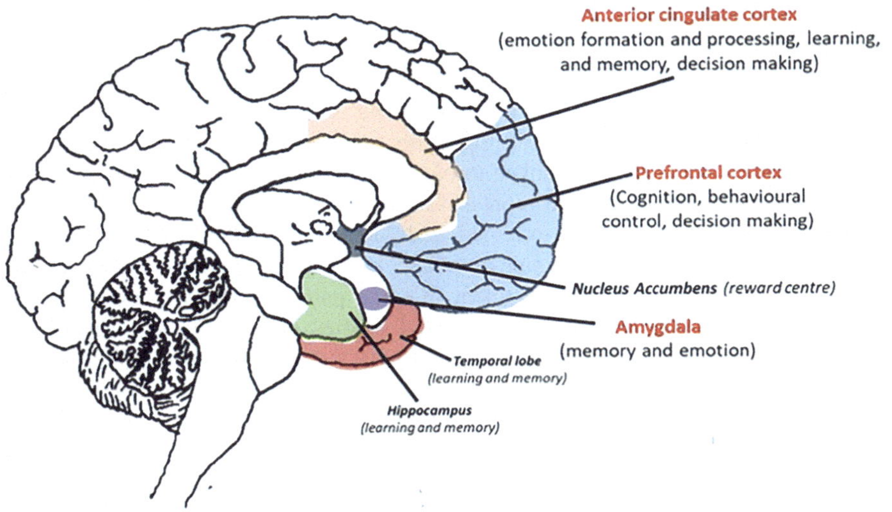

Fig. 3 Early insights into neural circuits modulating the neuropsychological effects of antidepressants using fMRI in human subjects and targeted brain manipulations in rodents. One of the benefits of using objective, neuropsychological tasks is that they can be performed alongside brain imaging studies in both patients and healthy volunteers. The studies suggest that the amygdala is particularly important in the modulation of affective biases. Changes in activity in the amygdala when participants are presented with cues that generate an affective bias also suggest that dysfunction within the amygdala or inputs that modulate amygdala reactivity may be important in the generation of negative biases in MDD. Both in the amygdala and in the prefrontal cortex, the remediation of altered activity changes correlates with antidepressant efficacy. Studies in rodents permit the use of more invasive methods to test these hypotheses. In a study in rats, lesions to the amygdala prevented the formation of a positive affective bias following treatment with an antidepressant consistent with its role in modulating the formation of a biased memory. Using drug infusions localised to the medial prefrontal cortex, the effects of the RAAD ketamine were localised to this region, demonstrating an important role for this brain region in modulating the affective bias during retrieval. Although further studies are needed, an 'affective bias circuit' may involve both the amygdala and the medial prefrontal cortex and possibly a direct modulatory role for medial prefrontal cortex inputs to the amygdala. Regions implicated from human imaging studies and rodent models are illustrated in red, whilst other regions potentially linked to affective biases are illustrated in black italics

reversing the impacts of negative affective biases and their impact on cognition, particularly learning and memory, which may have fuelled the development of low mood and the loss of interest in rewarding activities, the core symptoms of MDD. This neuropsychological model could also help explain the rapid and sustained antidepressant effects observed with ketamine and other potential RAADs, such as psilocybin (Stuart et al. 2015; Hinchcliffe et al. 2023). In contrast to conventional antidepressants, studies in rodents suggest that ketamine and psilocybin can remediate negative affective biases associated with past experiences and may also enable these experiences to be re-learnt with a more positive affective bias (Stuart et al. 2015; Hinchcliffe et al. 2023). The fact that these different classes of antidepressants interact with affective biases in unique ways may help explain the temporal differences in the onset of clinical benefits. RAAD may also be more effective in treatment-resistant populations as they are not dependant on new experiences. From a translational perspective, the ABT, as well as other models of affective biases developed for non-human animals, not only provides opportunities to use animals as predictive models in the development of new drugs but also offers the potential to investigate underlying mechanisms that may play a causal role in the development of mood disorders in a model with construct validity.

Acknowledgments ESJR has current research funding from the MRC, BBSRC, Wellcome Trust and academic research grants from Boehringer Ingelheim, COMPASS Pathways and Eli Lilly. ESJR has undertaken paid consultancy work for COMPASS Pathways and Pangea Botanica and contract research for COMPASS Pathways, SmallPharma and IRLab Therapeutics.

References

Abramson LY, Metalsky GI, Alloy LB (1989) Hopelessness depression: a theory-based subtype of depression. Psychol Rev 96(2):358

Admon R, Pizzagalli DA (2015) Dysfunctional reward processing in depression. Curr Opin Psychol 4:114–118

Amitai N, Young JW, Higa K, Sharp RF, Geyer MA, Powell SB (2014) Isolation rearing effects on probabilistic learning and cognitive flexibility in rats. Cogn Affect Behav Neurosci 14:388–406

Anderson MH, Hardcastle C, Munafò MR, Robinson ES (2012) Evaluation of a novel translational task for assessing emotional biases in different species. Cogn Affect Behav Neurosci 12:373–381

Anderson MH, Munafò MR, Robinson ES (2013) Investigating the psychopharmacology of cognitive affective bias in rats using an affective tone discrimination task. Psychopharmacology (Berl) 226:601–613

Anisman H, Matheson K (2005) Stress, depression, and anhedonia: caveats concerning animal models. Neurosci Biobehav Rev 29(4–5):525–546

Baldessarini RJ, Lau WK, Sim J, Sum MY, Sim K (2015) Duration of initial antidepressant treatment and subsequent relapse of major depression. J Clin Psychopharmacol 35(1):75–76

Bari A, Theobald DE, Caprioli D, Mar AC, Aidoo-Micah A, Dalley JW, Robbins TW (2010) Serotonin modulates sensitivity to reward and negative feedback in a probabilistic reversal learning task in rats. Neuropsychopharmacology 35(6):1290–1301

Barr AM, Phillips AG (1998) Chronic mild stress has no effect on responding by rats for sucrose under a progressive ratio schedule. Physiol Behav 64(5):591–597

Barrett FS, Doss MK, Sepeda ND, Pekar JJ, Griffiths RR (2020) Emotions and brain function are altered up to one month after a single high dose of psilocybin. Sci Rep 10(1):2214

Beck AT (1967) Depression: clinical, experimental and theoretical aspects

Beck A (1970) The core problem in depression. The cognitive triad. Depression. Theories and therapies

Beck AT (ed) (1979) Cognitive therapy of depression. Guilford Press

Beck AT, Clark DA (2015) Anxiety and depression: an information processing perspective. In: Anxiety and self-focused attention. Routledge, pp 41–54

Belzung C, Lemoine M (2011) Criteria of validity for animal models of psychiatric disorders: focus on anxiety disorders and depression. Biol Mood Anxiety Disord 1(1):1–14

Berman RM, Cappiello A, Anand A, Oren DA, Heninger GR, Charney DS, Krystal JH (2000) Antidepressant effects of ketamine in depressed patients. Biol Psychiatry 47(4):351–4

Berton O, Hahn CG, Thase ME (2012) Are we getting closer to valid translational models for major depression? Science 338(6103):75–79

Bodnoff SR, Suranyi-Cadotte B, Aitken DH, Quirion R, Meaney MJ (1988) The effects of chronic antidepressant treatment in an animal model of anxiety. Psychopharmacology (Berl) 95:298–302

Boku S, Nakagawa S, Toda H, Hishimoto A (2018) Neural basis of major depressive disorder: beyond monoamine hypothesis. Psychiatry Clin Neurosci 72(1):3–12

Broomfield NM, Davies R, MacMahon K, Ali F, Cross SM (2007) Further evidence of attention bias for negative information in late life depression. Int J Geriatr Psychiatry 22(3):175–180

Browning M, Kingslake J, Dourish CT, Goodwin GM, Harmer CJ, Dawson GR (2019) Predicting treatment response to antidepressant medication using early changes in emotional processing. Eur Neuropsychopharmacol 29(1):66–75

Buggy Y, Cornelius V, Wilton L, Shakir SA (2011) Risk of depressive episodes with rimonabant: a before and after modified prescription event monitoring study conducted in England. Drug Saf 34:501–509

Cao B, Park C, Subramaniapillai M, Lee Y, Iacobucci M, Mansur RB et al (2019) The efficacy of vortioxetine on anhedonia in patients with major depressive disorder. Front Psych 10:17

Carratalá-Ros C, Martínez-Verdú A, Olivares-García R, Salamone JD, Correa M (2023) Effects of the dopamine depleting agent tetrabenazine in tests evaluating different components of depressive-like behavior in mice: sex-dependent response to antidepressant drugs with SERT and DAT blocker profiles. Psychopharmacology (Berl):1–14

Castagné V, Moser P, Roux S, Porsolt RD (2010) Rodent models of depression: forced swim and tail suspension behavioral despair tests in rats and mice. Curr Protoc Pharmacol 49(1):5–8

Chan SW, Goodwin GM, Harmer CJ (2007) Highly neurotic never-depressed students have negative biases in information processing. Psychol Med 37(9):1281–1291

Chandra P, Hafizi S, Massey-Chase RM, Goodwin GM, Cowen PJ, Harmer CJ (2010) NK1 receptor antagonism and emotional processing in healthy volunteers. J Psychopharmacol 24(4):481–487

Christensen R, Kristensen PK, Bartels EM, Bliddal H, Astrup A (2007) Efficacy and safety of the weight-loss drug rimonabant: a meta-analysis of randomised trials. Lancet 370(9600): 1706–1713

Cipriani A, Furukawa TA, Salanti G, Chaimani A, Atkinson LZ, Ogawa Y et al (2018) Comparative efficacy and acceptability of 21 antidepressant drugs for the acute treatment of adults with major depressive disorder: a systematic review and network meta-analysis. Lancet 391(10128): 1357–1366

Contreras-Mora H, Rowland MA, Yohn SE, Correa M, Salamone JD (2018) Partial reversal of the effort-related motivational effects of tetrabenazine with the MAO-B inhibitor deprenyl (selegiline): implications for treating motivational dysfunctions. Pharmacol Biochem Behav 166:13–20

Corriger A, Pickering G (2019) Ketamine and depression: a narrative review. Drug Des Devel Ther:3051–3067

Costafreda SG, Brammer MJ, David AS, Fu CH (2008) Predictors of amygdala activation during the processing of emotional stimuli: a meta-analysis of 385 PET and fMRI studies. Brain Res Rev 58(1):57–70

Cryan JF, Valentino RJ, Lucki I (2005) Assessing substrates underlying the behavioral effects of antidepressants using the modified rat forced swimming test. Neurosci Biobehav Rev 29(4–5): 547–569

Der-Avakian A, D'Souza MS, Pizzagalli DA, Markou AM (2011) Development of a rodent probabilistic reward task used to objectively assess human anhedonia: implications for translational research

Der-Avakian A, D'Souza MS, Pizzagalli DA, Markou A (2013) Assessment of reward responsiveness in the response bias probabilistic reward task in rats: implications for cross-species translational research. Transl Psychiatry 3(8):e297–e297

Detke MJ, Johnson J, Lucki I (1997) Acute and chronic antidepressant drug treatment in the rat forced swimming test model of depression. Exp Clin Psychopharmacol 5(2):107
Dichter GS, Smoski MJ, Kampov-Polevoy AB, Gallop R, Garbutt JC (2010) Unipolar depression does not moderate responses to the sweet taste test. Depress Anxiety 27(9):859–863
Dickstein DP, Finger EC, Brotman MA, Rich BA, Pine DS, Blair JR, Leibenluft E (2010) Impaired probabilistic reversal learning in youths with mood and anxiety disorders. Psychol Med 40(7): 1089–1100
Dieterich A, Liu T, Samuels BA (2021) Chronic non-discriminatory social defeat stress reduces effort-related motivated behaviors in male and female mice. Transl Psychiatry 11(1):125
Disner SG, Beevers CG, Haigh EA, Beck AT (2011) Neural mechanisms of the cognitive model of depression. Nat Rev Neurosci 12(8):467–477
Disner SG, Shumake JD, Beevers CG (2017) Self-referential schemas and attentional bias predict severity and naturalistic course of depression symptoms. Cognit Emot 31(4):632–644
Dozois DJ, Beck AT (2008) Cognitive schemas, beliefs and assumptions. Risk factors in depression, pp 119–143
Drevets WC, Bhattacharya A, Furey ML (2020) The antidepressant efficacy of the muscarinic antagonist scopolamine: past findings and future directions. Adv Pharmacol 89:357–386
Duman RS (2022) Neural plasticity: consequences of stress and actions of antidepressant treatment. Dialogues Clin Neurosci
Dunlop BW, Mayberg HS (2022) Neuroimaging-based biomarkers for treatment selection in major depressive disorder. Dialogues Clin Neurosci
Elliott R, Zahn R, Deakin JF, Anderson IM (2011) Affective cognition and its disruption in mood disorders. Neuropsychopharmacology 36(1):153–182
Enkel T, Gholizadeh D, Bohlen V, Halbach O, Sanchis-Segura C, Hurlemann R, Spanagel R et al (2010) Ambiguous-cue interpretation is biased under stress-and depression-like states in rats. Neuropsychopharmacology 35(4):1008–1015
Fales CL, Barch DM, Rundle MM, Mintun MA, Mathews J, Snyder AZ, Sheline YI (2009) Antidepressant treatment normalizes hypoactivity in dorsolateral prefrontal cortex during emotional interference processing in major depression. J Affect Disord 112(1–3):206–211
Frazer A, Benmansour S (2002) Delayed pharmacological effects of antidepressants. Mol Psychiatry 7(1):S23–S28
Fu CH, Williams SC, Cleare AJ, Brammer MJ, Walsh ND, Kim J et al (2004) Attenuation of the neural response to sad faces in major depression by antidepressant treatment: a prospective, event-related functional magnetic resonance imaging study. Arch Gen Psychiatry 61(9): 877–889
Gartlehner G, Wagner G, Matyas N, Titscher V, Greimel J, Lux L et al (2017) Pharmacological and non-pharmacological treatments for major depressive disorder: review of systematic reviews. BMJ Open 7(6):e014912
Geddes JR, Carney SM, Davies C, Furukawa TA, Kupfer DJ, Frank E, Goodwin GM (2003) Relapse prevention with antidepressant drug treatment in depressive disorders: a systematic review. Lancet 361(9358):653–661
Geugies H, Mocking RJ, Figueroa CA, Groot PF, Marsman JBC, Servaas MN et al (2019) Impaired reward-related learning signals in remitted unmedicated patients with recurrent depression. Brain 142(8):2510–2522
Godlewska BR (2019) Cognitive neuropsychological theory: reconciliation of psychological and biological approaches for depression. Pharmacol Ther 197:38–51
Godlewska BR, Harmer CJ (2021) Cognitive neuropsychological theory of antidepressant action: a modern-day approach to depression and its treatment. Psychopharmacology (Berl) 238(5): 1265–1278
Gonda X, Pompili M, Serafini G, Carvalho AF, Rihmer Z, Dome P (2015) The role of cognitive dysfunction in the symptoms and remission from depression. Ann Gen Psychiatry 14(1):1–7

Goodwin GM, Aaronson ST, Alvarez O, Arden PC, Baker A, Bennett JC et al (2022) Single-dose psilocybin for a treatment-resistant episode of major depression. N Engl J Med 387(18): 1637–1648

Hales CA, Stuart SA, Anderson MH, Robinson ES (2014) Modelling cognitive affective biases in major depressive disorder using rodents. Br J Pharmacol 171(20):4524–4538

Hales CA, Bartlett JM, Arban R, Hengerer B, Robinson ESJ (2020) Role of the medial prefrontal cortex in the effects of rapid acting antidepressants on decision-making biases in rodents. Neuropsychopharmacology 45(13):2278–2288

Hamilton JP, Gotlib IH (2008) Neural substrates of increased memory sensitivity for negative stimuli in major depression. Biol Psychiatry 63(12):1155–1162

Hao Y, Ge H, Sun M, Gao Y (2019) Selecting an appropriate animal model of depression. Int J Mol Sci 20(19):4827

Harding EJ, Paul ES, Mendl M (2004) Cognitive bias and affective state. Nature 427(6972): 312–312

Hare BD, Ghosal S, Duman RS (2017) Rapid acting antidepressants in chronic stress models: molecular and cellular mechanisms. Chronic Stress 1:2470547017697317

Harmer CJ, Bhagwagar Z, Perrett DI, Völlm BA, Cowen PJ, Goodwin GM (2003a) Acute SSRI administration affects the processing of social cues in healthy volunteers. Neuropsychopharmacology 28(1):148–152

Harmer CJ, Hill SA, Taylor MJ, Cowen PJ, Goodwin GM (2003b) Toward a neuropsychological theory of antidepressant drug action: increase in positive emotional bias after potentiation of norepinephrine activity. Am J Psychiatry 160(5):990–992

Harmer CJ, Shelley NC, Cowen PJ, Goodwin GM (2004) Increased positive versus negative affective perception and memory in healthy volunteers following selective serotonin and norepinephrine reuptake inhibition. Am J Psychiatry 161(7):1256–1263

Harmer CJ, Goodwin GM, Cowen PJ (2009a) Why do antidepressants take so long to work? A cognitive neuropsychological model of antidepressant drug action. Br J Psychiatry 195(2): 102–108

Harmer CJ, O'Sullivan U, Favaron E, Massey-Chase R, Ayres R, Reinecke A et al (2009b) Effect of acute antidepressant administration on negative affective bias in depressed patients. Am J Psychiatry 166(10):1178–1184

Harmer CJ, Duman RS, Cowen PJ (2017) How do antidepressants work? New perspectives for refining future treatment approaches. Lancet Psychiatry 4(5):409–418

Harro J (2019) Animal models of depression: pros and cons. Cell Tissue Res 377(1):5–20

Hartling C, Metz S, Pehrs C, Scheidegger M, Gruzman R, Keicher C et al (2021) Comparison of four fMRI paradigms probing emotion processing. Brain Sci 11(5):525

Hasselmann H (2014) Scopolamine and depression: a role for muscarinic antagonism? CNS Neurol Dis-Drug Targets 13(4):673–683

Hershenberg R, Satterthwaite TD, Daldal A, Katchmar N, Moore TM, Kable JW, Wolf DH (2016) Diminished effort on a progressive ratio task in both unipolar and bipolar depression. J Affect Disord 196:97–100

Herzallah MM, Moustafa AA, Natsheh JY, Danoun OA, Simon JR, Tayem YI et al (2013) Depression impairs learning, whereas the selective serotonin reuptake inhibitor, paroxetine, impairs generalization in patients with major depressive disorder. J Affect Disord 151(2): 484–492

Hinchcliffe JK, Stuart SA, Mendl M, Robinson ES (2017) Further validation of the affective bias test for predicting antidepressant and pro-depressant risk: effects of pharmacological and social manipulations in male and female rats. Psychopharmacology (Berl) 234:3105–3116

Hinchcliffe JK, Bartlett JM, Thomas CW, Robinson ES (2023) Differential modulation of affective biases by psychedelics in a rodent affective bias test. J Psychopharmacol 37(8):G14

Hirschfeld RM (2000) History and evolution of the monoamine hypothesis of depression. J Clin Psychiatry 61(6):4–6

Hoffman KL (2016) New dimensions in the use of rodent behavioral tests for novel drug discovery and development. Expert Opin Drug Discovery 11(4):343–353
Holmes PV (2003) Rodent models of depression: reexamining validity without anthropomorphic inference. Crit Rev Neurobiol 15(2)
Horder J, Cowen PJ, Di Simplicio M, Browning M, Harmer CJ (2009) Acute administration of the cannabinoid CB1 antagonist rimonabant impairs positive affective memory in healthy volunteers. Psychopharmacology (Berl) 205:85–91
Huys QJ, Pizzagalli DA, Bogdan R, Dayan P (2013) Mapping anhedonia onto reinforcement learning: a behavioural meta-analysis. Biol Mood Anxiety Disord 3(1):1–16
Iadarola ND, Niciu MJ, Richards EM, Vande Voort JL, Ballard ED, Lundin NB et al (2015) Ketamine and other N-methyl-D-aspartate receptor antagonists in the treatment of depression: a perspective review. Ther Adv Chronic Dis 6(3):97–114
Ioannidis J (2008) Effectiveness of antidepressants: an evidence myth constructed from a thousand randomized trials? Philos Ethics Humanit Med 3(1):1–9
Jankovic J, Beach J (1997) Long-term effects of tetrabenazine in hyperkinetic movement disorders. Neurology 48(2):358–362
Jentsch MC, Van Buel EM, Bosker FJ, Gladkevich AV, Klein HC, Oude Voshaar RC et al (2015) Biomarker approaches in major depressive disorder evaluated in the context of current hypotheses. Biomark Med 9(3):277–297
Kangas BD, Wooldridge LM, Luc OT, Bergman J, Pizzagalli DA (2020) Empirical validation of a touchscreen probabilistic reward task in rats. Transl Psychiatry 10(1):285
Kenney C, Jankovic J (2006) Tetrabenazine in the treatment of hyperkinetic movement disorders. Expert Rev Neurother 6(1):7–17
Klawohn J, Joyner K, Santopetro N, Brush CJ, Hajcak G (2022) Depression reduces neural correlates of reward salience with increasing effort over the course of the progressive ratio task. J Affect Disord 307:294–300
Kraehenmann R, Preller KH, Scheidegger M, Pokorny T, Bosch OG, Seifritz E, Vollenweider FX (2015) Psilocybin-induced decrease in amygdala reactivity correlates with enhanced positive mood in healthy volunteers. Biol Psychiatry 78(8):572–581
Kuiper NA, Derry PA, MacDonald MR (1982) Self-reference and person perception in depression: a social cognition perspective. Integr Clin Soc Psychol 1982:79–103
Kunisato Y, Okamoto Y, Ueda K, Onoda K, Okada G, Yoshimura S et al (2012) Effects of depression on reward-based decision making and variability of action in probabilistic learning. J Behav Ther Exp Psychiatry 43(4):1088–1094
Lagisz M, Zidar J, Nakagawa S, Neville V, Sorato E, Paul ES et al (2020) Optimism, pessimism and judgement bias in animals: a systematic review and meta-analysis. Neurosci Biobehav Rev 118: 3–17
Lee JS, Mathews A, Shergill S, Yiend J (2016) Magnitude of negative interpretation bias depends on severity of depression. Behav Res Ther 83:26–34
Lemogne C, le Bastard G, Mayberg H, Volle E, Bergouignan L, Lehéricy S et al (2009) In search of the depressive self: extended medial prefrontal network during self-referential processing in major depression. Soc Cogn Affect Neurosci 4(3):305–312
Lemogne C, Mayberg H, Bergouignan L, Volle E, Delaveau P, Lehéricy S et al (2010) Self-referential processing and the prefrontal cortex over the course of depression: a pilot study. J Affect Disord 124(1–2):196–201
LeMoult J, Gotlib IH (2019) Depression: a cognitive perspective. Clin Psychol Rev 69:51–66
Leppänen JM (2006) Emotional information processing in mood disorders: a review of behavioral and neuroimaging findings. Curr Opin Psychiatry 19(1):34–39
Leppänen JM, Milders M, Bell JS, Terriere E, Hietanen JK (2004) Depression biases the recognition of emotionally neutral faces. Psychiatry Res 128(2):123–133
Levy MJ, Boulle F, Steinbusch HW, van den Hove DL, Kenis G, Lanfumey L (2018) Neurotrophic factors and neuroplasticity pathways in the pathophysiology and treatment of depression. Psychopharmacology (Berl) 235:2195–2220

Lewis LR, Benn A, Dwyer DM, Robinson ES (2019) Affective biases and their interaction with other reward-related deficits in rodent models of psychiatric disorders. Behav Brain Res 372: 112051

Liu RT, Kleiman EM, Nestor BA, Cheek SM (2015) The hopelessness theory of depression: a quarter-century in review. Clin Psychol Sci Pract 22(4):345

Ma Y (2015) Neuropsychological mechanism underlying antidepressant effect: a systematic meta-analysis. Mol Psychiatry 20(3):311–319

Mathews A, MacLeod C (2005) Cognitive vulnerability to emotional disorders. Annu Rev Clin Psychol 1:167–195

Mayberg HS, Lozano AM, Voon V, McNeely HE, Seminowicz D, Hamani C et al (2005) Deep brain stimulation for treatment-resistant depression. Neuron 45(5):651–660

McCabe C, Cowen PJ, Harmer CJ (2009) NK1 receptor antagonism and the neural processing of emotional information in healthy volunteers. Int J Neuropsychopharmacol 12(9):1261–1274

McCabe C, Mishor Z, Cowen PJ, Harmer CJ (2010) Diminished neural processing of aversive and rewarding stimuli during selective serotonin reuptake inhibitor treatment. Biol Psychiatry 67(5): 439–445

Mendl M, Burman OH, Parker RM, Paul ES (2009) Cognitive bias as an indicator of animal emotion and welfare: emerging evidence and underlying mechanisms. Appl Anim Behav Sci 118(3–4):161–181

Mendl M, Burman OH, Paul ES (2010) An integrative and functional framework for the study of animal emotion and mood. Proc R Soc B Biol Sci 277(1696):2895–2904

Mitchell AJ (2006) Depressed patients and treatment adherence. Lancet 367(9528):2041–2043

Montgomery SA, Baldwin DS, Blier P, Fineberg NA, Kasper S, Lader M et al (2007) Which antidepressants have demonstrated superior efficacy? A review of the evidence. Int Clin Psychopharmacol 22(6):323–329

Mukherjee D, Filipowicz AL, Vo K, Satterthwaite TD, Kable JW (2020) Reward and punishment reversal-learning in major depressive disorder. J Abnorm Psychol 129(8):810

Murphy FC, Michael A, Robbins TW, Sahakian BJ (2003) Neuropsychological impairment in patients with major depressive disorder: the effects of feedback on task performance. Psychol Med 33(3):455–467

Murphy SE, Norbury R, O'Sullivan U, Cowen PJ, Harmer CJ (2009) Effect of a single dose of citalopram on amygdala response to emotional faces. Br J Psychiatry 194(6):535–540

Murrough JW, Iosifescu DV, Chang LC, Al Jurdi RK, Green CE, Perez AM et al (2013) Antidepressant efficacy of ketamine in treatment-resistant major depression: a two-site randomized controlled trial. Am J Psychiatry 170(10):1134–1142

Nestler EJ, Hyman SE (2010) Animal models of neuropsychiatric disorders. Nat Neurosci 13(10): 1161–1169

Neville V, Nakagawa S, Zidar J, Paul ES, Lagisz M, Bateson M et al (2020) Pharmacological manipulations of judgement bias: a systematic review and meta-analysis. Neurosci Biobehav Rev 108:269–286

Nikiforuk A, Popik P (2009) Antidepressants alleviate the impact of reinforcer downshift. Eur Neuropsychopharmacol 19(1):41–48

Norbury R, Taylor MJ, Selvaraj S, Murphy SE, Harmer CJ, Cowen PJ (2009) Short-term antidepressant treatment modulates amygdala response to happy faces. Psychopharmacology (Berl) 206:197–204

Nunes EJ, Randall PA, Hart EE, Freeland C, Yohn SE, Baqi Y et al (2013) Effort-related motivational effects of the VMAT-2 inhibitor tetrabenazine: implications for animal models of the motivational symptoms of depression. J Neurosci 33(49):19120–19130

O'Leary OF, Cryan JF (2013) Towards translational rodent models of depression. Cell Tissue Res 354:141–153

Olausson P, Kiraly DD, Gourley SL, Taylor JR (2013) Persistent effects of prior chronic exposure to corticosterone on reward-related learning and motivation in rodents. Psychopharmacology (Berl) 225:569–577

Orchard F, Pass L, Reynolds S (2016) Associations between interpretation bias and depression in adolescents. Cogn Ther Res 40:577–583
Ormel J, Hollon SD, Kessler RC, Cuijpers P, Monroe SM (2022) More treatment but no less depression: the treatment-prevalence paradox. Clin Psychol Rev 91:102111
Papadatou-Pastou M, Miskowiak KW, Williams JMG, Harmer CJ, Reinecke A (2012) Acute antidepressant drug administration and autobiographical memory recall: a functional magnetic resonance imaging study. Exp Clin Psychopharmacol 20(5):364
Paul ES, Harding EJ, Mendl M (2005) Measuring emotional processes in animals: the utility of a cognitive approach. Neurosci Biobehav Rev 29(3):469–491
Pessoa L, Adolphs R (2010) Emotion processing and the amygdala: from a 'low road' to 'many roads' of evaluating biological significance. Nat Rev Neurosci 11(11):773–782
Phelps CE, Lumb BM, Donaldson LF, Robinson ES (2021) The partial saphenous nerve injury model of pain impairs reward-related learning but not reward sensitivity or motivation. Pain 162(3):956–966
Pigott HE, Leventhal AM, Alter GS, Boren JJ (2010) Efficacy and effectiveness of antidepressants: current status of research. Psychother Psychosom 79(5):267–279
Pizzagalli DA, Roberts AC (2022) Correction: prefrontal cortex and depression. Neuropsychopharmacology 47(2):609
Pizzagalli DA, Jahn AL, O'Shea JP (2005) Toward an objective characterization of an anhedonic phenotype: a signal-detection approach. Biol Psychiatry 57(4):319–327
Pizzagalli DA, Iosifescu D, Hallett LA, Ratner KG, Fava M (2008a) Reduced hedonic capacity in major depressive disorder: evidence from a probabilistic reward task. J Psychiatr Res 43(1): 76–87
Pizzagalli DA, Evins AE, Schetter EC, Frank MJ, Pajtas PE, Santesso DL, Culhane M (2008b) Single dose of a dopamine agonist impairs reinforcement learning in humans: behavioral evidence from a laboratory-based measure of reward responsiveness. Psychopharmacology (Berl) 196:221–232
Price JL, Drevets WC (2011) Neural circuits underlying the pathophysiology of mood disorders. Trends Cogn Sci 16(1):61–71
Pringle A, Browning M, Cowen PJ, Harmer CJ (2011a) A cognitive neuropsychological model of antidepressant drug action. Prog Neuropsychopharmacol Biol Psychiatry 35(7):1586–1592
Pringle A, McTavish SF, Williams C, Smith R, Cowen PJ, Harmer CJ (2011b) Short-term NK1 receptor antagonism and emotional processing in healthy volunteers. Psychopharmacology (Berl) 215:239–246
Pringle A, Parsons E, Cowen LG, McTavish SF, Cowen PJ, Harmer CJ (2012) Using an experimental medicine model to understand the antidepressant potential of the N-methyl-D-aspartic acid (NMDA) receptor antagonist memantine. J Psychopharmacol 26(11):1417–1423
Prouzeau D, Conejero I, Voyvodic PL, Becamel C, Abbar M, Lopez-Castroman J (2022) Psilocybin efficacy and mechanisms of action in major depressive disorder: a review. Curr Psychiatry Rep 24(10):573–581
Quan MN, Zhang N, Wang YY, Zhang T, Yang Z (2011) Possible antidepressant effects and mechanisms of memantine in behaviors and synaptic plasticity of a depression rat model. Neuroscience 182:88–97
Randall PA, Pardo M, Nunes EJ, López Cruz L, Vemuri VK, Makriyannis A et al (2012) Dopaminergic modulation of effort-related choice behavior as assessed by a progressive ratio chow feeding choice task: pharmacological studies and the role of individual differences. PloS One 7(10):e47934
Randall PA, Lee CA, Nunes EJ, Yohn SE, Nowak V, Khan B et al (2014) The VMAT-2 inhibitor tetrabenazine affects effort-related decision making in a progressive ratio/chow feeding choice task: reversal with antidepressant drugs. PloS One 9(6):e99320
Reed JL, Nugent AC, Furey ML, Szczepanik JE, Evans JW, Zarate CA Jr (2018) Ketamine normalizes brain activity during emotionally valenced attentional processing in depression. Neuroimage Clin 20:92–101

Rénéric JP, Lucki I (1998) Antidepressant behavioral effects by dual inhibition of monoamine reuptake in the rat forced swimming test. Psychopharmacology (Berl) 136:190–197

Richelson E (1996) Synaptic effects of antidepressants. J Clin Psychopharmacol 16(3):1S–7S

Rizvi SJ, Pizzagalli DA, Sproule BA, Kennedy SH (2016) Assessing anhedonia in depression: potentials and pitfalls. Neurosci Biobehav Rev 65:21–35

Robbins TW (1998) Homology in behavioural pharmacology: an approach to animal models of human cognition. Behav Pharmacol 9(7):509–519

Robinson ES (2018) Translational new approaches for investigating mood disorders in rodents and what they may reveal about the underlying neurobiology of major depressive disorder. Philos Trans R Soc B Biol Sci 373(1742):20170036

Robinson ESJ, Roiser JP (2016) Affective biases in humans and animals. Transl Neuropsychopharmacol:263–286

Roiser JP, Sahakian BJ (2013) Hot and cold cognition in depression. CNS Spectr 18(3):139–149

Roiser JP, Sahakian BJ (2017) Information processing in mood disorders. The Oxford handbook of mood disorders, p 179

Rosenblat JD, Leon-Carlyle M, Ali S, Husain MI, McIntyre RS (2023) Antidepressant effects of psilocybin in the absence of psychedelic effects. Am J Psychiatry 180(5):395–396

Rupprechter S, Stankevicius A, Huys QJ, Steele JD, Seriès P (2018) Major depression impairs the use of reward values for decision-making. Sci Rep 8(1):13798

Rychlik M, Bollen E, Rygula R (2017) Ketamine decreases sensitivity of male rats to misleading negative feedback in a probabilistic reversal-learning task. Psychopharmacology (Berl) 234: 613–620

Rygula R, Abumaria N, Havemann-Reinecke U, Rüther E, Hiemke C, Zernig G et al (2008) Pharmacological validation of a chronic social stress model of depression in rats: effects of reboxetine, haloperidol and diazepam. Behav Pharmacol 19(3):183–196

Rygula R, Szczech E, Papciak J, Nikiforuk A, Popik P (2014a) The effects of cocaine and mazindol on the cognitive judgement bias of rats in the ambiguous-cue interpretation paradigm. Behav Brain Res 270:206–212

Rygula R, Papciak J, Popik P (2014b) The effects of acute pharmacological stimulation of the 5-HT, NA and DA systems on the cognitive judgement bias of rats in the ambiguous-cue interpretation paradigm. Eur Neuropsychopharmacol 24(7):1103–1111

Rygula R, Noworyta-Sokolowska K, Drozd R, Kozub A (2018) Using rodents to model abnormal sensitivity to feedback in depression. Neurosci Biobehav Rev 95:336–346

Salamone JD, Cousins MS, Bucher S (1994) Anhedonia or anergia? Effects of haloperidol and nucleus accumbens dopamine depletion on instrumental response selection in a T-maze cost/ benefit procedure. Behav Brain Res 65(2):221–229

Sani G, Serra G, Kotzalidis GD, Romano S, Tamorri SM, Manfredi G et al (2012) The role of memantine in the treatment of psychiatric disorders other than the dementias: a review of current preclinical and clinical evidence. CNS Drugs 26:663–690

Scinska A, Sienkiewicz-Jarosz H, Kuran W, Ryglewicz D, Rogowski A, Wrobel E et al (2004) Depressive symptoms and taste reactivity in humans. Physiol Behav 82(5):899–904

Skandali N, Rowe JB, Voon V et al (2018) Dissociable effects of acute SSRI (escitalopram) on executive, learning and emotional functions in healthy humans. Neuropsychopharmacol 43 (13):2645–2651

Slaney C, Hinchcliffe JK, Robinson ES (2018) Translational shifts in preclinical models of depression: implications for biomarkers for improved treatments. Biomarkers Psychiatry:169–193

Slaney C, Perkins AM, Davis R, Penton-Voak I, Munafò MR, Houghton CJ, Robinson ES (2022) Objective measures of reward sensitivity and motivation in people with high v. low anhedonia. Psychol Med:1–9

Slattery DA, Cryan JF (2017) Modelling depression in animals: at the interface of reward and stress pathways. Psychopharmacology (Berl) 234:1451–1465

Steru L, Chermat R, Thierry B, Simon P (1985) The tail suspension test: a new method for screening antidepressants in mice. Psychopharmacology (Berl) 85:367–370

Stuart SA, Butler P, Munafo MR, Nutt DJ, Robinson ES (2013) A translational rodent assay of affective biases in depression and antidepressant therapy. Neuropsychopharmacology 38(9): 1625–1635

Stuart SA, Butler P, Munafò MR, Nutt DJ, Robinson ES (2015) Distinct neuropsychological mechanisms may explain delayed-versus rapid-onset antidepressant efficacy. Neuropsychopharmacology 40(9):2165–2174

Stuart SA, Wood CM, Robinson ESJ (2017) Using the affective bias test to predict drug-induced negative affect: implications for drug safety. Br J Pharmacol 174(19):3200–3210

Stuart SA, Hinchcliffe JK, Robinson ES (2019) Evidence that neuropsychological deficits following early life adversity may underlie vulnerability to depression. Neuropsychopharmacology 44(9):1623–1630

Taylor MJ (2007) Rapid onset of true antidepressant action. Curr Psychiatry Rep 9(6):475–479

Thomas JM, Higgs S, Dourish CT (2016) Test–retest reliability and effects of repeated testing and satiety on performance of an emotional test battery. J Clin Exp Neuropsychol 38(4):416–433

Tranter R, Bell D, Gutting P, Harmer C, Healy D, Anderson IM (2009) The effect of serotonergic and noradrenergic antidepressants on face emotion processing in depressed patients. J Affect Disord 118(1–3):87–93

Treadway MT, Buckholtz JW, Schwartzman AN, Lambert WE, Zald DH (2009) Worth the 'EEfRT'? The effort expenditure for rewards task as an objective measure of motivation and anhedonia. PloS One 4(8):e6598

Treadway MT, Bossaller NA, Shelton RC, Zald DH (2012) Effort-based decision-making in major depressive disorder: a translational model of motivational anhedonia. J Abnorm Psychol 121(3): 553

Underwood R, Tolmeijer E, Wibroe J, Peters E, Mason L (2021) Networks underpinning emotion: a systematic review and synthesis of functional and effective connectivity. Neuroimage 243: 118486

Warren MB, Pringle A, Harmer CJ (2015) A neurocognitive model for understanding treatment action in depression. Philos Trans R Soc B Biol Sci 370(1677):20140213

Wessa M, Lois G (2015) Brain functional effects of psychopharmacological treatment in major depression: a focus on neural circuitry of affective processing. Curr Neuropharmacol 13(4): 466–479

Wichers MC, Barge-Schaapveld DQCM, Nicolson NA, Peeters F, De Vries M, Mengelers R, Van Os J (2009) Reduced stress-sensitivity or increased reward experience: the psychological mechanism of response to antidepressant medication. Neuropsychopharmacology 34(4): 923–931

Wilkinson MP, Grogan JP, Mellor JR, Robinson ES (2020) Comparison of conventional and rapid-acting antidepressants in a rodent probabilistic reversal learning task. Brain Neurosci Adv 4: 2398212820907177

Williams JMG, Barnhofer T, Crane C, Herman D, Raes F, Watkins E, Dalgleish T (2007) Autobiographical memory specificity and emotional disorder. Psychol Bull 133(1):122

Williams LM, Korgaonkar MS, Song YC, Paton R, Eagles S, Goldstein-Piekarski A et al (2015) Amygdala reactivity to emotional faces in the prediction of general and medication-specific responses to antidepressant treatment in the randomized iSPOT-D trial. Neuropsychopharmacology 40(10):2398–2408

Willner P (1984) The validity of animal models of depression. Psychopharmacology (Berl) 83(1): 1–16

Willner P (1991) Methods for assessing the validity of animal models of human psychopathology. Anim Models Psychiatry I:1–23

Willner P (1997) Validity, reliability and utility of the chronic mild stress model of depression: a 10-year review and evaluation. Psychopharmacology (Berl) 134:319–329

Wright JH, Beck AT (1983) Cognitive therapy of depression: theory and practice. Psychiatr Serv 34(12):1119–1127

Yang XH, Huang J, Zhu CY, Wang YF, Cheung EF, Chan RC, Xie GR (2014) Motivational deficits in effort-based decision making in individuals with subsyndromal depression, first-episode and remitted depression patients. Psychiatry Res 220(3):874–882

Yohn SE, Reynolds S, Tripodi G, Correa M, Salamone JD (2018) The monoamine-oxidase B inhibitor deprenyl increases selection of high-effort activity in rats tested on a progressive ratio/chow feeding choice procedure: implications for treating motivational dysfunctions. Behav Brain Res 342:27–34

Yu CL, Liang CS, Yang FC, Tu YK, Hsu CW, Carvalho AF et al (2022) Trajectory of antidepressant effects after single-or two-dose administration of psilocybin: a systematic review and multivariate meta-analysis. J Clin Med 11(4):938

Zarate CA, Singh JB, Carlson PJ, Brutsche NE, Ameli R, Luckenbaugh DA et al (2006) A randomized trial of an N-methyl-D-aspartate antagonist in treatment-resistant major depression. Arch Gen Psychiatry 63(8):856–864

Zhang Z, Huang P, Li S, Liu Z, Zhang J, Liu Z (2022) Neural mechanisms underlying the processing of emotional stimuli in individuals with depression: an ALE meta-analysis study. Psychiatry Res 313:114598

Ketamine for Major Depressive Disorder

Sara Costi

Contents

Abstract Major Depressive Disorder (MDD) is a leading cause of disability worldwide. Conventional antidepressant treatment is characterised by a significant time to onset of therapeutic action (approximately 2 weeks) and fails to achieve a stable remission of symptoms in one-third of subjects with MDD. In the last 20 years the discovery of antidepressant effects of the N-methyl-d-aspartate (NMDA) receptor antagonist ketamine as a rapid acting (within hours) and sustained (up to 7 days) antidepressant has represented a major paradigm shift in the field.

The present chapter reviews the pharmacology, safety, and efficacy of ketamine as a novel therapeutic agent for MDD and specifically for subjects who did not

S. Costi (✉)
Department of Psychiatry, Warneford Hospital, University of Oxford, Oxford, UK

Oxford Health Foundation Trust, Warneford Hospital, Oxford, UK

Department of Psychiatry, Icahn School of Medicine at Mount Sinai, New York, NY, USA
e-mail: sara.costi@psych.ox.ac.uk

Curr Topics Behav Neurosci (2024) 66: 131–148
https://doi.org/10.1007/7854_2023_453
Published Online: 4 November 2023

respond to conventional antidepressant (treatment resistant depression). The impact of ketamine on suicidal ideation, the availability of brain biomarkers of ketamine treatment response and the association of ketamine and psychotherapy are considered.

Keywords Esketamine · Glutamate · Ketamine · Major depressive disorder · Suicide · Treatment resistant depression

1 Introduction

Major Depressive Disorder (MDD) is a chronic and debilitating condition characterised by social and occupational impairment, especially in severe cases (Ferrari et al. 2013). MDD is among the most common disorders affecting more than 264 million people worldwide (Federici et al. 2017). Subjects with MDD who fail to respond to at least two conventional antidepressant trials of adequate dose and duration may be classified as suffering from treatment resistant depression (TRD) (Rush et al. 2006). The conventional antidepressant medications show remission rates of 40–60% in randomised trials (Cipriani et al. 2007) and lower remission rates (20–40%) in naturalistic studies (Cuffel et al. 2003). The persistence of depressive symptoms is associated with a worsened quality of life for patients and caregivers, increased risk for suicide, and higher economic costs to society (Ferrari et al. 2013; Greenberg et al. 2015). Despite an urgent need for more efficient and rapidly acting antidepressants, the development of new drugs for MDD until the beginning of the twenty-first century has been almost exclusively based on medications targeting the monoamine system (Insel and Scolnick 2006).

2 Beyond the Monoaminergic Hypothesis of Depression

One of the first theory explaining the neurochemical imbalance associated with MDD is the monoamine hypothesis of depression developed approximately sixty years ago. This hypothesis suggests that depression is associated with lower functional activity of serotonin, norepinephrine, and dopamine neurons. First-generation antidepressants, including tricyclic antidepressants (TCA) and monoamine oxidase inhibitors (MAOIs), and second-generation drugs [selective serotonin reuptake inhibitors (SSRI), serotonin-norepinephrine reuptake inhibitors (SNRI), atypical antidepressants, and serotonin modulators], target mostly the monoamine neurotransmitters system and increase the levels of these mediators in the synaptic space. However, subsequent works suggest that an imbalance within the monoaminergic transmission is not a condition sufficient to regulate mood indicating that the monoamine role is mostly modulatory on other neurobiologic pathways associated with MDD (Heninger et al. 1996; Charney 1998). Moreover, the latency of response to antidepressant medications (approximately two weeks) and the percentage of MDD patients not responding to treatment (approximately one third (Fava and Davidson 1996)) provide additional support to the hypothesis that the monoaminergic theory of depression is incomplete in describing the neurobiological alterations contributing to the disorder.

The discovery of a rapid antidepressant effect of a sub-anesthetic dose of the non-competitive N-methyl-D-aspartate (NMDA) receptor antagonist ketamine as a mechanistically novel, rapidly acting antidepressant suggested the possibility of a transformative therapeutic breakthrough for refractory mood disorders (Berman et al. 2000). Placebo-controlled trials have provided strong evidence for the rapid-acting (within hours) and sustained (up to 7 days) antidepressant effects of a single administration of ketamine in TRD (Zarate et al. 2006). The rapid antidepressant effect ketamine contrasts with the delayed effect onset of currently approved antidepressants, which is particularly important in cases of suicidal ideation (Jick et al. 2004). Overall, the discovery of a rapid-acting antidepressant effects of ketamine led to a paradigm shift in the field of depression treatment and, due to its rapid action, provided a novel investigational tool to conduct mechanistic studies on behavioural changes such as clinical remission and relapse within hours and days of treatment, in striking contrast with the available antidepressant treatment that take days and week prior to exerting a clinically observable antidepressant effect. However, a critical but unresolved question concerns the exact mechanisms of action underlying ketamine's antidepressant response, and the longer-term feasibility and safety profile of ketamine for MDD, including the abuse potential of the drug, an intravenous (IV) route of administration, the potential for transient dissociative or psychotomimetic effects, and an uncertain safety profile for long-term treatment.

3 Clinical Effects of a Single Administration of a Sub-anaesthetic Dose of Ketamine

The first study testing the use of ketamine for depression was conceptualised as a trial to test glutamate synaptic alterations in the context of MDD using the response to ketamine as a probe and capitalised on previous preclinical work suggesting that NMDA receptor antagonism was associated with antidepressant effects (Pilc et al. 2013; Krystal et al. 2019). In line with this group's previous research on psychosis, ketamine was administered via IV infusion at the dosage of 0.5 mg/kg at a constant rate over 40 min. The study consisted of a placebo-controlled crossover trial conducted on a small sample ($n = 7$) and reported a marked decrease in depressive symptoms, as measured by the 25-item Hamilton Depression Rating Scale (HAM-D), after 240 min from the administration of the drug. The striking antidepressant response lasted within 72 h from the infusion with a return to baseline symptomatology in one to two weeks after infusion. Ketamine rapid and lasting antidepressant effects were temporally dissociated from the brief acute dissociative effects of the drug (Berman et al. 2000).

Findings from subsequent clinical trials of a single infusion of ketamine in depressed subjects utilising a crossover design showed similar results (Zarate et al. 2006; Diamond et al. 2014). Overall, a single infusion of ketamine in MDD appears to be well tolerated in the with response rates (defined as a reduction of ≥50% in depression severity scores from baseline) that varied between 50% (Berman et al. 2000) to 71% (Zarate et al. 2006) in clinical trials. These percentages appear higher in real-world settings with one third of TRD subjects achieving remission and approximately 50–75% of patients showing clinical response after a single dose (Wilkinson and Sanacora 2017).

4 How to Extend Ketamine Antidepressant Effect?

Given that the antidepressant effects of a single infusion of ketamine lasts approximately 7–10 days, subsequent research has focused on strategies to prolong the antidepressant effect of ketamine. Some of those studied the effect of repeated infusions of ketamine aiming at achieving a prolonged antidepressant effect (Rasmussen et al. 2013; Murrough et al. 2013; Shiroma et al. 2014). Murrough et al. (2013) reported six IV infusions of ketamine administered three times/weekly over 40 min at a 0.5 mg/kg dose in a sample of 24 medication-free TRD patients who had formerly displayed a positive response to a single IV ketamine infusion. Those who responded to a single IV ketamine dose were shown to maintain an antidepressant response until they received additional doses and for at least six days after each administration. The response duration of repeated ketamine infusions was greater than the duration for a single infusion (50% of patients retained remission for more than two weeks following the final ketamine infusion) with a median time to relapse of 18 days. Shiroma et al. replicated the findings in a sample of TRD subjects ($n = 14$) on a stable dose of antidepressant medications for at least two months prior to study onset. The authors reported 92% response rate and 67% remission rate after six-repeated ketamine infusions, suggesting that the administration of ketamine in subjects concomitantly treated with other antidepressants is safe (Shiroma et al. 2014).

Other studies tested the association with other agents (riluzole, lithium) to extend ketamine antidepressant effect. Riluzole reduces intrasynaptic levels of glutamate and is approved for the treatment of amyotrophic lateral sclerosis, but failed in demonstrating a difference in time to relapse when compared to placebo after an initial infusion with ketamine (Mathew et al. 2010; Ibrahim et al. 2012). Similarly, lithium appeared as a viable option to prolong ketamine antidepressant effect since it is used clinically as an adjunct treatment in TRD patients (Bschor 2014) and preclinical research suggested an overlapping mechanism of action between lithium and ketamine implicating the activation of the mammalian target of rapamycin (mTOR) signalling pathway and the inhibitory phosphorylation of glycogen synthase kinase (GSK)-3 within the prefrontal cortex (PFC) (Li et al. 2010; Beurel et al. 2015). Participants ($n = 34$) who showed at least an initial partial response to a single intravenous infusion of ketamine were randomised to lithium or placebo before receiving an additional three infusions of ketamine. After the ketamine treatments, participants remained on lithium or placebo during a double-blind continuation phase. No difference between lithium and placebo in continuing the acute antidepressant response to ketamine was observed (Costi et al. 2019).

5 Ketamine and Suicidality

Ketamine reduces all symptoms of depression, and in particular suicidal ideation (Krystal et al. 2013). Since 2000, several clinical trials have reported that ketamine has potential anti-suicidal properties (Price et al. 2014; Ballard et al. 2014; Murrough et al. 2015b) in patients with MDD. Considering ketamine's rapid antidepressant and anti-suicidal effects, ketamine appears as a potential pharmacological candidate for the stabilisation of patients at imminent risk of suicide. A meta-analysis conducted on patient-level data obtained for participants with baseline suicidal ideation who participated in a randomised clinical trial ($n = 167$) and received one infusion of ketamine showed that ketamine reduces suicidal ideation significantly more rapidly than control in both clinician-administered and self-report rating scales. Of note, the effect of ketamine on reduced suicidal ideation remained significant after controlling for improvement in severity of depressive symptoms and persisted for up to one week in more than 86% of subjects (Wilkinson et al. 2018). Similarly, repeat ketamine administrations have been shown to reduce suicidal ideation for up to 6 weeks (Phillips et al. 2020; Xiong et al. 2021). Of note, data from phase 3 studies and phase 2 study on intranasal esketamine in adults with MDD at imminent risk for suicide showed rapid improvement in depressive symptoms, but failed to detect significant improvement in suicidal ideation compared with placebo at 24 h and 25 days after administration (Canuso et al. 2019; Fu et al. 2020; Ionescu et al. 2021). In 2019, the Food and Drug Administration (FDA) and the European Medicines Agency (EMA) approved intranasal esketamine for the treatment of MDD with suicidal ideation or behaviour.

6 Ketamine Assisted Psychotherapy

Increasing evidence from clinical practice and research settings suggests that the combination of ketamine and psychotherapy for MDD could improve outcomes and delay relapse (Mathai et al. 2022; Kew et al. 2023). Combination strategies serve a variety of purposes and it well established that combining conventional antidepressants and psychotherapy improves outcomes and reduces relapse in MDD (Dunlop 2016; Karyotaki et al. 2016). However, the combination of ketamine and psychotherapy appears unique due to ketamine status as a medically available psychoactive drug. Several providers uses ketamine as a tool to facilitate psychological exploration in the context of the so-called ketamine-assisted psychotherapy (KAP) (Dore et al. 2019) during which the administrations of ketamine is highly variable and aims to achieve dose-dependent states of consciousness that represent different opportunities for therapeutic intervention, while other approaches focus on the use behavioural therapies outside of the period of acute drug effects (Wilkinson et al. 2017) capitalising on the notion that ketamine administration is associated with a period of enhanced neuroplasticity post treatment that could facilitate cognitive and behavioural interventions (Wilkinson et al. 2019). A recent meta-analysis suggests that the addition of psychotherapy to ketamine treatment is associated with largely positive outcomes (Kew et al. 2023). Although promising, the heterogeneity in psychotherapy approaches and treatment protocols limits the possibility to draw definitive conclusions on the benefits of combining ketamine and psychotherapy at the current time.

7 Ketamine Safety Profile

Data available from published trials suggest that overall, ketamine is well tolerated. Clinical trials, real-world case series, and a systematic review have established that most frequent side effects following ketamine administration are mild and transient, with onset occurring shortly after dosing, peaking within 2 h, and resolving within 1–24 h (Andrade 2017; Short et al. 2018).

Further, differences in percentage and severity of side effects occur as a function of dosage and formulation, route of administration, patient population, concomitant medications, and methodological aspects of studies with IV administration that tends to be associated with more severe side effects than other routes of administration (e.g. oral, SC, IM) (McIntyre et al. 2021).

The most commonly described side effects are acute psychiatric or dissociative side-effects have been described in the majority (up to 70%) of the studies. The most common acute psychiatric side-effects were anxiety, agitation or irritability, mood elevation, and unusual thoughts, while the most common psychotomimetic side-effects reported were dissociation, perceptual disturbance, derealisation, hallucinations, and depersonalisation. No long-term psychotomimetic side-effects have been reported. Another commonly described acute side effect of ketamine is a change in the cardiovascular status of patients, particularly hypertension and tachycardia. Other cardiovascular side-effects included palpitations, chest pain, chest-tightness, dizziness, and decreased blood pressure. Other less common side effects reported only in the short term include neurological (headache and dizziness) and cognitive side effects such as memory loss, difficulties in concentration, and confusion. The most frequently reported other side effects were blurred vision and nausea, while the incidence of serious adverse events and/or persistent medical sequelae in clinical trials is low (Caddy et al. 2015). Unfortunately, data concerning long-term safety is scarce and has been rarely assessed in randomised controlled trials. Thus, the general consensus in the field is that vigilance for safety and tolerability is a priority in the surveillance of patients, including monitoring for hepatic toxicity, cognitive changes, and genitourinary toxicity, as well as the potential for abuse liability (McIntyre et al. 2021; Ceban et al. 2021).

8 Pharmacology of Ketamine

8.1 *Pharmacokinetics and Route of Administration*

The pharmacokinetics and bioavailability of ketamine is strongly influenced by the route of administration. However, there have been few pharmacokinetic (PK) studies on ketamine in this therapeutic area and only a single population PK study (conducted on nine patients), each receiving a single treatment of ketamine, in which several metabolites apart from ketamine enantiomers were quantified (Zhao et al. 2012). Bioavailability is approximately 99% when given intravenously (IV), 93% when given intramuscularly (IM), 25–50% when given intranasally (IN), and 16–20% when administered orally, although peak effects occur rapidly regardless of administration method (Niciu et al. 2014). However, a recent study collected PK data from a double-blind, controlled, crossover study in 21 TRD subjects wherein ketamine was administered at a range of doses to each participant (0.1–0.5 mg/kg) using a multiple crossover design and three routes of drug administration (IV, subcutaneous, IM). This study estimated ketamine bioavailability after subcutaneous (SC) and IM administration approximately of 64% (Abuhelwa et al. 2022). Ketamine reaches its maximum plasma concentrations in less than 1 min with IV, 5–15 min with IM, and 30–60 min with oral administration. It is associated with a high systemic clearance (about 1 L/min) and short plasma half-life (approximately 1–3 h) (Niciu et al. 2014). The primary active metabolite of ketamine is norketamine, resulting from N-demethylation of ketamine. The relative plasma concentration of ketamine and norketamine can differ significantly as a result of oral (1:15) versus intravenous administration (1:1) (Yanagihara et al. 2003).

8.2 *Ketamine Molecular Mechanism of Action*

The mechanisms through which ketamine produces its antidepressant effects are still not completely known. At a molecular level, ketamine is a non-competitive NMDA receptor (NMDAR) antagonist that blocks the ion permeation pore located inside the NMDAR. Preclinical and clinical data suggest that this feature of ketamine is critical in generating the intracellular signalling and synaptic changes that produce antidepressant action. However, other signalling pathways may be involved in ketamine antidepressant action (Williams et al. 2018). One theory capitalises on work showing that ketamine can rapidly activate mammalian target of rapamycin (mTOR) signalling, increase synapse-related proteins and the number of new dendritic spines in the prefrontal cortex (Li et al. 2010). An infusion of the mTOR inhibitor rapamycin into the prefrontal cortex (PFC) inhibited the molecular and behavioural effects of ketamine, which corroborated the hypothesis that mTOR signalling mediates the fast-acting antidepressant effects through increased dendric spine numbers. Other authors (Moghaddam et al. 1997) also suggested that a limited function of NMDARs reduces the inhibitory tone of PFC neurons. Consequentially, the antidepressant action of ketamine can be explained as a preferential block of NMDARs on inhibitory interneurons, which remove the inhibition from glutamatergic neurons, increases glutamate levels that activate postsynaptic α-amino-3-hydroxy-5-methyl-4-isoxazolepropionic acid (AMPA) receptors, and determine the downstream mTOR activity and the subsequential increase in spine numbers (Gerhard et al. 2020).

Another theory for ketamine rapid antidepressant effect suggests that ketamine may impact intracellular signalling via blockage of NMDAR that are involved in a specific form of neurotransmission. The effects of ketamine on NMDAR are dependent on the basal state of the receptor at the time of drug administration. For active NMDARs, ketamine leads to augmented glutamate release and increased excitatory activity. For inactive NMDARs, ketamine leads to the inhibition of eukaryotic elongation factor (eEF2) kinase, with subsequent up-regulation of the brain-derived neurotrophic factor (BDNF), a key protein that activates signalling that induce synaptic trafficking of AMPARs with synaptic potentiation in the hippocampus. These molecular and synaptic changes in the hippocampus are required for the rapid antidepressant effects of ketamine (Autry et al. 2011; Suzuki et al. 2017; Kavalali and Monteggia 2020).

Of note, ketamine antidepressant effects extend beyond its half-life. It is possible that the active metabolites derived from ketamine metabolisation, may contribute to the prolonged pharmacological action. Another possibility is that the intracellular signal transduction (and synaptic changes) associated with ketamine rapid antidepressant action elicits further downstream signalling responsible for the sustained effects. For example, the prefrontal cortex (PFC) has been proposed as a key brain area for ketamine antidepressant response: ketamine can reduce dendric spine loss and increase new spines formation in the PFC of mice following chronic unpredictable stress (Li et al. 2011) via a synaptogenesis process driven by BDNF. Notably, prevention of early signalling involved in the rapid antidepressant action of ketamine also blocks its prolonged effects (Autry et al. 2011; Ma et al. 2017).

There has also been interest in a possible opioid mechanism of action of ketamine. Ketamine has a modest affinity for μ opioid receptors and in patients with TRD, pre-treatment with the opiate antagonist, naltrexone, blocked the antidepressant (but not the dissociative) effects of ketamine (Williams et al. 2018). Work in animals indicates that opioid mechanisms are necessary, but not sufficient, to explain the antidepressant action of ketamine (Klein et al. 2020).

9 Brain-Imaging Biomarkers of Ketamine Antidepressant Effects

9.1 Resting State Functional Magnetic Resonance Imaging (fMRI) Studies

Several studies have investigated the acute ketamine effects on brain activity and connectivity, mostly 24 h following the drug administration, corresponding to ketamine antidepressant peak effects. Abdallah and colleagues showed that brain connectivity in MDD subjects 24-h post-ketamine administration was characterised by increased connectivity in the PFC and reduced global brain connectivity in the left cerebellum. Further, depressed patients who responded to ketamine showed increased global brain connectivity in the bilateral caudate, the right lateral PFC and the left middle temporal gyrus (Abdallah et al. 2017). Another study looked at the delayed effects of ketamine, compared to placebo, 2 days and 10 days after the ketamine administration on depressed and healthy subjects. Compared to placebo, there were small differences in the connectivity between the DMN and the insula for the two groups 48 h post drug administration that normalised at day 10 (Evans et al. 2018). No significant changes were observed in terms of global brain connectivity at the level of PFC between the ketamine and placebo groups in MDD patients (Kraus et al. 2020). At the level of striatal connectivity, 2 days after ketamine administration, depressed patients showed increased connectivity between the different striatal regions and areas of the PFC, anterior cingulate cortex (ACC) and orbitofrontal cortex (OFC). Interestingly, increased connectivity between the dorsal caudate and the PFC also correlated with a reduction in anhedonia in the MDD group at 2 days and 10 days after ketamine (Mkrtchian et al. 2021).

A study by Vasavada et al. explored the effects of repeated ($n = 4$) ketamine infusions, compared to placebo, on the connectivity of the amygdala and the hippocampus with the Central Executive Network (CEN), the Default Monde Network (DMN) and the salience network (SN) in a sample of depressed subjects ($n = 44$). Compared to baseline, ketamine increased right amygdala connectivity to the right CEN, decreased amygdala connectivity to the left CEN 24 h after the first infusion, which increased after the fourth infusion compared to 24 h after the first infusion, and decreased left amygdala connectivity to the SN. Decreased left amygdala to SN connectivity 24 h after the first infusion predicted improvements in anxiety at the fourth infusion, while ketamine increased right hippocampus connectivity to left CEN and this change at 24 h after the first infusion predicted an improvement in anhedonia severity at the fourth infusion (Vasavada et al. 2021).

9.2 Positron Emission Tomography (PET) Studies

Other works have explored the effects of ketamine in depressed individuals and have used [^{18}F]-fluorodeoxyglucose (FDG) positron emission tomography (PET) imaging to assess changes in glucose metabolism. The FDG PET provides a proxy of glutamatergic transmission because it allows the measurement of glucose metabolism by glial uptake of glucose in response to glutamate release from neurons. A sample of 20 unmedicated TRD participants underwent positron emission tomography (PET) to measure regional cerebral glucose metabolism at baseline (one to three days before treatment), and following a single ketamine administration (120 min post-infusion). Relative to baseline, normalised metabolism decreased significantly following ketamine in the right habenula and increased in the right amygdala. No significant correlations were observed between change in metabolism and change in depressive severity score (Carlson et al. 2013). In a double-blind, randomised, crossover trial, twenty-one subjects with bipolar disorder in a depressed state underwent FDG PET imaging two hours after receiving both placebo and ketamine infusions. Compared to placebo, subjects who showed the greatest clinical improvement had the largest metabolic increase after ketamine and significantly lower glucose metabolism in the left hippocampus following the ketamine administration (Nugent et al. 2014).

9.3 Task-Based fMRI Studies

fMRI studies have also focused on investigating the effect of ketamine on functional systems of reward, emotion, and cognitive control. Despite these studies provide data showing that ketamine alters depression-relevant neural circuits, because of the differences in tasks and neural targets, only partial consensus has been reached overall.

Depression is associated with anhedonia (reduced ability to experience pleasure) and a dysfunctional reward processing at the brain-circuit level. A study conducted in a sample of 14 depressed subjects who received a single infusion of ketamine and completed a fMRI scan using a variant of the monetary incentive delay task (MID) at baseline and 24-h post infusion, showed reduced subgenual (sg)ACC activation to positive incentives, but not negative feedback (Morris et al. 2020). In another study, a small sample of MDD subjects ($n = 10$) underwent fMRI during a modified monetary incentive delay task at baseline, one day and seven days after ketamine infusion and showed increased orbitofrontal activation during the anticipatory phase of reward 24 h post ketamine, that persisted up to 1 week after ketamine injection. Increased ventral striatum and orbitofrontal cortex activity involved in reward processing persisted seven days post-infusion (Sterpenich et al. 2019).

Several studies have used emotion recognition task to test the effects of a single ketamine infusion on emotion processing related networks. In the study described above by Sterpenich et al. (2019), the authors used also a task wherein participants rated the emotional valence of the pictures displayed and they reported decreased amygdala and insula responses during negatively valanced pictures 24-h post ketamine while the insula and dorsal ACC response decreased 7 days post ketamine. Further, the ventral tegmental area (VTA) appeared more active for negative picture ratings compared to positive picture ratings at baseline and this appeared to be reversed seven days following ketamine (Sterpenich et al. 2019).

Another double-blind, placebo-controlled, crossover study of 33 unmedicated participants with MDD and 26 healthy controls explored how ketamine affected fMRI activation during an attentional bias dot probe task with emotional face stimuli across multiple time points (baseline, and one to three days after each infusion). Compared to placebo, ketamine affected fMRI signal during emotionally valenced attentional processing (angry and happy faces) within the PFC and dorsal ACC. Overall, MDD participants brain activity post-ketamine resembled brain activity in healthy controls post-placebo, suggesting a normalisation of the well-established MDD hyperactive function within the amygdala, insula, and dorsal ACC regions (Stuhrmann et al. 2011; Reed et al. 2018).

The same group extended this work using an implicit and explicit facial recognition task and showed reduced activity post-ketamine in frontomedial regions and posteriorly in parietal and occipital areas (Reed et al. 2019). These findings in participants with MDD included some areas forming part of the default mode network (DMN) such as the PFC, posterior cingulate, and precuneus, suggesting that ketamine's mechanism of action in MDD may be specifically associated with a decreased activity in these regions during emotional processing.

Conversely, another group used two separate emotion perception tasks during fMRI to examine the neural effects of ketamine in 20 medication-free subjects with TRD. One task used happy and neutral facial expressions and the other used sad and neutral facial expressions. While no effect was observed during the sad faces task, neural responses to positive faces were selectively increased within the right caudate and greater connectivity of the right caudate during positive emotion and this was associated with improvement in depression severity following ketamine (Murrough et al. 2015a).

One study investigated emotion recognition following multiple ketamine infusions in 27 TRD participants and in 17 subjects undergoing electroconvulsive therapy (ECT) using a fMRI face-matching task probed negative- and positive-valence systems at baseline and after treatment. Both treatments were associated with a decrease in amygdala reactivity for positive and negative emotional face processing. Changes for positive faces in the inferior parietal cortex correlated with overall symptom improvement, while changes in frontal regions correlated with anxiety for negative faces, and anhedonia for positive faces (Loureiro et al. 2020).

10 Conclusion

The discovery of the rapid antidepressant effects of ketamine has stimulated a substantial research effort into the role of NMDA receptor antagonist drugs in the treatment of depression. This has produced an important shift away from the prevailing focus on monoamine mechanisms. Thus far, however, no other NMDA receptor antagonist or glutamatergic agent has been demonstrated to have an antidepressant effect equivalent to that of ketamine.

Ketamine has become widely used clinically. Concern about possible long-term adverse effects of ketamine, for example, on cognitive function or the urinary bladder, mandates careful follow-up and surveillance of all patients receiving longer-term ketamine treatment. It will also be important to determine the role of psychotherapy in maintaining the therapeutic response to ketamine and decreasing the need for repeated ketamine treatments. Currently, the various psychotherapies that have been studied are too disparate to allow clear recommendations.

The mechanism of action of ketamine in producing rapid antidepressant effects is intriguing and a matter of intensive research. Conceivably, these studies might lead to medications that impact directly on neuroplasticity without the need for an intervening neurotransmitter mechanism. Functional imaging reveals effects of ketamine alter the balance of neural networks implicated in the pathophysiology of depression. As well as helping to elucidate the action of ketamine at a systems level these changes could in the future provide useful predictors of those depressed patients most likely to respond to ketamine treatment.

References

Abdallah CG, Averill LA, Collins KA, Geha P, Schwartz J, Averill C, DeWilde KE, Wong E, Anticevic A, Tang CY, Iosifescu DV, Charney DS, Murrough JW (2017) Ketamine treatment and global brain connectivity in major depression. Neuropsychopharmacology 42:1210–1219. https://doi.org/10.1038/npp.2016.186

Abuhelwa AY, Somogyi AA, Loo CK, Glue P, Barratt DT, Foster DJR (2022) Population pharmacokinetics and pharmacodynamics of the therapeutic and adverse effects of ketamine in patients with treatment-refractory depression. Clin Pharmacol Ther 112:720–729. https://doi.org/10.1002/cpt.2640

Andrade C (2017) Ketamine for depression, 4: in what dose, at what rate, by what route, for how long, and at what frequency? J Clin Psychiatry 78:e852–e857. https://doi.org/10.4088/JCP.17f11738

Autry AE, Adachi M, Nosyreva E, Na ES, Los MF, Cheng P, Kavalali ET, Monteggia LM (2011) NMDA receptor blockade at rest triggers rapid behavioural antidepressant responses. Nature 475:91–95. https://doi.org/10.1038/nature10130

Ballard ED, Ionescu DF, Vande Voort JL, Niciu MJ, Richards EM, Luckenbaugh DA, Brutsché NE, Ameli R, Furey ML, Zarate CA (2014) Improvement in suicidal ideation after ketamine infusion: relationship to reductions in depression and anxiety. J Psychiatr Res 58:161–166. https://doi.org/10.1016/j.jpsychires.2014.07.027

Berman RM, Cappiello A, Anand A, Oren DA, Heninger GR, Charney DS, Krystal JH (2000) Antidepressant effects of ketamine in depressed patients. Biol Psychiatry 47:351–354. https://doi.org/10.1016/s0006-3223(99)00230-9

Beurel E, Grieco SF, Jope RS (2015) Glycogen synthase kinase-3 (GSK3): regulation, actions, and diseases. Pharmacol Ther 148:114–131. https://doi.org/10.1016/j.pharmthera.2014.11.016

Bschor T (2014) Lithium in the treatment of major depressive disorder. Drugs 74:855–862. https://doi.org/10.1007/s40265-014-0220-x

Caddy C, Amit BH, McCloud TL, Rendell JM, Furukawa TA, McShane R, Hawton K, Cipriani A (2015) Ketamine and other glutamate receptor modulators for depression in adults. Cochrane Database Syst Rev:CD011612. https://doi.org/10.1002/14651858.CD011612.pub2

Canuso CM, Singh JB, Fedgchin M, Alphs L, Lane R, Lim P, Pinter C, Hough D, Sanacora G, Manji H, Drevets WC (2019) Efficacy and safety of intranasal esketamine for the rapid reduction of symptoms of depression and suicidality in patients at imminent risk for suicide: results of a double-blind, randomized, placebo-controlled study. Focus (Am Psychiatr Publ) 17: 55–65. https://doi.org/10.1176/appi.focus.17105

Carlson PJ, Diazgranados N, Nugent AC, Ibrahim L, Luckenbaugh DA, Brutsche N, Herscovitch P, Manji HK, Zarate CA, Drevets WC (2013) Neural correlates of rapid antidepressant response to ketamine in treatment-resistant unipolar depression: a preliminary positron emission tomography study. Biol Psychiatry 73:1213–1221. https://doi.org/10.1016/j.biopsych.2013.02.008

Ceban F, Rosenblat JD, Kratiuk K, Lee Y, Rodrigues NB, Gill H, Subramaniapillai M, Nasri F, Lui LMW, Lipsitz O, Kumar A, Lee JG, Chau EH, Cao B, Lin K, Ho RC, Mansur RB, Swainson J, McIntyre RS (2021) Prevention and management of common adverse effects of ketamine and esketamine in patients with mood disorders. CNS Drugs 35:925–934. https://doi.org/10.1007/s40263-021-00846-5

Charney DS (1998) Monoamine dysfunction and the pathophysiology and treatment of depression. J Clin Psychiatry 59(Suppl 14):11–14

Cipriani A, Geddes JR, Furukawa TA, Barbui C (2007) Metareview on short-term effectiveness and safety of antidepressants for depression: an evidence-based approach to inform clinical practice. Can J Psychiatry 52:553–562. https://doi.org/10.1177/070674370705200903

Costi S, Soleimani L, Glasgow A, Brallier J, Spivack J, Schwartz J, Levitch CF, Richards S, Hoch M, Stade EC, Welch A, Collins KA, Feder A, Iosifescu DV, Charney DS, Murrough JW (2019) Lithium continuation therapy following ketamine in patients with treatment resistant unipolar depression: a randomized controlled trial. Neuropsychopharmacology 44:1812–1819. https://doi.org/10.1038/s41386-019-0365-0

Cuffel BJ, Azocar F, Tomlin M, Greenfield SF, Busch AB, Croghan TW (2003) Remission, residual symptoms, and nonresponse in the usual treatment of major depression in managed clinical practice. J Clin Psychiatry 64:397–402. https://doi.org/10.4088/jcp.v64n0406

Diamond PR, Farmery AD, Atkinson S, Haldar J, Williams N, Cowen PJ, Geddes JR, McShane R (2014) Ketamine infusions for treatment resistant depression: a series of 28 patients treated weekly or twice weekly in an ECT clinic. J Psychopharmacol (Oxf) 28:536–544. https://doi.org/10.1177/0269881114527361

Dore J, Turnipseed B, Dwyer S, Turnipseed A, Andries J, Ascani G, Monnette C, Huidekoper A, Strauss N, Wolfson P (2019) Ketamine assisted psychotherapy (KAP): patient demographics, clinical data and outcomes in three large practices administering ketamine with psychotherapy. J Psychoactive Drugs 51:189–198. https://doi.org/10.1080/02791072.2019.1587556

Dunlop BW (2016) Evidence-based applications of combination psychotherapy and pharmacotherapy for depression. Focus (Madison) 14:156–173. https://doi.org/10.1176/appi.focus.20150042

Evans JW, Szczepanik J, Brutsché N, Park LT, Nugent AC, Zarate CA (2018) Default mode connectivity in major depressive disorder measured up to 10 days after ketamine administration. Biol Psychiatry 84:582–590. https://doi.org/10.1016/j.biopsych.2018.01.027

Fava M, Davidson KG (1996) Definition and epidemiology of treatment-resistant depression. Psychiatr Clin North Am 19:179–200. https://doi.org/10.1016/S0193-953X(05)70283-5

Federici S, Bracalenti M, Meloni F, Luciano JV (2017) World Health Organization disability assessment schedule 2.0: an international systematic review. Disabil Rehabil 39:2347–2380. https://doi.org/10.1080/09638288.2016.1223177

Ferrari AJ, Charlson FJ, Norman RE, Patten SB, Freedman G, Murray CJL, Vos T, Whiteford HA (2013) Burden of depressive disorders by country, sex, age, and year: findings from the global burden of disease study 2010. PLoS Med 10:e1001547. https://doi.org/10.1371/journal.pmed.1001547

Fu D-J, Ionescu DF, Li X, Lane R, Lim P, Sanacora G, Hough D, Manji H, Drevets WC, Canuso CM (2020) Esketamine nasal spray for rapid reduction of major depressive disorder symptoms in patients who have active suicidal ideation with intent: double-blind, Randomized Study (ASPIRE I). J Clin Psychiatry. https://doi.org/10.4088/JCP.19m13191

Gerhard DM, Pothula S, Liu R-J, Wu M, Li X-Y, Girgenti MJ, Taylor SR, Duman CH, Delpire E, Picciotto M, Wohleb ES, Duman RS (2020) GABA interneurons are the cellular trigger for ketamine's rapid antidepressant actions. J Clin Invest 130:1336–1349. https://doi.org/10.1172/JCI130808

Greenberg PE, Fournier A-A, Sisitsky T, Pike CT, Kessler RC (2015) The economic burden of adults with major depressive disorder in the United States (2005 and 2010). J Clin Psychiatry 76:155–162. https://doi.org/10.4088/JCP.14m09298

Heninger GR, Delgado PL, Charney DS (1996) The revised monoamine theory of depression: a modulatory role for monoamines, based on new findings from monoamine depletion experiments in humans. Pharmacopsychiatry 29:2–11. https://doi.org/10.1055/s-2007-979535

Ibrahim L, Diazgranados N, Franco-Chaves J, Brutsche N, Henter ID, Kronstein P, Moaddel R, Wainer I, Luckenbaugh DA, Manji HK, Zarate CA (2012) Course of improvement in depressive symptoms to a single intravenous infusion of ketamine vs add-on riluzole: results from a 4-week, double-blind, placebo-controlled study. Neuropsychopharmacology 37:1526–1533. https://doi.org/10.1038/npp.2011.338

Insel TR, Scolnick EM (2006) Cure therapeutics and strategic prevention: raising the bar for mental health research. Mol Psychiatry 11:11–17. https://doi.org/10.1038/sj.mp.4001777

Ionescu DF, Fu D-J, Qiu X, Lane R, Lim P, Kasper S, Hough D, Drevets WC, Manji H, Canuso CM (2021) Esketamine nasal spray for rapid reduction of depressive symptoms in patients with major depressive disorder who have active suicide ideation with intent: results of a phase 3, double-blind, randomized study (ASPIRE II). Int J Neuropsychopharmacol 24:22–31. https://doi.org/10.1093/ijnp/pyaa068

Jick H, Kaye JA, Jick SS (2004) Antidepressants and the risk of suicidal behaviors. JAMA 292: 338–343. https://doi.org/10.1001/jama.292.3.338

Karyotaki E, Smit Y, Holdt Henningsen K, Huibers MJH, Robays J, de Beurs D, Cuijpers P (2016) Combining pharmacotherapy and psychotherapy or monotherapy for major depression? A meta-analysis on the long-term effects. J Affect Disord 194:144–152. https://doi.org/10.1016/j.jad.2016.01.036

Kavalali ET, Monteggia LM (2020) Targeting homeostatic synaptic plasticity for treatment of mood disorders. Neuron 106:715–726. https://doi.org/10.1016/j.neuron.2020.05.015

Kew BM, Porter RJ, Douglas KM, Glue P, Mentzel CL, Beaglehole B (2023) Ketamine and psychotherapy for the treatment of psychiatric disorders: systematic review. BJPsych Open 9: e79. https://doi.org/10.1192/bjo.2023.53

Klein ME, Chandra J, Sheriff S, Malinow R (2020) Opioid system is necessary but not sufficient for antidepressive actions of ketamine in rodents. Proc Natl Acad Sci U S A 117:2656–2662

Kraus C, Mkrtchian A, Kadriu B, Nugent AC, Zarate CA, Evans JW (2020) Evaluating global brain connectivity as an imaging marker for depression: influence of preprocessing strategies and placebo-controlled ketamine treatment. Neuropsychopharmacology 45:982–989. https://doi.org/10.1038/s41386-020-0624-0

Krystal JH, Sanacora G, Duman RS (2013) Rapid-acting glutamatergic antidepressants: the path to ketamine and beyond. Biol Psychiatry 73:1133–1141. https://doi.org/10.1016/j.biopsych.2013.03.026

Krystal JH, Abdallah CG, Sanacora G, Charney DS, Duman RS (2019) Ketamine: a paradigm shift for depression research and treatment. Neuron 101:774–778. https://doi.org/10.1016/j.neuron.2019.02.005

Li N, Lee B, Liu R-J, Banasr M, Dwyer JM, Iwata M, Li X-Y, Aghajanian G, Duman RS (2010) mTOR-dependent synapse formation underlies the rapid antidepressant effects of NMDA antagonists. Science 329:959–964. https://doi.org/10.1126/science.1190287

Li N, Liu R-J, Dwyer JM, Banasr M, Lee B, Son H, Li X-Y, Aghajanian G, Duman RS (2011) Glutamate N-methyl-D-aspartate receptor antagonists rapidly reverse behavioral and synaptic deficits caused by chronic stress exposure. Biol Psychiatry 69:754–761. https://doi.org/10.1016/j.biopsych.2010.12.015

Loureiro JRA, Leaver A, Vasavada M, Sahib AK, Kubicki A, Joshi S, Woods RP, Wade B, Congdon E, Espinoza R, Narr KL (2020) Modulation of amygdala reactivity following rapidly acting interventions for major depression. Hum Brain Mapp 41:1699–1710. https://doi.org/10.1002/hbm.24895

Ma Z, Zang T, Birnbaum SG, Wang Z, Johnson JE, Zhang C-L, Parada LF (2017) TrkB dependent adult hippocampal progenitor differentiation mediates sustained ketamine antidepressant response. Nat Commun 8:1668. https://doi.org/10.1038/s41467-017-01709-8

Mathai DS, Mora V, Garcia-Romeu A (2022) Toward synergies of ketamine and psychotherapy. Front Psychol 13:868103. https://doi.org/10.3389/fpsyg.2022.868103

Mathew SJ, Murrough JW, aan het Rot M, Collins KA, Reich DL, Charney DS (2010) Riluzole for relapse prevention following intravenous ketamine in treatment-resistant depression: a pilot randomized, placebo-controlled continuation trial. Int J Neuropsychopharmacol 13:71–82. https://doi.org/10.1017/S1461145709000169

McIntyre RS, Rosenblat JD, Nemeroff CB, Sanacora G, Murrough JW, Berk M, Brietzke E, Dodd S, Gorwood P, Ho R, Iosifescu DV, Lopez Jaramillo C, Kasper S, Kratiuk K, Lee JG, Lee Y, Lui LMW, Mansur RB, Papakostas GI, Subramaniapillai M, Thase M, Vieta E, Young AH, Zarate CA, Stahl S (2021) Synthesizing the evidence for ketamine and esketamine in treatment-resistant depression: an international expert opinion on the available evidence and implementation. Am J Psychiatry 178:383–399. https://doi.org/10.1176/appi.ajp.2020.20081251

Mkrtchian A, Evans JW, Kraus C, Yuan P, Kadriu B, Nugent AC, Roiser JP, Zarate CA (2021) Ketamine modulates fronto-striatal circuitry in depressed and healthy individuals. Mol Psychiatry 26:3292–3301. https://doi.org/10.1038/s41380-020-00878-1

Moghaddam B, Adams B, Verma A, Daly D (1997) Activation of glutamatergic neurotransmission by ketamine: a novel step in the pathway from NMDA receptor blockade to dopaminergic and cognitive disruptions associated with the prefrontal cortex. J Neurosci 17:2921–2927. https://doi.org/10.1523/JNEUROSCI.17-08-02921.1997

Morris LS, Costi S, Tan A, Stern ER, Charney DS, Murrough JW (2020) Ketamine normalizes subgenual cingulate cortex hyper-activity in depression. Neuropsychopharmacology 45:975–981. https://doi.org/10.1038/s41386-019-0591-5

Murrough JW, Perez AM, Pillemer S, Stern J, Parides MK, aan het Rot M, Collins KA, Mathew SJ, Charney DS, Iosifescu DV (2013) Rapid and longer-term antidepressant effects of repeated ketamine infusions in treatment-resistant major depression. Biol Psychiatry 74:250–256. https://doi.org/10.1016/j.biopsych.2012.06.022

Murrough JW, Collins KA, Fields J, DeWilde KE, Phillips ML, Mathew SJ, Wong E, Tang CY, Charney DS, Iosifescu DV (2015a) Regulation of neural responses to emotion perception by ketamine in individuals with treatment-resistant major depressive disorder. Transl Psychiatry 5: e509. https://doi.org/10.1038/tp.2015.10

Murrough JW, Soleimani L, DeWilde KE, Collins KA, Lapidus KA, Iacoviello BM, Lener M, Kautz M, Kim J, Stern JB, Price RB, Perez AM, Brallier JW, Rodriguez GJ, Goodman WK, Iosifescu DV, Charney DS (2015b) Ketamine for rapid reduction of suicidal ideation: a randomized controlled trial. Psychol Med 45:3571–3580. https://doi.org/10.1017/S0033291715001506

Niciu MJ, Luckenbaugh DA, Ionescu DF, Guevara S, Machado-Vieira R, Richards EM, Brutsche NE, Nolan NM, Zarate CA (2014) Clinical predictors of ketamine response in treatment-resistant major depression. J Clin Psychiatry 75:e417–e423. https://doi.org/10.4088/JCP.13m08698

Nugent AC, Diazgranados N, Carlson PJ, Ibrahim L, Luckenbaugh DA, Brutsche N, Herscovitch P, Drevets WC, Zarate CA (2014) Neural correlates of rapid antidepressant response to ketamine in bipolar disorder. Bipolar Disord 16:119–128. https://doi.org/10.1111/bdi.12118

Phillips JL, Norris S, Talbot J, Hatchard T, Ortiz A, Birmingham M, Owoeye O, Batten LA, Blier P (2020) Single and repeated ketamine infusions for reduction of suicidal ideation in treatment-resistant depression. Neuropsychopharmacology 45:606–612. https://doi.org/10.1038/s41386-019-0570-x

Pilc A, Wierońska JM, Skolnick P (2013) Glutamate-based antidepressants: preclinical psychopharmacology. Biol Psychiatry 73:1125–1132. https://doi.org/10.1016/j.biopsych.2013.01.021

Price RB, Iosifescu DV, Murrough JW, Chang LC, Al Jurdi RK, Iqbal SZ, Soleimani L, Charney DS, Foulkes AL, Mathew SJ (2014) Effects of ketamine on explicit and implicit suicidal cognition: a randomized controlled trial in treatment-resistant depression. Depress Anxiety 31:335–343. https://doi.org/10.1002/da.22253

Rasmussen KG, Lineberry TW, Galardy CW, Kung S, Lapid MI, Palmer BA, Ritter MJ, Schak KM, Sola CL, Hanson AJ, Frye MA (2013) Serial infusions of low-dose ketamine for major depression. J Psychopharmacol (Oxf) 27:444–450. https://doi.org/10.1177/0269881113478283

Reed JL, Nugent AC, Furey ML, Szczepanik JE, Evans JW, Zarate CA (2018) Ketamine normalizes brain activity during emotionally valenced attentional processing in depression. Neuroimage Clin 20:92–101. https://doi.org/10.1016/j.nicl.2018.07.006

Reed JL, Nugent AC, Furey ML, Szczepanik JE, Evans JW, Zarate CA (2019) Effects of ketamine on brain activity during emotional processing: differential findings in depressed versus healthy control participants. Biol Psychiatry Cogn Neurosci Neuroimaging 4:610–618. https://doi.org/10.1016/j.bpsc.2019.01.005

Rush AJ, Trivedi MH, Wisniewski SR, Nierenberg AA, Stewart JW, Warden D, Niederehe G, Thase ME, Lavori PW, Lebowitz BD, McGrath PJ, Rosenbaum JF, Sackeim HA, Kupfer DJ, Luther J, Fava M (2006) Acute and longer-term outcomes in depressed outpatients requiring one or several treatment steps: a STAR*D report. Am J Psychiatry 163:1905–1917. https://doi.org/10.1176/ajp.2006.163.11.1905

Shiroma PR, Johns B, Kuskowski M, Wels J, Thuras P, Albott CS, Lim KO (2014) Augmentation of response and remission to serial intravenous subanesthetic ketamine in treatment resistant depression. J Affect Disord 155:123–129. https://doi.org/10.1016/j.jad.2013.10.036

Short B, Fong J, Galvez V, Shelker W, Loo CK (2018) Side-effects associated with ketamine use in depression: a systematic review. Lancet Psychiatry 5:65–78. https://doi.org/10.1016/S2215-0366(17)30272-9

Sterpenich V, Vidal S, Hofmeister J, Michalopoulos G, Bancila V, Warrot D, Dayer A, Desseilles M, Aubry J-M, Kosel M, Schwartz S, Vutskits L (2019) Increased reactivity of the mesolimbic reward system after ketamine injection in patients with treatment-resistant major depressive disorder. Anesthesiology 130:923–935. https://doi.org/10.1097/ALN.0000000000002667

Stuhrmann A, Suslow T, Dannlowski U (2011) Facial emotion processing in major depression: a systematic review of neuroimaging findings. Biol Mood Anxiety Disord 1:10. https://doi.org/10.1186/2045-5380-1-10

Suzuki K, Nosyreva E, Hunt KW, Kavalali ET, Monteggia LM (2017) Effects of a ketamine metabolite on synaptic NMDAR function. Nature 546:E1–E3. https://doi.org/10.1038/nature22084

Vasavada MM, Loureiro J, Kubicki A, Sahib A, Wade B, Hellemann G, Espinoza RT, Congdon E, Narr KL, Leaver AM (2021) Effects of serial ketamine infusions on corticolimbic functional connectivity in major depression. Biol Psychiatry Cogn Neurosci Neuroimaging 6:735–744. https://doi.org/10.1016/j.bpsc.2020.06.015

Wilkinson ST, Sanacora G (2017) Considerations on the off-label use of ketamine as a treatment for mood disorders. JAMA 318:793–794. https://doi.org/10.1001/jama.2017.10697

Wilkinson ST, Wright D, Fasula MK, Fenton L, Griepp M, Ostroff RB, Sanacora G (2017) Cognitive behavior therapy may sustain antidepressant effects of intravenous ketamine in treatment-resistant depression. Psychother Psychosom 86:162–167. https://doi.org/10.1159/000457960

Wilkinson ST, Ballard ED, Bloch MH, Mathew SJ, Murrough JW, Feder A, Sos P, Wang G, Zarate CA, Sanacora G (2018) The effect of a single dose of intravenous ketamine on suicidal ideation: a systematic review and individual participant data meta-analysis. Am J Psychiatry 175:150–158. https://doi.org/10.1176/appi.ajp.2017.17040472

Wilkinson ST, Holtzheimer PE, Gao S, Kirwin DS, Price RB (2019) Leveraging neuroplasticity to enhance adaptive learning: the potential for synergistic somatic-behavioral treatment combinations to improve clinical outcomes in depression. Biol Psychiatry 85:454–465. https://doi.org/10.1016/j.biopsych.2018.09.004

Williams NR, Heifets BD, Blasey C, Sudheimer K, Pannu J, Pankow H, Hawkins J, Birnbaum J, Lyons DM, Rodriguez CI, Schatzberg AF (2018) Attenuation of antidepressant effects of ketamine by opioid receptor antagonism. Am J Psychiatry 175:1205–1215. https://doi.org/10.1176/appi.ajp.2018.18020138

Xiong J, Lipsitz O, Chen-Li D, Rosenblat JD, Rodrigues NB, Carvalho I, Lui LMW, Gill H, Narsi F, Mansur RB, Lee Y, McIntyre RS (2021) The acute antisuicidal effects of single-dose intravenous ketamine and intranasal esketamine in individuals with major depression and bipolar disorders: a systematic review and meta-analysis. J Psychiatr Res 134:57–68. https://doi.org/10.1016/j.jpsychires.2020.12.038

Yanagihara Y, Ohtani M, Kariya S, Uchino K, Hiraishi T, Ashizawa N, Aoyama T, Yamamura Y, Yamada Y, Iga T (2003) Plasma concentration profiles of ketamine and norketamine after administration of various ketamine preparations to healthy Japanese volunteers. Biopharm Drug Dispos 24:37–43. https://doi.org/10.1002/bdd.336

Zarate CA, Singh JB, Carlson PJ, Brutsche NE, Ameli R, Luckenbaugh DA, Charney DS, Manji HK (2006) A randomized trial of an N-methyl-D-aspartate antagonist in treatment-resistant major depression. Arch Gen Psychiatry 63:856–864. https://doi.org/10.1001/archpsyc.63.8.856

Zhao X, Venkata SLV, Moaddel R, Luckenbaugh DA, Brutsche NE, Ibrahim L, Zarate CA, Mager DE, Wainer IW (2012) Simultaneous population pharmacokinetic modelling of ketamine and three major metabolites in patients with treatment-resistant bipolar depression. Br J Clin Pharmacol 74:304–314. https://doi.org/10.1111/j.1365-2125.2012.04198.x

Psilocybin and Other Classic Psychedelics in Depression

D. J. Nutt, J. M. Peill, B. Weiss, K. Godfrey, R. L. Carhart-Harris, and D. Erritzoe

Contents

Abstract Psychedelic drugs such as psilocybin and ketamine are returning to clinical research and intervention across several disorders including the treatment of depression. This chapter focusses on psychedelics that specifically target the 5-HT_{2A} receptor such as psilocybin and DMT. These produce plasma-concentration

D. J. Nutt (✉), J. M. Peill, B. Weiss, K. Godfrey, and D. Erritzoe
Centres for Neuropsychopharmacology & Psychedelic Research, Division of Psychiatry, Department of Brain Sciences, Imperial College London, London, UK
e-mail: d.nutt@imperial.ac.uk

R. L. Carhart-Harris
Centres for Neuropsychopharmacology & Psychedelic Research, Division of Psychiatry, Department of Brain Sciences, Imperial College London, London, UK

Psychedelics Division, Neuroscape, University of California San Francisco, San Francisco, CA, USA

Curr Topics Behav Neurosci (2024) 66: 149–174
https://doi.org/10.1007/7854_2023_451
Published Online: 14 November 2023

related psychological effects such as hallucinations and out of body experiences, insightful and emotional breakthroughs as well as mystical-type experiences. When coupled with psychological support, effects can produce a rapid improvement in mood among people with depression that can last for months. In this chapter, we summarise the scientific studies to date that explore the use of psychedelics in depressed individuals, highlighting key clinical, psychological and neuroimaging features of psychedelics that may account for their therapeutic effects. These include alterations in brain entropy that disrupt fixed negative ruminations, a period of post-treatment increased cognitive flexibility, and changes in self-referential psychological processes. Finally, we propose that the brain mechanisms underlying the therapeutic effect of serotonergic psychedelics might be distinct from those underlying classical serotonin reuptake-blocking antidepressants.

Keywords Antidepressants · Connectivity · Depression · Dimethyltryptamine (DMT) · Electroencephalograhy (EEG) · Entropy · Lysergic acid diethylamide (LSD) · Magnetic resonance imaging (MRI) · Modularity · Neuroimaging · Psilocybin · Psychedelics · Serotonin · Serotonin 2A receptor (5-HT_{2A}R)

1 Introduction

Psychedelic therapy with psilocybin and other serotonergic drugs such as Lysergic Acid Diethylamide (LSD) and Dimethyltryptamine (DMT) is arguably the most exciting new development in the treatment of depression since the discovery of the monoamine-reuptake blocking antidepressant agents in the late 1950s. This method offers a radical new approach to treatment where traditional remedies have proven ineffective – for a review see (Morgese et al. 2023), but often it works faster and with just a single treatment (Galvão-Coelho et al. 2020; G. M. Goodwin et al. 2022a, b; R. R. Griffiths et al. 2016; F. Palhano-Fontes et al. 2019a, b; S. Ross et al. 2016a, b) (Table 1 shows the studies performed for psilocybin (and other classic psychedelics) in primary depressive disorders). In this chapter, we explore the growing research base on the brain mechanisms that might explain the activity of psychedelic agents in depression.

2 Basic Facts

The receptor target for serotonergic psychedelics is the serotonin 5-HT_{2A} receptor where these drugs act as agonists or partial agonists. Their impact is dose and plasma concentration related – positron emission tomography (PET) studies show that that the impact of psychedelics on the brain is linked to their proportional 5-HT_{2A} receptor occupation (Madsen et al. 2019). Moreover, blocking the 5-HT_{2A} receptor

Table 1 Comprehensive list of clinical studies investigating classic psychedelic use in the treatment of primary depressive disorders or related disorders including within- and between-condition Cohen's *d* effect size estimates for Psychedelic Interventions

Unique study/ sample	Paper	Intervention condition	Comparison condition	Subjects per group	Diagnosis	Measure	Timepoint	Within effect size	Between effect size
Psilocybin									
1	Carhart-Harris et al. (2021)	COMP360 psilocybin 25 mg/25 mg (2 doses, 3 weeks apart) + psychological support	COMP360 psilocybin 1 mg/1 mg + psychological support with daily escitalopram for 6 weeks	Psilo: 30 Escit: 29	MDD	QIDS-SR-16	3 weeks	2.05	0.50
	Daws et al. (2022)	Two studies: 1. See doses from paper above 2. See study 4	1. As above 2. None	1. Psilo: 22, Escit: 21 2. 16	1. TRD 2. MDD	–	–	–	–
2	Davis et al. (2021a, b)	20 mg AND 30 mg/70 kg (2 doses) + supportive psychotherapy	Wait list (immediate or delayed dosing)	24	MDD	GRID-HAM-D	1 week	2.30	2.50
	Gukasyan et al. (2022)					GRID-HAM-D	1, 3, 6, 12 months	2.3, 2.0, 2.6, 2.4	–
3	Goodwin et al. (2022a, b)[a]	COMP360 10/25 mg (1 dose) + psychological support	COMP360 psilocybin (1 mg) + psychological support	79/75/79 (1 mg/ 10 mg/ 25 mg)	TRD	MADRS	3 weeks	2.22	1.13
4	Carhart-Harris et al. (2016a, b)	10/25 mg (2 doses) + psychological support	None	12	TRD	QIDS-SR-16	1 week	2.70	–
	Mertens et al. (2020)			19		–	–	–	–

(continued)

Table 1 (continued)

Unique study/ sample	Paper	Intervention condition	Comparison condition	Subjects per group	Diagnosis	Measure	Timepoint	Within effect size	Between effect size
	Erritzoe et al. (2018)			20		–	–	–	–
	Roseman et al. (2018)			20		–	–	–	–
	Carhart-Harris et al. (2018a, b)			20		QIDS-SR-16	3, 6 months	1.5, 1.4	–
				16		–	–	–	–
5	Griffiths et al. (2016)[b]	22 or 30 mg/70 kg (1 dose) + psychological support	1 or 3 mg/70 kg psilocybin + psychological support	51 (crossover)	Cancer-related anxiety and depression	GRID-HAM-D	5 weeks	1.41–1.53	0.42
6	Ross et al. (2016a, b)[b]	0.3 mg/kg + psychotherapy	Niacin (250 mg) + psychotherapy	29 (crossover)	Cancer-related anxiety and depression	BDI	6 weeks	0.93–1.12	1.07
7	Sloshower et al. (2023)	0.3 mg/kg (1 dose) + psychotherapy	Inactive placebo	Psilo: 17 Placebo: 19	MDD	GRID-HAM-D	1 week	1.99	1.16
LSD									
1	Holze et al. (2023)	200 mcg (2 doses) + talking psychotherapy	Inactive placebo + talking psychotherapy	42 (crossover)	Anxiety with and without life threatening illness	STAI-G	16 weeks	0.31	0.87

2	Grof et al. (1973)	200–500 mcg	DPT (60-105 mg)	LSD: 44 DPT: 19	Terminal cancer patients	–	–	–	–
DMT									
1	D'Souza et al. (2022)	0.3 mg/kg (dose escalation) + psychological support	0.1 mg/kg + psychological support	MDD:7 HC:3	TRD	HAM-D	1 day	0.68	0.53
Ayahuasca									
1	Zeifman et al. (2021)	Ayahuasca (2.2 ml/kg, 1 dose) containing: • DMT (0.8 mg/ml) • Harmine (0.21 mg/ml) + psychoeducation	None	15	MDD	MADRS-SI	1, 7, 14, 21 days	1.71, 1.65, 1.61, 1.85	–
	Sanches et al. (2016)		None	17	MDD	MADRS	–	–	–
2	Osório et al. (2015) [a]	Ayahuasca (2.2 ml/kg, 1 dose) containing: • DMT (0.8 mg/ml) • Harmine (0.21 mg/ml) + psychoeducation	None	6	MDD	HAM-D	1 week	1.64	–
3	Palhano-Fontes et al. (2019b)	Ayahuasca ((1 ml/kg, 1 dose), single-dose) containing: • N,N DMT (0.36 mg/ml) • Harmine (1.86 mg/ml) • Harmaline (0.24 mg/ml) • Tetraharmaline (1.20 mg/ml) + psychological support	Placebo liquid containing water, yeast, citric acid, zinc sulfate, caramel colorant + psychological support	Aya: 14 Placebo: 15	TRD	MADRS	1 week	2.90	1.49
	Galvão-Coelho et al. (2020)			Aya:14 Placebo:14 HC: 45	TRD	–	–	–	–

Note: Within-condition and between-condition effect size estimates were calculated in cases where estimates were not presented in the article. Within-condition effects were calculated by dividing the difference between pre-test and post-test scores by the pooled standard deviation. Between-condition effects were calculated by dividing the difference between the experimental group's pre-post difference and the comparison group's pre-post difference by the pooled standard deviation of post-test scores. Where post-test standard deviations were not presented, pre-test standard deviations were used

[a] Cohen's *d*s were calculated using just baseline standard deviation

[b] Indicates that within-condition effect size estimates are provided for each of the two cross-over groups post-psilocybin

with selective antagonists, e.g. ketanserin, blocks the psychedelic effects (Preller et al. 2018; Quednow et al. 2012).

The psychedelic brain is profoundly altered as compared with the brain in normal consciousness (R. L. Carhart-Harris and K. J. Friston 2019) as revealed by measures such as EEG/MEG synchronicity (decreased) and fMRI connectivity and EEG/MEG/fMRI entropy (increased) (Carhart-Harris 2018; Carhart-Harris et al. 2014; Lebedev et al. 2016; Viol et al. 2017). The impact of these on depression outcomes is discussed later. With the serotonergic psychedelics these changes occur due to disruption of the normal integrative functioning of layer 5 pyramidal cells in the cortex which express a high density of 5-HT_{2A} receptors (S. D. Muthukumaraswamy et al. 2013a, b). Some other drugs, e.g. ketamine and salvia, can produce similar increases in cortical entropy through different pharmacological mechanisms and with different subjective experiences (Doss et al. 2020; Farnes et al. 2020; Roy et al. 2021; M. M. Schartner et al. 2017a, b).

3 Clinical Effects

Over the last two decades, a substantial body of clinical research has demonstrated impressive therapeutic efficacy for psilocybin therapy and other psychedelic drug interventions in the treatment of depression. These clinical interventions vary in terms of dosage, drug, and psychotherapeutic aspects, but generally have included three stages of care including the preparation of the patient for the unique psychoactive experience of psychedelics (Preparation), administration of psychedelic compounds (Dosing experience), and post-dose sessions devoted to processing the meaning of the psychedelic experience and enacting desired psycho-behavioural changes (Integration).

To date, the antidepressant efficacy of such interventions has been shown across 12 published clinical studies (six examining psilocybin), wherein within-condition pre-post Cohen's *d* effect size estimates have ranged from 1.12 (Stephen Ross et al. 2016a, b) to 2.90 (Fernanda Palhano-Fontes et al. 2019a, b) (see Table 1). In all but one double-blind randomised control trial (DB-RCT), the psychedelic treatment condition has outperformed the comparison condition, with between-condition Cohen's *d* effect size estimates, comparing a psychedelic therapy condition to a placebo (or quasi-placebo, e.g. 1 mg psilocybin) condition, ranging from 0.42 (Roland R. Griffiths et al. 2016) to 1.49 (Fernanda Palhano-Fontes et al. 2019a, b) (see Table 1). Notably, these estimates exceed placebo-controlled effects of specific psychotherapies ($d = 0.25$) (Cuijpers et al. 2014) and pharmacotherapies ($d = 0.30$) (Cipriani et al. 2018; Khan and Brown 2015), and generally exceed estimates of efficacy produced by combined psycho- and pharmacotherapy (Cuijpers et al. 2009). Collectively, clinical trials have shown efficacy across a number of variants of depression including Major Depressive Disorder (MDD) (Robin Carhart-Harris et al. 2021; A. K. Davis et al. 2021a, b; Sloshower et al. 2023), treatment resistant MDD (Carhart-Harris et al. 2016a, b; Guy M. Goodwin et al. 2022a, b; Osório et al.

2015; Fernanda Palhano-Fontes et al. 2019a, b; Sanches et al. 2016), and cancer-related depression (Roland R. Griffiths et al. 2016; Stephen Ross et al. 2016a, b). Furthermore, they have demonstrated a rapid antidepressant onset that is likely to be extremely valuable to depressed patients, and especially those bearing suicidal risk where rapid alleviation of depressive symptomatology is of great importance.

So far, only one study (Robin Carhart-Harris et al. 2021) has compared Psilocybin Therapy to a gold-standard antidepressant treatment, namely a combined treatment containing escitalopram pharmacotherapy, a common selective serotonin reuptake inhibitor (SSRI), and psychological support (as well as components of psychotherapy). Of note, this combination has shown incremental efficacy over pharmacotherapy alone (Cuijpers et al. 2009). Although the two treatments did not significantly differ on the primary outcome measure, Psilocybin Therapy showed evidence of superior efficacy on 14 of 16 outcome measures assessed in the trial, as well as evidence of remediating core symptom components of depression, including *depressed mood*, *anhedonia*, and *sexual dysfunction* (Weiss et al. under review).

SSRIs share some properties in common with Psilocybin Therapy, namely serotonergic modulation, but psilocybin and other psychedelic compounds differ in key respects, namely in directly agonising the 5-HT_{2A} receptor both extra- and intra-cellularly (Vargas et al. 2023). These properties may be responsible for signs of outperforming SSRIs in key areas of weakness, including potentially delayed onset (Quitkin et al. 1987; Taylor et al. 2006), reduced limbic responsiveness to reward (McCabe et al. 2010), and sexual dysfunction (Cascade et al. 2009; Clayton et al. 2007).

Some scientists have expressed scepticism toward modern psychedelic trials, noting that the dramatic cognitive-emotional effects of psychedelics combined with their recognisable psychoactive profile (at times referred to as the 'Michael Pollan effect') alert participants to their actual treatment condition (see references (Muthukumaraswamy et al. 2022; Sloshower et al. 2023) for evidence of high blind-breaking rate). Because adequately 'blinding' participants to the psychedelic condition is important for ensuring the placebo response is not magnified within the experimental treatment condition, such problems are particularly threatening to the validity of psychedelic-related clinical trials. Nevertheless, a recent study has shown that pre-trial expectancy (linked to the placebo effect) was predictive of enhanced therapeutic efficacy in the Psilocybin Therapy condition (Szigeti et al. 2023); however, this study was not equipped to rule out the absolute effect of expectancy on outcomes, e.g. by manipulating or otherwise controlling expectancy as a variable. Nevertheless, it seems possible that expectancy effects are in fact part of the therapeutic mechanism of psychedelics which interact in a complex way with the direct pharmacological consequences of treatment (Butler et al. 2022).

Mitigation strategies have been proposed to improve the validity of psychedelic-related clinical trials (e.g. recruitment of low expectancy patients, periodic monitoring of blind-breaking) (Aday et al. 2022; Muthukumaraswamy et al. 2022). However, other criteria besides placebo-controlled efficacy may be equally important in determining Psilocybin Therapy's clinical utility. One such criterion involves the duration of positive therapeutic changes. So far, two clinical trials have demonstrated impressive sustained response from Psilocybin Therapy at 6 months (within-condition $d = 1.4$) (Robin L. Carhart-Harris et al. 2018a, b) and 12 months (within-condition $d = 2.4$) (Gukasyan et al. 2022), with the caveat that both studies were open-label and more susceptible to exhibiting placebo response. Other criteria include the rate at which psychedelic therapies are utilised among depressed individuals and the rate of adherence in continuing with psychedelic therapies.

With respect to utilisation, current rates of mental health service uptake are low in part due to perceptions of stigma and low treatment efficacy, among others (Abuse 2021; Cuijpers et al. 2010). Psychedelics appear to be increasing in popularity as media coverage, conference attendance, and interest group participation expands. Although psychedelics may merely be within an early hype cycle, it is conceivable that psychedelic therapies will be better utilised by depressed patients than conventional mental health treatments. With respect to acceptability and adherence, on average, 17.5% of patients drop out from antidepressant psychotherapy (Cooper and Conklin 2015; Swift and Greenberg 2014), and, due in part to treatment duration, onerous side-effects, and treatment resistance, depression is associated with lower medication adherence (Grenard et al. 2011). Because psychedelic therapies are time-limited in comparison to other therapies and rapid in their onset, they may offer more effective relief to patients who would otherwise drop out or discontinue. Indeed, psychedelic therapies may function like similar therapies such as MDMA-assisted therapy, which is associated with significantly lower dropout than gold-standard therapies for PTSD (Mithoefer et al. 2019). There is also some suggestive evidence that psychedelic therapies may produce greater utilisation of other mental health services, given observed increases in patients' motivation to invest in change behaviour (Nielson et al. 2018; Teixeira et al. 2022; Watts et al. 2017).

3.1 Psychological Processes

Research on key causal mechanisms underlying Psilocybin Therapy's antidepressant effect have identified a number of important acute psychological experiences. Perhaps the strongest clinical evidence has been found for the effects of so-called 'mystical experiences' (Barrett et al. 2015; Kangaslampi 2023) on antidepressive response (i.e. depression related to terminal cancer) (Roland R. Griffiths et al. 2016; Stephen Ross et al. 2016a, b). The construct of mystical experience is multifaceted, referring to states of consciousness containing four factors, namely: (1) mysticism involving experiences of undifferentiated unity with a larger whole that can be deemed 'sacred', (2) positive mood states, e.g. awe, (3) ineffability, and (4) non-ordinary perceptions of space and time (Barrett et al. 2015; Stace 1960). Notably, mystical experience has shown evidence of mediating the effect of treatment condition on depressive outcome while controlling for the intensity of the drug experience (Roland R. Griffiths et al. 2016), suggesting some therapeutic specificity to the mystical experience itself beyond mere generic drug effects. Mystical experience, and a related construct, oceanic boundlessness (Studerus et al. 2010), have also demonstrated moderate-to-large associations with antidepressant response ($r = -0.41$, (A. K. Davis et al. 2021a, b); $r = -0.50$, (Roseman et al. 2018)). Further clinical evidence has been found for role of other key acute experiences including *ego dissolution*, describing the loss of one's sense of self and self-world boundary during the psychedelic states (Nour et al. 2016; Weiss et al. 2021), *emotional breakthrough*, the emotional-cognitive process of confronting challenging emotional material and reaching resolution (Roseman et al. 2019; Weiss et al. under review) and the *emotional impact of music* on antidepressant response (Kaelen et al. 2018; Weiss et al. under review).

Outside of clinical trials, the relevance of acute experiences to positive mental health outcomes post-psychedelic use has found additional support. Reappraisal (therapeutic shifts in maladaptive core beliefs) (Agin-Liebes et al. 2022), acute and post-acute insight (Alan K. Davis et al. 2021a, b; Peill et al. 2022), psychological flexibility (the ability to manage negative psychological material while advancing toward goals) (Agin-Liebes et al. 2022; Close et al. 2020; Zeifman et al. 2020) and communitas (experiences of perceived togetherness) (Kettner et al. 2021) have all demonstrated associations with improvements in negative emotionality and well-being.

Extra-pharmacological factors are also considered critical to the therapeutic benefits of psychedelic therapy, perhaps owing to heightened environmental sensitivity within the psychedelic state (Carhart-Harris et al. 2015; Hartogsohn 2016; Kaelen et al. 2015). Such factors include psychological support within the preparation, dosing, and integration stages of therapy, therapeutic alliance (Murphy et al. 2022), music-listening (Belser et al. 2017; Kaelen et al. 2018; Noorani et al. 2018; Watts et al. 2017) (see (O'Callaghan et al. 2020) for review), individual characteristics such as trait absorption, a cognitive disposition to be immersed in one's internal mental landscape (Lifshitz et al. 2019), and so-called 'set and setting', describing one's mental state when entering a psychedelic state along with features of the environment that impact it (Adamson and Metzner 1988; Hartogsohn 2016; Hyde 1960; Johnson et al. 2008; Metzner et al. 1965). Experimental designs will ultimately be necessary that manipulate the presence or absence of these properties within ethical limits. However, correlational work has shown that individuals undergoing Psilocybin Therapy who report greater therapeutic alliance (Murphy et al. 2022) and positive emotional response to music (Kaelen et al. 2018) tend to show better antidepressant response. In addition, early experimental research (Hartogsohn 2016; Hyde 1960) observed a number of setting-related dimensions that were associated with adverse acute experiences, including (1) interpersonal attitudes of nonacceptance (versus acceptance) to participants' behaviour or feelings, (2) the rigidity (versus flexibility) of goals (e.g. undergoing tests), and (3) unfamiliarity (versus familiarity) of the environment. More recent work has furthermore linked pre-dose psychological states of preoccupation, apprehensiveness, and confusion to adverse acute experiences (Aday et al. 2021).

These factors are not only important for potentiating positive outcomes for patients, but equally for minimising risk of negative outcomes and serious adverse events. In a large observational survey study of psychedelic users ($N = 1{,}993$), 7.6% of participants reported seeking treatment for enduring psychological symptoms following psychedelic use (Carbonaro et al. 2016). The incidence of symptoms appears to be significantly lower in controlled drug administration studies, namely 0.90% of 110 participants in the Vollenweider laboratory in Switzerland, and 0.90% of 250 participants in the Johns Hopkins laboratory (Yaden et al. 2022). However, risk may be greater in a depressed population. A recently published clinical trial of Psilocybin Therapy for treatment-resistant depression (Guy M Goodwin et al. 2022a, b) reported a 6.3% and 8% occurrence of serious adverse events including suicidal thoughts/behaviour and self-injury in the 25 mg and 10 mg conditions compared with 1.3% in the 1 mg control group. This evidence speaks to the importance of heeding the heightened sensitivity induced by psychedelic substances and the extra-pharmacological factors that modulate their effects as well as a possible nocebo effect when the hoped-for clinical benefits do not occur. Adverse experiences under psychedelics may be even more concerning when considering psychedelics' effect on plasticity (Lepow et al. 2021) and learning (Carhart-Harris et al. 2022), as such experiences could disproportionately exacerbate mental health problems in the longer-term.

4 Brain Mechanisms

4.1 Magnetic Resonance Imaging

Neuroimaging offers a number of techniques/methods to take an objective look at the effects of psychedelics. Magnetic resonance imaging (MRI) is one of the most used modalities to study psychedelics in treatment studies. Observations in neuronal activation can point to differences between patient groups, e.g. depression and controls, or in patients before and after the drug. Paired with tasks in the scanner (task-based MRI), we can also observe brain activity changes related to cognitive/behavioural function. Here, we will explore the changes in MRI measures in those who have taken psychedelics and how it may be different in patient populations that exhibit depressive-like symptomatology.

Recent research suggests that brain functions are carried out by a group of defined networks. These are areas of the brain that functionally match-up and aid in performing tasks, internalising/externalising thoughts and integrating our senses of the external world. For example, some are involved in high-level cognition and sensory processing, e.g. default mode network and visual network respectively. Psychedelics reduce 'functional connectivity' within these networks in the acute state (on drug) (Carhart-Harris et al. 2016a, b; Mason et al. 2021; Müller et al. 2018; Roseman et al. 2014; Timmermann et al. 2023). More specifically, research has demonstrated that during certain task-based fMRI, following psilocybin, there were decreases in functional connectivity within the amygdala (a key centre involved with emotion processing) in people with depression (Mertens et al. 2020). This change

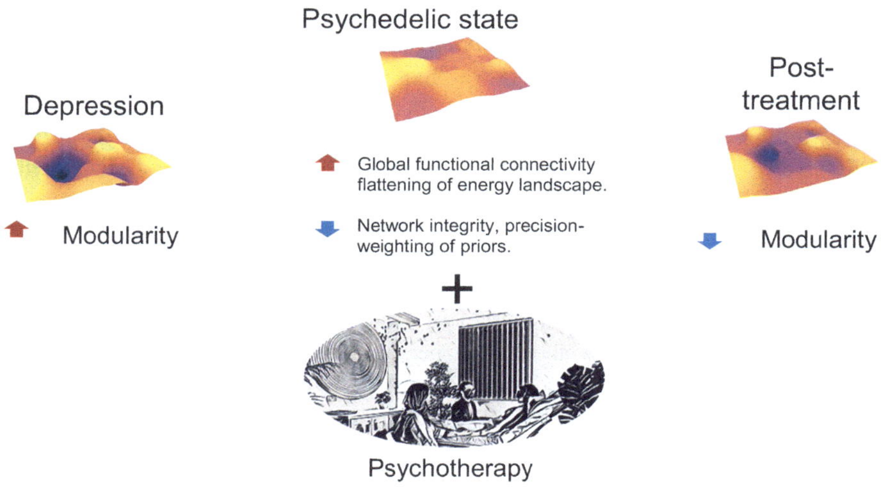

Fig. 1 Model of the action of psychedelic therapy in the treatment of depression, taken from (Girn et al. 2023). The steep valleys and peaks in depression demonstrate a higher level of modularity and segregation of brain networks which is flattened during psychedelic treatment. Post-psychedelic assisted psychotherapy treatment leads to a slightly flatter landscape than before treatment. This relates to a change in prior beliefs, perceptions and rigid thought patterns

was associated with alterations to rumination (a key characteristic of depression characterised by repetitive negative thinking). Given that patients with depression and anxiety generally exhibit emotion regulation problems, it is interesting that psychedelics produce functional connectivity changes in the amygdala.

In contrast to the drop in within-network connectivity, there is increased between-network connectivity in the acute psychedelic state (on-drug). An increase in functional connectivity globally (throughout the brain) and between networks appears consistently reported during classic psychedelic administration (Carhart-Harris et al. 2016a, b; Müller et al. 2018; Preller et al. 2020; Roseman et al. 2014; Tagliazucchi et al. 2016; Timmermann et al. 2023). Interestingly, this seems to be a larger increase when compared to other psychoactive agents such as MDMA (Roseman et al. 2014), suggesting that there is increased cross-talk between usually segregated systems in the brain during psychedelic experiences; possibly giving rise to phenomena such as synaesthesia (Petri et al. 2014).

During the decades of development and learning the brain undergoes from babyhood it becomes more efficient in its functioning. Most processing becomes concentrated in the brain regions that are responsible for such functions, e.g. the visual cortex communicates mostly within itself, the auditory cortex similarly. This is known as the small-world brain network concept, and is responsible for the remarkable consistency of brain functioning and the exceptional energy efficiency of the brain as a computing system. However, efficiency can come at a cost if maladaptive thoughts and behaviours become overly encoded and entrenched as in depression. People can get locked into thought loops from which they cannot escape (see Fig. 1). The increased cross-talk seen under psychedelics may give rise to a similar brain state seen in early childhood when connectivity is maximal, before education and learning drive small-world brain network segregation.

This increased cross-talk or connectivity in the brain under psychedelics has been shown during the psychedelic trip in healthy control volunteers but not depressed individuals. We presume brain changes during the psychedelic experience in patients are the same as in controls and suspect that this increased connectivity is a critical aspect of the recovery process, but this is yet to be investigated. We believe that this recovery of cross-talk that occurs during the experience allows people to break free from entrenched negative thought loops. For example, increased connectivity may allow participants to re-connect with long-forgotten memories of events and/or decisions enabling revisions about their perspective/s leading to therapeutic outcome. Also, the increased connectivity allows patients to come up with new ways of thinking about their past and future – something we call ‘finding new solutions to old problems’.

Another analytical take on MRI data involves decomposing the brain into distinct regions known as ‘modules’. Each module contains areas of the brain that have high levels of connectivity to each other when compared to other regions of the brain. Note – brain areas are with higher ‘connectivity’ does not necessarily mean they have more ‘connections’ but rather they seem to work together functionally. In depression, there tends to be a higher level of ‘modularity’ (or a higher level of connectivity within-modules) and so a higher level of *segregation* (Li et al. 2022; Ye et al. 2015). This may form the basis for repetitive thought patterns and ‘grooves’ of thought processes we see in depression. In a recent study, it was demonstrated that the depressed group exhibited decreases in modularity following psychedelic-assisted psychotherapy which was reflective of improvements in depressive-like symptoms (Daws et al. 2022) – see Fig. 1.

Although thus far we can see that MRI is producing some exciting results within the world of psychiatry and psychedelics, much of the current research is based on a limited number of datasets (Drummond et al. 2022). Furthermore, most of the acute state work has been performed in participants without any mental health difficulties. This suggests that we must look to perform more neuroimaging research within this area to see if the results are reliably replicated and whether hypotheses about changing connectivity driving antidepressant action can be supported.

4.2 Electroencephalography

Another potential brain mechanism which may provide insights into psychedelic-assisted therapy are what is known as neural oscillations. Neural oscillations are rhythmic patterns of electrical activity in the brain that are thought to play a critical role in various cognitive and affective processes. These are measurable using electroencephalography (EEG) or magnetoencephalography (MEG). EEG is a non-invasive technique that measures the electrical activity of the brain using electrodes placed on the scalp. It has several advantages over other neuroimaging methods, including high temporal resolution and portability, making it a useful tool for investigating the effects of psychedelics on brain function in naturalistic settings. It provides a means for investigating the acute effects of psychedelics in clinical therapeutic trials as it can be implemented with little effect to the set and setting, an advantage over MRI for which acute studies are generally limited to healthy experienced users.

Previous studies in healthy individuals have shown that psychedelics can modulate neural oscillations. Broadband desynchronisation has been found following the administration of LSD, psilocybin, DMT, and the DMT-containing brew ayahuasca (Carhart-Harris et al. 2016a, b; Timmermann et al. 2023) (While these studies show some divergences, the evidence overwhelmingly indicates that serotonergic psychedelics decrease alpha (8–13 Hz) power and these reductions correlate temporally with the intensity of the psychedelic experience (Timmermann et al. 2023; Timmermann et al. 2019). Desynchronisation in EEG spectral power in patients during therapy dosing sessions has not yet been reported on in the literature, however in future this may elucidate differences in response to therapy across participants, stratifying them based on these neural oscillation markers of psychedelic action.

In addition to broadband desynchronisation, recent studies involving psychedelics have shown that LSD, DMT, psilocybin (as well as ketamine) elicit increased levels of signal diversity compared to placebo, as determined by Lempel-Ziv complexity (LZ), a data compressibility algorithm that is closely related to informational entropy (Michael M. Schartner et al. 2017a, b; Timmermann et al. 2023; Timmermann et al. 2019). Increased brain entropy under psychedelics is consistent with the Entropic Brain Hypothesis (Carhart-Harris et al. 2014), which posits that the richness of conscious experience is underpinned by richness in spontaneous brain activity, characterised by measures of entropy (or disorder). In clinical studies, the extent to which these drugs acutely increase signal diversity or entropy in the brain could be hypothesised to map onto meaningfulness of experience, the degree of psychological insight or some subsequent therapeutic benefit reported by that participant. If this is observed, it would open up possibilities around personalised dose adjustment based on real-time brain activity to reach a threshold of signal diversity which may correlate with increased likelihood of response.

As with fMRI (above) EEG has been used to investigate the effects of psychedelics on functional connectivity and studies have shown that psychedelics can increase functional connectivity between brain regions involved in emotional processing and self-referential processing (Carhart-Harris et al. 2012; Palhano-Fontes et al. 2015). Understanding the effects of psychedelics on functional connectivity in clinical populations utilising EEG is an important research area as it may provide insight into the neural mechanisms underlying their therapeutic effects.

Furthermore, EEG recorded prior to psychedelic assisted therapy can be used to investigate individual differences in brain function that may predict treatment response. Recent studies have shown that baseline EEG activity can predict the likelihood of experiencing positive therapeutic effects following psychedelic-assisted therapy. For example, individuals with greater baseline alpha power in the frontal cortex and greater baseline theta power in the temporal cortex have been shown to have a greater likelihood of experiencing positive drug effects (Suresh D. Muthukumaraswamy et al. 2013a, b; Tagliazucchi et al. 2021), This may carry over to clinical trials EEG used to identify biomarkers that predict the therapeutic outcomes or risk of adverse reactions to psychedelics.

To conclude, EEG is a low cost imaging tool that can be used for exploring the neural mechanisms involved in the therapeutic effects of psychedelic-assisted therapy. By examining neural oscillations and functional connectivity, and by identifying biomarkers that predict treatment response and the risk of adverse reactions, future EEG research has the potential to enhance our understanding of the therapeutic potential of psychedelics, as well as to improve the safety and efficacy of psychedelic-assisted therapy.

4.3 Computational Aspects of Psychedelics in Depression

Current ideas regarding the mechanisms of altered consciousness with psychedelics centre on the concept of the brain as an inference-making machine whereby learning through experience generates hypothesises about the world – so called priors – that are then tested and updated by reference to external perceptions (R. L. Carhart-Harris and K. Friston 2019; Carhart-Harris et al. 2014).

Hierarchical predictive processing is a relatively new concept that looks to explain the human brain's beliefs, perception and conscious experience of the external world. This model suggests that higher-level cognitive processes (or top processes) integrate our sensory inputs (bottom processes) allowing interpretation of our conscious experience (Clark 2015, 2016; Friston 2010; Friston and Kiebel 2009; Hohwy and Seth 2020).

It has been suggested that psychedelics disrupt this regulation of top-down control, thus giving rise to more bottom-up flow of neuronal processes. Through producing *entropy* (i.e. the degree to which neural activation patterns cannot be predicted by prior states of activation, disorder) in the higher-level control areas of the brain, psychedelics are thought to relax the rigidity of the predictions, which can themselves exert a constraining influence over mental experience. This is theorised to allow an opening of the mind to new sensory experience, which may extend to different higher-order beliefs about oneself and one's world. In a therapeutic context, such potential for revising beliefs could conceivably guide reappraisal of rigid maladaptive beliefs underlying depression and mechanistically explain the effects of psychedelic antidepressant therapy. Relaxation of priors (or models about the world) in people with rigid thought patterns may help to re-calibrate and revise the models that have become heavily entrenched (Carhart-Harris et al. 2022).

Due to a relaxing of priors and a higher level of sensory input during the psychedelic experience, it has been suggested that the psychedelic state provides increased 'richness' of self-consciousness (Carhart-Harris 2018; Carhart-Harris et al. 2014; Lebedev et al. 2016); leading to experiences such as emotional breakthroughs, insightful experiences and mystical-type encounters. In line with this, increases in the personality trait openness following psychedelic-assisted psychotherapy (Erritzoe et al. 2019; MacLean et al. 2011) in depressed individuals suggests a relaxing of higher-level priors/beliefs enabling an increase in openness to external experience or changes to internal beliefs or rigid thought patterns. This current theory is paving the way for modern neuropsychiatry to understand how psychedelics are affecting the brain on a computational level.

5 Current Research Questions

5.1 Duration of Trip for Psychedelics: Why Does This Matter?

There are profound differences in the duration of action of serotonin psychedelics. LSD and mescaline are long lasting with effects for up to a day and so; often requiring overnight stays in clinical studies. Oral Psilocybin lasts approximately 4–6 h. Other psychedelics, like DMT (i.v. or smoked/intranasal) or 5-MEO-DMT (i.v./im/intranasal) can produce very short altered experiences (just 5–10 min). A key question is: are each equally effective for depression? If short acting psychedelics such as DMT and 5-MEO-DMT prove just as effective for treating depression, this would offer a cost-benefit when compared to longer lasting psychedelics such as psilocybin or LSD. In a recent press release, a single dose of intravenous DMT demonstrated anti-depressant effects similar to those seen with a single dose of psilocybin (https://smallpharma.com/press-releases/positive-topline-results-from-phase-iia-trial-of-spl026-in-major-depressive-disorder/). If the peer-reviewed articles reveal similar results, this would be groundbreaking in the psychedelic community. An alternative approach, being considered by some companies, is utilising a 5-HT_{2A} receptor antagonist such as ketanserin (Becker et al. 2023) which can terminate an LSD trip after a given time, so shortening the treatment period in the clinic. Going forward, if psychedelics are integrated into health services; cost-benefit analyses will be crucial in delivering them as a viable treatment option. Overall, a shortened experience with similar antidepressant effects will be favourable.

5.2 Optimal Number of Treatments

Some studies have found powerful antidepressant effects of just one psychedelic experience (Galvão-Coelho et al. 2020; G. M. Goodwin et al. 2022a, b; Zeifman et al. 2021) . In the recent psilocybin-escitalopram paper [28] our group found a 2nd dose 3 weeks after the first had little extra immediate benefit. A recent press release revealed that two doses of DMT is no more favourable than one dose for MDD symptoms (https://smallpharma.com/press-releases/positive-topline-results-from-phase-iia-trial-of-spl026-in-major-depressive-disorder/). Therefore, the question remains as to the optimal number of doses for psychedelic treatment. Published studies directly comparing one dose to multiple doses would be gladly received by the community. Future studies may look to perform multiple doses over the course of months, e.g. 5 doses over 2.5 months to observe the effect of multiple drug treatments in therapeutic outcomes for participants. Unfortunately, generally due to lack of funding, studies are unable to execute such large scale treatment methods which explains why most groups have performed 1–2 doses thus far.

5.3 Prior Antidepressant Treatment

As SSRI treatment has been shown to attenuate the response of individuals to psilocybin [in rodent studies also] most trials have taken people off SSRIs and related medications before the psychedelic dosing sessions. In our recent trial (R. Carhart-Harris et al. 2021), we found that patients who had been withdrawn from SSRIs close to the trial start date did not have as beneficial outcomes as those having been withdrawn for many months, even though they had a similar psychedelic experiences. One hypothesis for this difference is that SSRI treatment immediately prior to psychedelics may lead to neuronal adaptive changes that impede the psychedelic plasticity effects [35]. It might be possible to overcome this attenuation by giving a larger dose of psychedelics to those who have been treated with SSRIs and demonstrated an inadequate response. For example, one team increased doses of LSD in people with Generalised Anxiety Disorder (GAD) treated with SSRIs (Holze et al. 2023).

5.4 Psychedelics and Prophylaxis

There is almost no work on whether psychedelics could be used as a prophylactic treatment, although see (Kočárová et al. 2021) for a review on this idea. Perhaps the initial antidepressant response to psychedelics could alter the brain to make other prophylactic interventions such as an SSRI or a mood stabiliser (e.g. lithium) effective (even where they were not before).

Whilst most patients gain significant benefits from a single psilocybin treatment there is a great deal of individual variation in the duration of effects of psilocybin on patients with depression (R. L. Carhart-Harris et al. 2018a, b). Some patients seem to have a long lasting – perhaps permanent – recovery whereas others may improve for a few months but the depression then slowly recurs. One alternative option is spacing treatments over the course of months or years to help prevent recurrence of depressive symptoms.

5.5 *Can Psychedelics Be Developed that Treat Depression Without Having a Psychedelic Effect?*

In rodents, scientists have used molecules that work in models of depression. These molecules do not produce the classic head twitch response [a proxy for psychedelic effects] (Olson 2022). If proven efficacious in humans this would be an appealing alternative approach for patients who do not wish to undergo full-dose psychedelic therapy sessions.

Another option could be microdosing; using very low sub-[or almost sub-] detectable amounts of psilocybin or LSD over longer periods to elevate mood or creativity. Although there are no controlled studies so far, the evidence is mixed (Murphy et al. 2023; Szigeti et al. 2021), with one study demonstrating acute improvements in mood and energy levels. Currently, more studies are being conducted in Australia (Woke pharmaceuticals; https://wokeph.com) and Imperial College London.

5.6 *Legality of Psychedelics*

We also know that the legal status of psychedelics – as illegal Schedule 1 drugs under the UN Conventions grossly exaggerates their harms and risks and is not evidence based (Nutt et al. 2013). It has led to a common perception that these drugs are addictive whereas there is growing evidence that they are not. This growing body of evidence has been instrumental in the Australian government decision to re-schedule psilocybin for treatment-resistant depression as of July 2023. https://www.tga.gov.au/resources/publication/scheduling-decisions-final/notice-final-decision-amend-or-not-amend-current-poisons-standard-june-2022-acms-38-psilocybine-and-mdma/re-scheduling-psilocybin-and-mdma-poisons-standard-questions-and-answers.

6 Final Thoughts

Psychedelic drugs offer a new way to treat depression that has significant potential advantages over current approaches. These include speed of onset (just one or two treatment sessions with implications for fewer longer-term adverse effects) and efficacy where other treatments have failed. They offer a paradigm shift in the treatment of depression and give psychiatrists and patients a new treatment option. Although, as described in this article, there are still many research questions to be answered, and controlled drugs regulations to be overcome, it seems likely that they will become established therapies within a decade. As well as the different acute effects of psychedelics from other antidepressant medicines and other physical

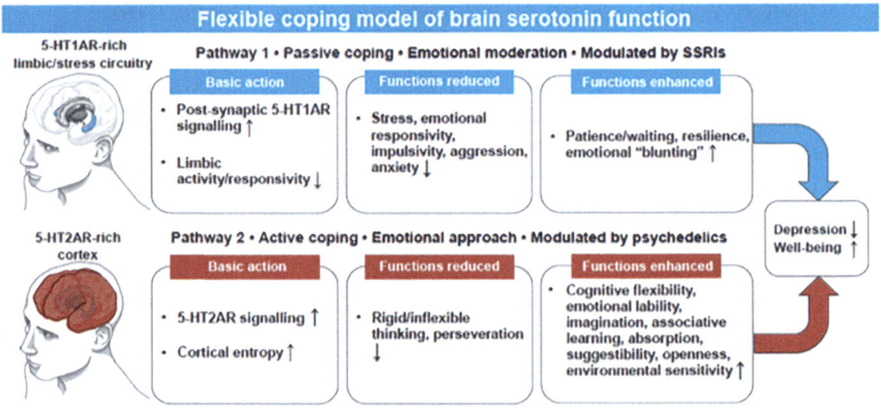

Fig. 2 Two different ways antidepressant treatments work. Model of the action of SSRIs vs. Psychedelics, taken from (Carhart-Harris and Nutt 2017). SSRIs are proposed to elicit their action via the 5-HT_{1A}R-rich limbic circuitry, decreasing limbic responsivity ultimately leading to decreases in depressive symptom severity. Proposed psychedelic action is via 5-HT_{2A}-rich cortex, increasing flexible thinking and decreasing rigidity ultimately leading to decreases in depression and increases in well-being

treatments, psychedelic appear to have a distinct pharmacology and biology to other serotonin-acting antidepressants. These differences are conceptualised in the hypothesis paper (Carhart-Harris and Nutt 2017) from which Fig. 2 is taken.

The therapeutic effects of SSRIs are described elsewhere in this book but in essence they are presumed to involve increased synaptic levels of serotonin, which through the inhibitory 5-HT_{1A} receptors dampens down the limbic hyper-reactivity found in depression as a result of chronic stress (Harmer et al. 2017). This allows the patients to engage with behavioural processes to overcome their anxiety and engage in rewarding social/recreational interactions and so become over time less depressed. Psychedelics work through cortical 5-HT_{2A} receptors to immediately disrupt longstanding canalised thinking process such as negative depressive ruminations. In many cases this escape is associated with major cognitive changes; revisions of their life-experiences and self-worth providing a new way of thinking about, and engaging with, the world (see Ref. Watts et al. 2017 for examples of patients reports).

The BDI was selected for Ross et al. (2016a, b) because it was considered by the authors to have better psychometric properties than the study's primary depression measure. For Ross et al. (2016a, b) and Griffiths et al. (2016), only pre-cross-over data was used to calculate the between-condition effect size. For Davis et al. (2021a, b), only pre-delayed treatment data was used to calculate the between-condition effect size, and data from both the immediate treatment and delayed treatment groups were used to calculate the within-condition effect size. Sloshower et al. (2023) was a within-subject, fixed-order study, such that the between-condition effect size estimate represents the difference between outcome scores at 1 week post-psilocybin dosing and 4 weeks post-placebo dosing. D'Souza et al. (2022) was also a within-subject, fixed-order study, such that the between-condition effect size estimate represents the difference between outcome scores at 1 day post 0.3 mg/kg DMT dosing and 1 day post-placebo (.1 mg/kg DMT) dosing. For D'Souza et al. (2022), post 0.3 mg/kg DMT HAM-D scores were from 6 subjects versus 7, as 1 subject dropped out.

Beck Depression Inventory (BDI) (Beck et al. 1961), Dipropyltryptamine (DPT), GRID-Hamilton Depression Rating Scale (GRID-HAM-D) (Team 2003), Hamilton Depression Rating Scale (HAM-D) (Hamilton 1960) Healthy control (HC), Montgomery-Asberg Depression Rating Scale (MADRS) (Montgomery and Åsberg 1979), MADRS–Suicidal Index (MADRS-SI) (Zeifman et al. 2021), Major Depressive Disorder (MDD), Quick Inventory of Depressive Symptomology (QIDS) (Rush et al. 2003), State-Trait Anxiety Inventory – Global (STAI-G) (Spielberger 1983), Treatment Resistant depression (TRD).

References

Abuse S (2021) Key substance use and mental health indicators in the United States: results from the 2020 National Survey on Drug Use and Health (HHS Publication No. PEP21-07-01-003, NSDUH Series H-56)

Adamson S, Metzner R (1988) The nature of the MDMA experience and its role in healing, psychotherapy and spiritual practice. Revision 10(4):59–72

Aday JS, Davis AK, Mitzkovitz CM, Bloesch EK, Davoli CC (2021) Predicting reactions to psychedelic drugs: a systematic review of states and traits related to acute drug effects. ACS Pharmacol Transl Sci 4(2):424–435

Aday JS, Heifets BD, Pratscher SD, Bradley E, Rosen R, Woolley JD (2022) Great expectations: recommendations for improving the methodological rigor of psychedelic clinical trials. Psychopharmacology (Berl):1–22

Agin-Liebes G, Zeifman R, Luoma JB, Garland EL, Campbell WK, Weiss B (2022) Prospective examination of the therapeutic role of psychological flexibility and cognitive reappraisal in the ceremonial use of ayahuasca. J Psychopharmacol 36(3):295–308

Barrett FS, Johnson MW, Griffiths RR (2015) Validation of the revised mystical experience questionnaire in experimental sessions with psilocybin. J Psychopharmacol 29(11):1182–1190

Beck AT, Ward CH, Mendelson M, Mock J, Erbaugh J (1961) An inventory for measuring depression. Arch Gen Psychiatry 4(6):561–571

Becker AM, Klaiber A, Holze F, Istampoulouoglou I, Duthaler U, Varghese N, Eckert A, Liechti ME (2023) Ketanserin reverses the acute response to LSD in a randomized, double-blind, placebo-controlled, crossover study in healthy participants. Int J Neuropsychopharmacol 26(2):97–106. https://doi.org/10.1093/ijnp/pyac075

Belser AB, Agin-Liebes G, Swift TC, Terrana S, Devenot N, Friedman HL, Guss J, Bossis A, Ross S (2017) Patient experiences of psilocybin-assisted psychotherapy: an interpretative phenomenological analysis. J Humanist Psychol 57(4):354–388

Butler M, Jelen L, Rucker J (2022) Expectancy in placebo-controlled trials of psychedelics: if so, so what? Psychopharmacology (Berl) 239(10):3047–3055. https://doi.org/10.1007/s00213-022-06221-6. PMCID: PMC9481484. PMID: 36063208

Carbonaro TM, Bradstreet MP, Barrett FS, MacLean KA, Jesse R, Johnson MW, Griffiths RR (2016) Survey study of challenging experiences after ingesting psilocybin mushrooms: acute and enduring positive and negative consequences. J Psychopharmacol 30(12):1268–1278

Carhart-Harris RL, Friston K (2019) REBUS and the anarchic brain: toward a unified model of the brain action of psychedelics. Pharmacol Rev 71(3):316–344

Carhart-Harris R, Giribaldi B, Watts R, Baker-Jones M, Murphy-Beiner A, Murphy R, Martell J, Blemings A, Erritzoe D, Nutt DJ (2021) Trial of psilocybin versus escitalopram for depression. N Engl J Med 384(15):1402–1411. https://doi.org/10.1056/NEJMoa2032994

Carhart-Harris RL (2018) The entropic brain-revisited. Neuropharmacology 142:167–178

Carhart-Harris RL, Nutt DJ (2017) Serotonin and brain function: a tale of two receptors. J Psychopharmacol 31(9):1091–1120. https://doi.org/10.1177/0269881117725915

Carhart-Harris RL, Leech R, Williams T, Erritzoe D, Abbasi N, Bargiotas T, Hobden P, Sharp D, Evans J, Feilding A (2012) Implications for psychedelic-assisted psychotherapy: functional magnetic resonance imaging study with psilocybin. Br J Psychiatry 200(3):238–244

Carhart-Harris RL, Leech R, Hellyer PJ, Shanahan M, Feilding A, Tagliazucchi E, Chialvo DR, Nutt D (2014) The entropic brain: a theory of conscious states informed by neuroimaging research with psychedelic drugs. Front Hum Neurosci 20

Carhart-Harris RL, Kaelen M, Whalley M, Bolstridge M, Feilding A, Nutt DJ (2015) LSD enhances suggestibility in healthy volunteers. Psychopharmacology (Berl) 232:785–794

Carhart-Harris RL, Bolstridge M, Rucker J, Day CM, Erritzoe D, Kaelen M, Bloomfield M, Rickard JA, Forbes B, Feilding A (2016a) Psilocybin with psychological support for treatment-resistant depression: an open-label feasibility study. Lancet Psychiatry 3(7):619–627

Carhart-Harris RL, Muthukumaraswamy S, Roseman L, Kaelen M, Droog W, Murphy K, Tagliazucchi E, Schenberg EE, Nest T, Orban C (2016b) Neural correlates of the LSD experience revealed by multimodal neuroimaging. Proc Natl Acad Sci 113(17):4853–4858

Carhart-Harris RL, Bolstridge M, Day C, Rucker J, Watts R, Erritzoe D, Kaelen M, Giribaldi B, Bloomfield M, Pilling S (2018a) Psilocybin with psychological support for treatment-resistant depression: six-month follow-up. Psychopharmacology (Berl) 235(2):399–408

Carhart-Harris RL, Bolstridge M, Day CMJ, Rucker J, Watts R, Erritzoe DE, Kaelen M, Giribaldi B, Bloomfield M, Pilling S, Rickard JA, Forbes B, Feilding A, Taylor D, Curran HV, Nutt DJ (2018b) Psilocybin with psychological support for treatment-resistant depression: six-month follow-up. Psychopharmacology (Berl) 235(2):399–408. https://doi.org/10.1007/s00213-017-4771-x

Carhart-Harris R, Chandaria S, Erritzoe D, Gazzaley A, Girn M, Kettner H, Mediano P, Nutt D, Rosas F, Roseman L (2022) Canalization and plasticity in psychopathology. Neuropharmacology:109398

Cascade E, Kalali AH, Kennedy SH (2009) Real-world data on SSRI antidepressant side effects. Psychiatry (Edgmont) 6(2):16

Cipriani A, Furukawa TA, Salanti G, Chaimani A, Atkinson LZ, Ogawa Y, Leucht S, Ruhe HG, Turner EH, Higgins JP (2018) Comparative efficacy and acceptability of 21 antidepressant drugs for the acute treatment of adults with major depressive disorder: a systematic review and network meta-analysis. Focus 16(4):420–429

Clark A (2015) Surfing uncertainty: prediction, action, and the embodied mind. Oxford University Press

Clark A (2016) Attention alters predictive processing. Behav Brain Sci 39

Clayton A, Kornstein S, Prakash A, Mallinckrodt C, Wohlreich M (2007) Psychology: changes in sexual functioning associated with duloxetine, escitalopram, and placebo in the treatment of patients with major depressive disorder. J Sex Med 4(4):917–929

Close JB, Hajien EC, Watts R, Roseman L, Carhart-Harris RL (2020) Psychedelics and psychological flexibility–results of a prospective web-survey using the acceptance and action questionnaire II. J Contextual Behav Sci

Cooper AA, Conklin LR (2015) Dropout from individual psychotherapy for major depression: A meta-analysis of randomized clinical trials. Clin Psychol Rev 40:57–65

Cuijpers P, Dekker J, Hollon SD, Andersson G (2009) Adding psychotherapy to pharmacotherapy in the treatment of depressive disorders in adults: a meta-analysis. J Clin Psychiatry 70(9):401

Cuijpers P, van Straten A, Warmerdam L, van Rooy MJ (2010) Recruiting participants for interventions to prevent the onset of depressive disorders: Possible ways to increase participation rates. BMC Health Serv Res 10(1):1–6

Cuijpers P, Turner E, Mohr D, Hofmann S, Andersson G, Berking M, Coyne J (2014) Comparison of psychotherapies for adult depression to pill placebo control groups: a meta-analysis. Psychol Med 44(4):685–695

Davis AK, Barrett FS, May DG, Cosimano MP, Sepeda ND, Johnson MW, Finan PH, Griffiths RR (2021a) Effects of psilocybin-assisted therapy on major depressive disorder: a randomized clinical trial. JAMA Psychiatry 78(5):481–489. https://doi.org/10.1001/jamapsychiatry.2020.3285

Davis AK, Barrett FS, So S, Gukasyan N, Swift TC, Griffiths RR (2021b) Development of the psychological insight questionnaire among a sample of people who have consumed psilocybin or LSD. J Psychopharmacol 35(4):437–446

Daws RE, Timmermann C, Giribaldi B, Sexton JD, Wall MB, Erritzoe D, Roseman L, Nutt D, Carhart-Harris R (2022) Increased global integration in the brain after psilocybin therapy for depression. Nat Med 28(4):844–851. https://doi.org/10.1038/s41591-022-01744-z

Doss MK, May DG, Johnson MW, Clifton JM, Hedrick SL, Prisinzano TE, Griffiths RR, Barrett FS (2020) The acute effects of the atypical dissociative hallucinogen Salvinorin A on functional connectivity in the human brain. Sci Rep 10(1):16392. https://doi.org/10.1038/s41598-020-73216-8

Drummond E, McCulloch W, Knudsen GM, Barrett FS, Doss MK, Carhart-Harris RL, Rosas FE, Deco G, Kringelbach ML, Preller KH (2022) Psychedelic resting-state neuroimaging: a review and perspective on balancing replication and novel analyses. Neurosci Biobehav Rev:104689

D'Souza DC, Syed SA, Flynn LT, Safi-Aghdam H, Cozzi NV, Ranganathan M (2022) Exploratory study of the dose-related safety, tolerability, and efficacy of dimethyltryptamine (DMT) in healthy volunteers and major depressive disorder. Neuropsychopharmacology 47(10):1854–1862. https://doi.org/10.1038/s41386-022-01344-y. Epub 2022 Jun 3. PMID: 35660802. PMCID: PMC9372173

Erritzoe D, Smith J, Fisher PM, Carhart-Harris R, Frokjaer VG, Knudsen GM (2019) Recreational use of psychedelics is associated with elevated personality trait openness: exploration of associations with brain serotonin markers. J Psychopharmacol 33(9):1068–1075

Erritzoe D, Roseman L, NourMM MLK, Kaelen M, Nutt DJ, Carhart-Harris RL (2018) Effects of psilocybin therapy on personality structure. Acta Psychiatr Scand 138(5):368–378. https://doi.org/10.1111/acps.12904. Epub 2018 Jun 19. PMID: 29923178. PMCID: PMC6220878

Farnes N, Juel BE, Nilsen AS, Romundstad LG, Storm JF (2020) Increased signal diversity/complexity of spontaneous EEG, but not evoked EEG responses, in ketamine-induced psychedelic state in humans. PloS One 15(11):e0242056. https://doi.org/10.1371/journal.pone.0242056

Friston K (2010) The free-energy principle: a unified brain theory? Nat Rev Neurosci 11(2):127–138

Friston K, Kiebel S (2009) Predictive coding under the free-energy principle. Philos Trans R Soc B Biol Sci 364(1521):1211–1221

Galvão-Coelho NL, de Menezes Galvão AC, de Almeida RN, Palhano-Fontes F, Campos Braga I, Lobão Soares B, Maia-de-Oliveira JP, Perkins D, Sarris J, de Araujo DB (2020) Changes in inflammatory biomarkers are related to the antidepressant effects of Ayahuasca. J Psychopharmacol 34(10):1125–1133. https://doi.org/10.1177/0269881120936486

Girn M, Rosas FE, Daws RE, Gallen CL, Gazzaley A, Carhart-Harris RL (2023) A complex systems perspective on psychedelic brain action. Trends Cogn Sci

Goodwin GM, Aaronson ST, Alvarez O, Arden PC, Baker A, Bennett JC, Bird C, Blom RE, Brennan C, Brusch D (2022a) Single-dose psilocybin for a treatment-resistant episode of major depression. N Engl J Med 387(18):1637–1648

Goodwin, G. M., Aaronson, S. T., Alvarez, O., Arden, P. C., Baker, A., Bennett, J. C., Bird, C., Blom, R. E., Brennan, C., Brusch, D., Burke, L., Campbell-Coker, K., Carhart-Harris, R.,

Cattell, J., Daniel, A., DeBattista, C., Dunlop, B. W., Eisen, K., Feifel, D., . . . Malievskaia, E. (2022b). Single-dose psilocybin for a treatment-resistant episode of major depression. N Engl J Med, 387(18), 1637–1648. https://doi.org/10.1056/NEJMoa2206443

Grenard JL, Munjas BA, Adams JL, Suttorp M, Maglione M, McGlynn EA, Gellad WF (2011) Depression and medication adherence in the treatment of chronic diseases in the United States: a meta-analysis. J Gen Intern Med 26:1175–1182

Griffiths RR, Johnson MW, Carducci MA, Umbricht A, Richards WA, Richards BD, Cosimano MP, Klinedinst MA (2016) Psilocybin produces substantial and sustained decreases in depression and anxiety in patients with life-threatening cancer: a randomized double-blind trial. J Psychopharmacol 30(12):1181–1197. https://doi.org/10.1177/0269881116675513

Grof S, Goodman LE, Richards WA, Kurland AA (1973) LSD-assisted psychotherapy in patients with terminal cancer. Int Pharmacopsychiatry 8(3):129–144. https://doi.org/10.1159/000467984

Gukasyan N, Davis AK, Barrett FS, Cosimano MP, Sepeda ND, Johnson MW, Griffiths RR (2022) Efficacy and safety of psilocybin-assisted treatment for major depressive disorder: prospective 12-month follow-up. J Psychopharmacol 36(2):151–158

Hamilton M (1960) A rating scale for depression. J Neurol Neurosurg Psychiatry 23(1):56

Harmer CJ, Duman RS, Cowen PJ (2017) How do antidepressants work? New perspectives for refining future treatment approaches. Lancet Psychiatry 4(5):409–418. https://doi.org/10.1016/s2215-0366(17)30015-9

Hartogsohn I (2016) Set and setting, psychedelics and the placebo response: an extra-pharmacological perspective on psychopharmacology. J Psychopharmacol 30(12):1259–1267

Hohwy J, Seth A (2020) Predictive processing as a systematic basis for identifying the neural correlates of consciousness. Philos Mind Sci 1(II)

Holze F, Gasser P, Müller F, Dolder PC, Liechti ME (2023) Lysergic acid diethylamide-assisted therapy in patients with anxiety with and without a life-threatening illness: a randomized, double-blind, placebo-controlled phase II study. Biol Psychiatry 93(3):215–223. https://doi.org/10.1016/j.biopsych.2022.08.025

Hyde RW (1960) Psychological and social determinants of drug action. The dynamics of psychiatric drug therapy, pp 297–315

Johnson MW, Richards WA, Griffiths RR (2008) Human hallucinogen research: guidelines for safety. J Psychopharmacol 22(6):603–620

Kaelen M, Barrett FS, Roseman L, Lorenz R, Family N, Bolstridge M, Curran HV, Feilding A, Nutt DJ, Carhart-Harris R (2015) LSD enhances the emotional response to music. Psychopharmacology (Berl) 232(19):3607–3614

Kaelen M, Giribaldi B, Raine J, Evans L, Timmerman C, Rodriguez N, Roseman L, Feilding A, Nutt D, Carhart-Harris R (2018) The hidden therapist: evidence for a central role of music in psychedelic therapy. Psychopharmacology (Berl) 235(2):505–519

Kangaslampi S (2023) Association between mystical-type experiences under psychedelics and improvements in well-being or mental health – a comprehensive review of the evidence. J Psychedelic Stud

Kettner H, Rosas F, Timmermann C, Kärtner L, Carhart-Harris RL, Roseman L (2021) Psychedelic communitas: intersubjective experience during psychedelic group sessions predicts enduring changes in psychological wellbeing and social connectedness. Front Pharmacol:234

Khan A, Brown WA (2015) Antidepressants versus placebo in major depression: an overview. World Psychiatry 14(3):294–300

Kočárová R, Horáček J, Carhart-Harris R (2021) Does psychedelic therapy have a transdiagnostic action and prophylactic potential? Front Psych 12:661233. https://doi.org/10.3389/fpsyt.2021.661233

Lebedev AV, Kaelen M, Lövdén M, Nilsson J, Feilding A, Nutt DJ, Carhart-Harris RL (2016) LSD-induced entropic brain activity predicts subsequent personality change. Hum Brain Mapp 37(9):3203–3213

Lepow L, Morishita H, Yehuda R (2021) Critical period plasticity as a framework for psychedelic-assisted psychotherapy. Front Neurosci:1165
Li N, Jin D, Wei J, Huang Y, Xu J (2022) Functional brain abnormalities in major depressive disorder using a multiscale community detection approach. Neuroscience 501:1–10
Lifshitz M, van Elk M, Luhrmann TM (2019) Absorption and spiritual experience: a review of evidence and potential mechanisms. Conscious Cogn 73:102760
MacLean KA, Johnson MW, Griffiths RR (2011) Mystical experiences occasioned by the hallucinogen psilocybin lead to increases in the personality domain of openness. J Psychopharmacol 25(11):1453–1461
Madsen MK, Fisher PM, Burmester D, Dyssegaard A, Stenbæk DS, Kristiansen S, Johansen SS, Lehel S, Linnet K, Svarer C (2019) Psychedelic effects of psilocybin correlate with serotonin 2A receptor occupancy and plasma psilocin levels. Neuropsychopharmacology 44(7):1328–1334
Mason N, Kuypers K, Reckweg J, Müller F, Tse D, Da Rios B, Toennes S, Stiers P, Feilding A, Ramaekers J (2021) Spontaneous and deliberate creative cognition during and after psilocybin exposure. Transl Psychiatry 11(1):209
McCabe C, Mishor Z, Cowen PJ, Harmer CJ (2010) Diminished neural processing of aversive and rewarding stimuli during selective serotonin reuptake inhibitor treatment. Biol Psychiatry 67(5): 439–445
Mertens LJ, Wall MB, Roseman L, Demetriou L, Nutt DJ, Carhart-Harris RL (2020) Therapeutic mechanisms of psilocybin: changes in amygdala and prefrontal functional connectivity during emotional processing after psilocybin for treatment-resistant depression. J Psychopharmacol 34(2):167–180. https://doi.org/10.1177/0269881119895520
Metzner R, Litwin G, Weil G (1965) The relation of expectation and mood to psilocybin reactions: a questionnaire study. Psychedelic Rev 5:3–39
Mithoefer MC, Feduccia AA, Jerome L, Mithoefer A, Wagner M, Walsh Z, Hamilton S, Klosinski B, Emerson A, Doblin R (2019) MDMA-assisted psychotherapy for treatment of PTSD: study design and rationale for phase 3 trials based on pooled analysis of six phase 2 randomized controlled trials. Psychopharmacology (Berl) 236(9):2735–2745
Montgomery SA, Åsberg M (1979) A new depression scale designed to be sensitive to change. Br J Psychiatry 134(4):382–389
Morgese MG, Chrobak AA, Jiang C, Maroteaux L, Romay-Tallon R, Sowa-Kućma M, Stachowicz K, Sowa-Kućma M (2023) The treatment of depression searching for new ideas. The pharmacotherapy of depression-searching for new mechanisms and drug interactions. Basic and clinical research, 16648714, p 153
Müller F, Dolder PC, Schmidt A, Liechti ME, Borgwardt S (2018) Altered network hub connectivity after acute LSD administration. Neuroimage Clin 18:694–701. https://doi.org/10.1016/j.nicl.2018.03.005
Murphy R, Kettner H, Zeifman R, Giribaldi B, Kartner L, Martell J, Read T, Murphy-Beiner A, Baker-Jones M, Nutt D (2022) Therapeutic alliance and rapport modulate responses to psilocybin assisted therapy for depression. Front Pharmacol 3819
Murphy RJ, Sumner R, Evans W, Ponton R, Ram S, Godfrey K, Forsyth A, Cavadino A, Krishmamuthy V, Smith T, Hoeh NR, Menkes DB, Muthukumaraswamy S (2023) Acute mood-elevating properties of microdosed LSD in healthy volunteers: a home-administered randomised controlled trial. Biol Psychiatry. https://doi.org/10.1016/j.biopsych.2023.03.013
Muthukumaraswamy SD, Carhart-Harris RL, Moran RJ, Brookes MJ, Williams TM, Errtizoe D, Sessa B, Papadopoulos A, Bolstridge M, Singh KD (2013a) Broadband cortical desynchronization underlies the human psychedelic state. J Neurosci 33(38):15171–15183
Muthukumaraswamy SD, Carhart-Harris RL, Moran RJ, Brookes MJ, Williams TM, Errtizoe D, Sessa B, Papadopoulos A, Bolstridge M, Singh KD, Feilding A, Friston KJ, Nutt DJ (2013b) Broadband cortical desynchronization underlies the human psychedelic state. J Neurosci 33(38): 15171–15183. https://doi.org/10.1523/JNEUROSCI.2063-13.2013
Muthukumaraswamy S, Forsyth A, Sumner RL (2022) The challenges ahead for psychedelic 'medicine'. Aust N Z J Psychiatry:00048674221081763

Nielson EM, May DG, Forcehimes AA, Bogenschutz MP (2018) The psychedelic debriefing in alcohol dependence treatment: illustrating key change phenomena through qualitative content analysis of clinical sessions. Front Pharmacol 9:132

Noorani T, Garcia-Romeu A, Swift TC, Griffiths RR, Johnson MW (2018) Psychedelic therapy for smoking cessation: qualitative analysis of participant accounts. J Psychopharmacol 32(7): 756–769

Nour MM, Evans L, Nutt D, Carhart-Harris RL (2016) Ego-dissolution and psychedelics: validation of the ego-dissolution inventory (EDI). Front Hum Neurosci 269

Nutt DJ, King LA, Nichols DE (2013) Effects of schedule I drug laws on neuroscience research and treatment innovation. Nat Rev Neurosci 14(8):577–585. https://doi.org/10.1038/nrn3530

O'Callaghan C, Hubik DJ, Dwyer J, Williams M, Ross M (2020) Experience of music used with psychedelic therapy: a rapid review and implications. J Music Ther 57(3):282–314

Olson DE (2022) Biochemical mechanisms underlying psychedelic-induced neuroplasticity. Biochemistry 61(3):127–136. https://doi.org/10.1021/acs.biochem.1c00812

Osório F d L, Sanches RF, Macedo LR, Dos Santos RG, Maia-de-Oliveira JP, Wichert-Ana L, De Araujo DB, Riba J, Crippa JA, Hallak JE (2015) Antidepressant effects of a single dose of ayahuasca in patients with recurrent depression: a preliminary report. Braz J Psychiatry 37:13–20

Palhano-Fontes F, Andrade KC, Tofoli LF, Santos AC, Crippa JA, Hallak JE, Ribeiro S, de Araujo DB (2015) The psychedelic state induced by ayahuasca modulates the activity and connectivity of the default mode network. PloS One 10(2):e0118143. https://doi.org/10.1371/journal.pone.0118143

Palhano-Fontes F, Barreto D, Onias H, Andrade KC, Novaes MM, Pessoa JA, Mota-Rolim SA, Osório FL, Sanches R, Dos Santos RG (2019a) Rapid antidepressant effects of the psychedelic ayahuasca in treatment-resistant depression: a randomized placebo-controlled trial. Psychol Med 49(4):655–663

Palhano-Fontes F, Barreto D, Onias H, Andrade KC, Novaes MM, Pessoa JA, Mota-Rolim SA, Osório FL, Sanches R, Dos Santos RG, Tófoli LF, de Oliveira G, Silveira MY, Riba J, Santos FR, Silva-Junior AA, Alchieri JC, Galvão-Coelho NL, Lobão-Soares B, Hallak JEC, Arcoverde E, Maia-de-Oliveira JP, Araújo DB (2019b) Rapid antidepressant effects of the psychedelic ayahuasca in treatment-resistant depression: a randomized placebo-controlled trial. Psychol Med 49(4):655–663. https://doi.org/10.1017/S0033291718001356. Epub 2018 Jun 15. PMID: 29903051. PMCID: PMC6378413

Palhano-Fontes, F., Barreto, D., Onias, H., Andrade, K. C., Novaes, M. M., Pessoa, J. A., Mota-Rolim, S. A., Osório, F. L., Sanches, R., Dos Santos, R. G., Tófoli, L. F., de Oliveira Silveira, G., Yonamine, M., Riba, J., Santos, F. R., Silva-Junior, A. A., Alchieri, J. C., Galvão-Coelho, N. L., Lobão-Soares, B., . . . Araújo, D. B. (2019b). Rapid antidepressant effects of the psychedelic ayahuasca in treatment-resistant depression: a randomized placebo-controlled trial. Psychol Med, 49(4), 655–663. https://doi.org/10.1017/s0033291718001356

Peill JM, Trinci KE, Kettner H, Mertens LJ, Roseman L, Timmermann C, Rosas FE, Lyons T, Carhart-Harris RL (2022) Validation of the psychological insight scale: a new scale to assess psychological insight following a psychedelic experience. J Psychopharmacol 36(1):31–45

Petri G, Expert P, Turkheimer F, Carhart-Harris R, Nutt D, Hellyer PJ, Vaccarino F (2014) Homological scaffolds of brain functional networks. J R Soc Interface 11(101):20140873

Preller KH, Burt JB, Ji JL, Schleifer CH, Adkinson BD, Stämpfli P, Seifritz E, Repovs G, Krystal JH, Murray JD, Vollenweider FX, Anticevic A (2018) Changes in global and thalamic brain connectivity in LSD-induced altered states of consciousness are attributable to the 5-HT2A receptor. eLife 7. https://doi.org/10.7554/eLife.35082

Preller KH, Duerler P, Burt JB, Ji JL, Adkinson B, Stämpfli P, Seifritz E, Repovš G, Krystal JH, Murray JD (2020) Psilocybin induces time-dependent changes in global functional connectivity. Biol Psychiatry 88(2):197–207

Quednow BB, Kometer M, Geyer MA, Vollenweider FX (2012) Psilocybin-induced deficits in automatic and controlled inhibition are attenuated by ketanserin in healthy human volunteers. Neuropsychopharmacology 37(3):630–640

Quitkin FM, Rabkin JD, Markowitz JM, Stewart JW, McGrath PJ, Harrison W (1987) Use of pattern analysis to identify true drug response: a replication. Arch Gen Psychiatry 44(3): 259–264

Roseman L, Leech R, Feilding A, Nutt DJ, Carhart-Harris RL (2014) The effects of psilocybin and MDMA on between-network resting state functional connectivity in healthy volunteers. Front Hum Neurosci 8:204

Roseman L, Nutt DJ, Carhart-Harris RL (2018) Quality of acute psychedelic experience predicts therapeutic efficacy of psilocybin for treatment-resistant depression. Front Pharmacol 8:974

Roseman L, Haijen E, Idialu-Ikato K, Kaelen M, Watts R, Carhart-Harris R (2019) Emotional breakthrough and psychedelics: validation of the emotional breakthrough inventory. J Psychopharmacol 33(9):1076–1087

Ross S, Bossis A, Guss J, Agin-Liebes G, Malone T, Cohen B, Mennenga SE, Belser A, Kalliontzi K, Babb J (2016a) Rapid and sustained symptom reduction following psilocybin treatment for anxiety and depression in patients with life-threatening cancer: a randomized controlled trial. J Psychopharmacol 30(12):1165–1180

Ross S, Bossis A, Guss J, Agin-Liebes G, Malone T, Cohen B, Mennenga SE, Belser A, Kalliontzi K, Babb J, Su Z, Corby P, Schmidt BL (2016b) Rapid and sustained symptom reduction following psilocybin treatment for anxiety and depression in patients with life-threatening cancer: a randomized controlled trial. J Psychopharmacol 30(12):1165–1180. https://doi.org/10.1177/0269881116675512

Roy AV, Thai M, Klimes-Dougan B, Westlund Schreiner M, Mueller BA, Albott CS, Lim KO, Fiecas M, Tye SJ, Cullen KR (2021) Brain entropy and neurotrophic molecular markers accompanying clinical improvement after ketamine: preliminary evidence in adolescents with treatment-resistant depression. J Psychopharmacol 35(2):168–177. https://doi.org/10.1177/0269881120928203

Rush AJ, Trivedi MH, Ibrahim HM, Carmody TJ, Arnow B, Klein DN, Markowitz JC, Ninan PT, Kornstein S, Manber R (2003) The 16-item quick inventory of depressive symptomatology (QIDS), clinician rating (QIDS-C), and self-report (QIDS-SR): a psychometric evaluation in patients with chronic major depression. Biol Psychiatry 54(5):573–583

Sanches RF, de Lima Osório F, Dos Santos RG, Macedo LR, Maia-de-Oliveira JP, Wichert-Ana L, de Araujo DB, Riba J, Crippa JAS, Hallak JE (2016) Antidepressant effects of a single dose of ayahuasca in patients with recurrent depression: a SPECT study. J Clin Psychopharmacol 36(1): 77–81

Schartner MM, Carhart-Harris RL, Barrett AB, Seth AK, Muthukumaraswamy SD (2017a) Increased spontaneous MEG signal diversity for psychoactive doses of ketamine, LSD and psilocybin. Sci Rep 7(1):1–12

Schartner MM, Carhart-Harris RL, Barrett AB, Seth AK, Muthukumaraswamy SD (2017b) Increased spontaneous MEG signal diversity for psychoactive doses of ketamine, LSD and psilocybin. Sci Rep 7:46421. https://doi.org/10.1038/srep46421

Sloshower J, Skosnik PD, Safi-Aghdam H, Pathania S, Syed S, Pittman B, D'Souza DC (2023) Psilocybin-assisted therapy for major depressive disorder: an exploratory placebo-controlled, fixed-order trial. J Psychopharmacol:02698811231154852

Spielberger CD (1983) State-trait anxiety inventory for adults

Stace WT (1960) Mysticism and philosophy. Philosophy 37(140)

Studerus E, Gamma A, Vollenweider FX (2010) Psychometric evaluation of the altered states of consciousness rating scale (OAV). PloS One 5(8):e12412

Swift JK, Greenberg RP (2014) A treatment by disorder meta-analysis of dropout from psychotherapy. J Psychother Integr 24(3):193

Szigeti B, Kartner L, Blemings A, Rosas F, Feilding A, Nutt DJ, Carhart-Harris RL, Erritzoe D (2021) Self-blinding citizen science to explore psychedelic microdosing. eLife 10. https://doi.org/10.7554/eLife.62878

Szigeti B, Nutt D, Carhart-Harris R, Erritzoe D (2023) The difference between 'placebo group' and 'placebo control': a case study in psychedelic icrodosing. Sci Rep 13. https://doi.org/10.1038/s41598-023-34938-7

Tagliazucchi E, Roseman L, Kaelen M, Orban C, Muthukumaraswamy SD, Murphy K, Laufs H, Leech R, McGonigle J, Crossley N (2016) Increased global functional connectivity correlates with LSD-induced ego dissolution. Curr Biol 26(8):1043–1050

Tagliazucchi E, Zamberlan F, Cavanna F, de la Fuente L, Romero C, Sanz Perl Y, Pallavicini C (2021) Baseline power of theta oscillations predicts mystical-type experiences induced by DMT in a natural setting. Front Psych 12:720066

Taylor MJ, Freemantle N, Geddes JR, Bhagwagar Z (2006) Early onset of selective serotonin reuptake inhibitor antidepressant action: systematic review and meta-analysis. Arch Gen Psychiatry 63(11):1217–1223. https://doi.org/10.1001/archpsyc.63.11.1217

Team, D. R. S. S. (2003) GRID-HAMD-17, GRID-HAMD-21 structured interview guide. In: International society for CNS drug development San Diego

Teixeira PJ, Johnson MW, Timmermann C, Watts R, Erritzoe D, Douglass H, Kettner H, Carhart-Harris RL (2022) Psychedelics and health behaviour change. J Psychopharmacol 36(1):12–19

Timmermann C, Roseman L, Schartner M, Milliere R, Williams LT, Erritzoe D, Muthukumaraswamy S, Ashton M, Bendrioua A, Kaur O (2019) Neural correlates of the DMT experience assessed with multivariate EEG. Sci Rep 9(1):1–13

Timmermann C, Roseman L, Haridas S, Rosas FE, Luan L, Kettner H, Martell J, Erritzoe D, Tagliazucchi E, Pallavicini C, Girn M, Alamia A, Leech R, Nutt DJ, Carhart-Harris RL (2023) Human brain effects of DMT assessed via EEG-fMRI. Proc Natl Acad Sci U S A 120(13): e2218949120. https://doi.org/10.1073/pnas.2218949120

Vargas MV, Dunlap LE, Dong C, Carter SJ, Tombari RJ, Jami SA, Cameron LP, Patel SD, Hennessey JJ, Saeger HN (2023) Psychedelics promote neuroplasticity through the activation of intracellular 5-HT2A receptors. Science 379(6633):700–706

Viol A, Palhano-Fontes F, Onias H, de Araujo DB, Viswanathan G (2017) Shannon entropy of brain functional complex networks under the influence of the psychedelic Ayahuasca. Sci Rep 7(1):1–13

Watts R, Day C, Krzanowski J, Nutt D, Carhart-Harris R (2017) Patients' accounts of increased "connectedness" and "acceptance" after psilocybin for treatment-resistant depression. J Humanist Psychol 57(5):520–564

Weiss B, Miller JD, Carter NT, Keith Campbell W (2021) Examining changes in personality following shamanic ceremonial use of ayahuasca. Sci Rep 11(1):1–15

Weiss B et al (under review) Unique psychological mechanisms in the treatment of major depressive disorder underlying psilocybin therapy versus escitalopram treatment. Int J Ment Heal Addict

Yaden DB, Earp BD, Griffiths RR (2022) Ethical issues regarding nonsubjective psychedelics as standard of care. Camb Q Healthc Ethics 31(4):464–471

Ye M, Yang T, Qing P, Lei X, Qiu J, Liu G (2015) Changes of functional brain networks in major depressive disorder: a graph theoretical analysis of resting-state fMRI. PloS One 10(9): e0133775

Zeifman RJ, Wagner AC, Watts R, Kettner H, Mertens LJ, Carhart-Harris RL (2020) Post-psychedelic reductions in experiential avoidance are associated with decreases in depression severity and suicidal ideation. Front Psych:782

Zeifman RJ, Singhal N, Dos Santos RG, Sanches RF, de Lima Osório F, Hallak JEC, Weissman CR (2021) Rapid and sustained decreases in suicidality following a single dose of ayahuasca among individuals with recurrent major depressive disorder: results from an open-label trial. Psychopharmacology (Berl) 238(2):453–459. https://doi.org/10.1007/s00213-020-05692-9

Microbiota-Gut-Brain Axis and Antidepressant Treatment

David Antoine Lachmansingh, Aonghus Lavelle, John F. Cryan, and Gerard Clarke

Contents

Abstract In the treatment of depressive disorders, conventional antidepressant therapy has been the mainstay of clinical management, along with well-established nonpharmacological interventions such as various kinds of psychotherapy. Over the last 2 decades, there has been considerable interest in the role of the gastrointestinal

D. A. Lachmansingh
APC Microbiome Ireland, University College Cork, Cork, Ireland

Department of Psychiatry and Neurobehavioural Science, University College Cork, Cork, Ireland

Department of Anatomy and Neuroscience, University College Cork, Cork, Ireland

A. Lavelle and J. F. Cryan (✉)
APC Microbiome Ireland, University College Cork, Cork, Ireland

Department of Anatomy and Neuroscience, University College Cork, Cork, Ireland
e-mail: j.cryan@ucc.ie

G. Clarke
APC Microbiome Ireland, University College Cork, Cork, Ireland

Department of Psychiatry and Neurobehavioural Science, University College Cork, Cork, Ireland

Curr Topics Behav Neurosci (2024) 66: 175–216
https://doi.org/10.1007/7854_2023_449
Published Online: 15 November 2023

system and its microbiota on brain function, behavior, and mental health. Components of what is referred to as the microbiota-gut-brain axis have been uncovered, and further research has elicited functional capabilities such as "gut-brain modules." Some studies have found associations with compositional alterations of gut microbiota in patients with depressive disorders and individuals experiencing symptoms of depression. Regarding the pathogenesis and neurobiology of depression itself, there appears to be a multifactorial contribution, in addition to the theories involving deficits in catecholaminergic and monoamine neurotransmission. Interestingly, there is evidence to suggest that antidepressants may play a role in modulating the gut microbiota, thereby possibly having an impact on the microbiota-gut-brain axis in this manner. The development of prebiotics, probiotics, and synbiotics has led to studies investigating not only their impact on the microbiota but also their therapeutic value in mental health. These psychobiotics have the potential to be used as therapeutic adjuncts in the treatment of depression. Regarding future directions, and in an attempt to further understand the role of the microbiota-gut-brain axis in depression, more studies such as those involving fecal microbiota transplantation will be required. In addition to recent findings, it is also suggested that more research will have to be undertaken to elicit whether specific strains of gut organisms are linked to depression. In terms of further investigation of the therapeutic potential of prebiotics, probiotics, and synbiotics as adjuncts to antidepressant treatment, we also expect there to be more research targeting specific microorganisms, as well as a strong focus on the effects of specific prebiotic fibers from an individualized (personalized) point of view.

Keywords Antidepressants · Depression · Fecal microbiota transplantation · Gut microbiota · Microbiota-gut-brain axis · Prebiotics · Probiotics · Psychobiotics · Synbiotics

1 Introduction

Antidepressant treatment options have long focused on targeting neurotransmitter systems, with monoamine-based mechanisms of action dominating the conceptual landscape in this field. The emergence of the gut microbiota as a potential key regulator of brain function and behavior is an exciting new development that may have implications from both neurobiological and clinical management perspectives. In this chapter, we discuss the preclinical and clinical observations that have shaped this narrative and consider the options for therapeutic targeting of the gut microbiota for the treatment of depression. We also outline the implications arising from the impact of current antidepressant drugs on the gut microbiota.

2 The Gut Microbiota

Researchers across the biological sciences and the behavioral and clinical neurosciences have become increasingly aware of the health implications of various body sites, including the skin surface, the nasal and oral pharynx, and all sampled parts of the alimentary canal. Examples of types of organisms within the gastrointestinal (GI) tract include bacteria, fungi, viruses, and archaea (Matijašić et al. 2020; Lurie-Weinberger and Gophna 2015), which collectively are known as the gut microbiota (Butler et al. 2019a, b). Previously, such microorganisms were thought to have had a commensal relationship with humans. However, as time has progressed and more research has been conducted, it is now clearer that this relationship is mutualistic. Many experts in the field consider the gut microbiota to have coevolved with the host, leading to its proposed role as a virtual endocrine organ in regulating numerous critical functions with implications in health and disease (Clarke et al. 2014; Bordenstein and Theis 2015; Sun et al. 2022).

The microbiome plays a critical role in the development and maturation of both the digestive and immune systems of infants. Bacterial colonization of the gastrointestinal (GI) tract of infants not only aids in improved digestion but also stimulates the development of lymphoid organs and structures (e.g., gut-associated lymphoid tissue and mucosal-associated lymphoid tissue), as well as additionally regulating the innate and adaptive immune systems (Hooper et al. 2012) while reinforcing colonization resistance against gut pathogens (Nunez et al. 2021; Phillips-Farfán et al. 2021).

The gut microbiota is known to break down complex carbohydrates, including naturally occurring plant-derived polysaccharides, resulting in the production of short-chain fatty acids (SCFAs) (O'Riordan et al. 2022; Sun et al. 2022). Microbial species in our gut also contribute to the production of other essential metabolites and neurochemicals, as well as modify the availability of their precursors, imperative for gut-brain axis signaling and many aspects of the central nervous system (CNS) function (Valles-Colomer et al. 2019). Additionally, this community of organisms within our GI tract detoxifies exogenous chemicals and is itself subject to modification by drugs that are intended to target the host physiology, especially xenobiotics (Clarke et al. 2019). Certain bacterial species influence the metabolism of medications such as digoxin, chemotherapeutic agents, and anti-inflammatory drugs used in systemic illnesses (Collins and Patterson 2020), while others aid in the elimination of toxins such as azo dyes (Cussotto et al. 2019).

Another beneficial effect of the gut microbiota is providing defense against pathogenic species and infections. It does this by supporting the intestinal epithelial barrier via immunological means, as well as interacting synergistically with the immune system on a local and systemic basis (Wiertsema et al. 2021). The microbiota is also thought to play a role in synthesizing certain vitamins, such as some vitamers of vitamin B; however, the exact quantities and contribution as part of daily human intake is unknown (Pham et al. 2021). The ability to biotransform molecules such as bile acids is also an important function of gut microbial

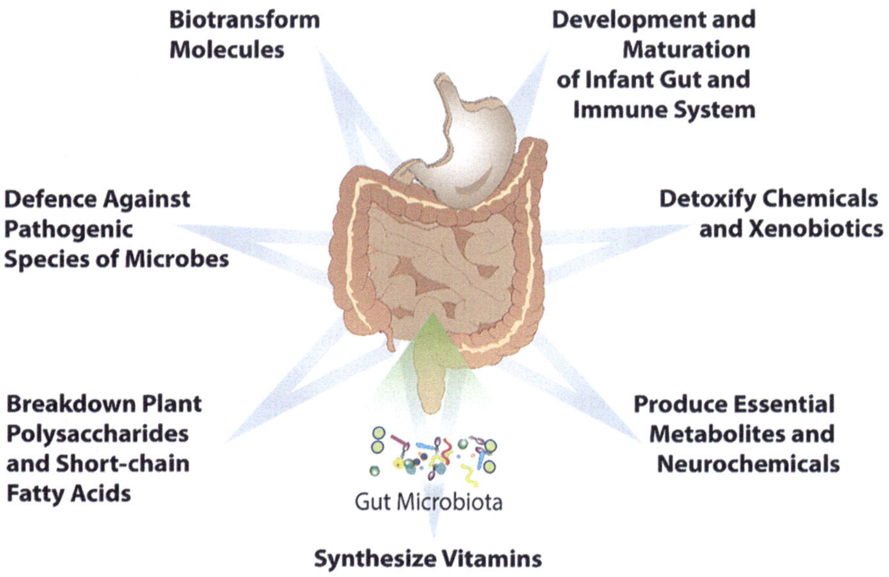

Fig. 1 Salient functions of the gut microbiota. These include synthesizing vitamins, providing defense against pathogenic species of microorganisms, development and maturation of the infant gut immune system, and others indicated in the diagram

organisms. Bile acids are deconjugated after their use in the GI tract by enzymes produced by a broad range of microbes. Unconjugated bile acids have several important biological functions and are under investigation for their potentially neuroprotective properties (Spichak et al. 2021).

Figure 1 above outlines the salient functions of the gut microbiota. It is important to, however, appreciate that aspects of this collection of organisms can be structurally divided into ecological functional units, referred to as guilds (Wu et al. 2021a, b). Guilds are groups of bacteria that portray consistent coabundant behavior and are likely to cooperate toward the same ecological function.

3 Microbiota-Gut-Brain Axis

The microbiota-gut-brain axis includes the CNS, the GI tract, and its associated microorganisms, along with key elements in this communication pathway, such as the hypothalamic-pituitary-adrenal (HPA) axis, the immune system, the enteric nervous system, and the autonomic nervous system (Cryan et al. 2019). There are also humoral routes of communication between the gut and the brain – a route available to bacterial components such as peptidoglycan and lipoteichoic acid and microbial metabolites such as those produced from tryptophan, SCFAs, and others, to name just a few (Generoso et al. 2020). Communication between the brain and gut is bidirectional and has been described as an "ancient, evolutionary conserved, biomolecular language" (Oleskin and Shenderov 2019; Järbrink-Sehgal and Andreasson 2020).

The intestinal barrier is the interface between the GI lumen and the host. This consists of an epithelial layer as well as a mucus gel layer containing antimicrobial peptides and immunoglobulins, with some differences between the small intestine and the large intestine – for example, the latter having a double-layered mucus lining (Kelly et al. 2015; Hansson 2019). Functions of the intestinal barrier include the prevention of pathogens and luminal toxic substances from entering the body, as well as regulation of the absorption of water, nutrients, and electrolytes into the circulation (Kelly et al. 2015). The integrity of the intestinal barrier may be influenced by the gut microbiota, the disruption of which may result in systemic inflammation, including the release of cytokines. Systemically elevated cytokines may in turn relay signals to the brain through blood circulation and via specific transport mechanisms at the blood-brain barrier (Pan et al. 2011). Substances produced by the gut microbiota may also reach the brain in a similar "endocrine" fashion, i.e., via the bloodstream. Gut microorganisms and their metabolites may stimulate vagal afferent neuronal fibers located beneath the intestinal epithelium, resulting in more direct neural signals being transmitted to the CNS (Fülling et al. 2019; Berding et al. 2021).

As alluded to earlier, the axis is bidirectional, and the brain may in turn influence the gut and the gut microbiome by similar "endocrine" as well as neuronal pathways facilitated by means of the vagus nerve. Such influence may also involve higher-order behaviors relating to health and food choices (Järbrink-Sehgal and Andreasson 2020), such as exercise, food cravings, and the binge consumption of alcohol (Alcock et al. 2014; Carbia et al. 2023). Disruptions in brain-gut-microbiota communication appear to influence eating behaviors, which may lead to a preference for consuming high-calorie food, which in turn leads to compositional alterations of the gut microbiota (Gupta et al. 2020).

4 Microbiota-Gut-Brain Axis, Brain Function, and Behavior

Stress has a notable effect on gastrointestinal health as well as the gut microbiota composition (Dinan and Cryan 2012). This indicates the potential for an emotional health component to this "top-down" aspect of communication between the brain and gut.

The microbiota may play a role in mood regulation. Preclinical research has shown that germ-free mice exhibited reduced depressive-like behavior in contrast to mice with a conventional microbiota (Bastiaanssen et al. 2019; Zheng et al. 2016). However, it should be appreciated that there may be difficulty in interpreting these results as germ-free mice are more prone to exhibiting increased locomotion (Heijtz et al. 2011). Additionally, it was found that probiotics and prebiotics (see Sect. 9) reduced depressive-like behavior in rodent models (Bastiaanssen et al. 2019).

However, in terms of clinical research, there have been inconsistent findings in studies investigating the microbiota and depression. This has been the case in studies examining global indices such as microbial diversity (Nikolova et al. 2021; McGuinness et al. 2022). Some studies have failed to show reduced alpha diversity within the gut microbiota in patients with major depressive disorder (MDD) in comparison to healthy controls, while others have reported a reduction in said diversity (Simpson et al. 2021). A decrease in the relative abundance of *Faecalibacterium* in depressed patients has been observed, and its presence was negatively correlated with the severity of the illness (Bastiaanssen et al. 2019). Other studies reported lower relative abundances of *Bifidobacterium* and *Lactobacillus* within the gut microbiota of depressed individuals (Bastiaanssen et al. 2019). Additionally, the genera *Coprococcus* appears to be depleted in patients suffering from depression (Valles-Colomer et al. 2019). On the other hand, some studies have found an association between some types of gut microbial taxa and depressive symptoms (Radjabzadeh et al. 2022; Bosch et al. 2022). These will be discussed later in Sects. 5 and 6.

There is evidence to suggest that the microbiota plays a role in cognitive function. Certain strains of germ-free mice have been shown to display impaired short-term recognition and working memory (Gareau et al. 2010; Luczynski et al. 2016). Further evidence that the microbiota plays a role in cognitive function can be inferred from results from probiotic studies. In a study conducted by Kim and colleagues, it was found that supplementation with a probiotic containing *Bifidobacteria* in older adults resulted in improved cognitive function and mood while at the same time being associated with a reduction of *Eubacterium* and Clostridiales organisms (Kim et al. 2020). This was also correlated with an increase in brain-derived neurotrophic factor (BDNF), a growth factor that has a physiological role in neuronal maturation, synaptic plasticity, and synapse formation and is important in CNS functions such as learning and memory (Björkholm and Monteggia 2016). Deficits in BDNF signaling and activity have been associated with depression (Colucci-D'Amato et al. 2020) (see Sect. 7).

The HPA axis (see Fig. 2) is a key pillar in brain-gut-microbiota communication. Cortisol produced as part of this pathway is thought to contribute to altered gut microbiota composition and gut permeability (Moloney et al. 2015; Misiak et al. 2020). Chronic elevations of cortisol may lead to resistance to cortisol-based immunosuppression and reduced immunosuppressive effects on inflammatory cells. This resistance may result in there being more circulating inflammatory cytokines, with some immune cells remaining activated in the process (Morey et al. 2015). In animal studies, it was found that the species *Enterococcus faecalis* reduced corticosterone levels and restored social activity in germ-free mice following social stress (Wu et al. 2021a, b). Such findings suggest that the microbiota may influence the HPA axis and social behavior via particular neuronal circuits in the brain. Alterations in gut microbiota composition and function may lead to the release of cytokines and other molecules, which can cross the blood-brain barrier (BBB) and activate the

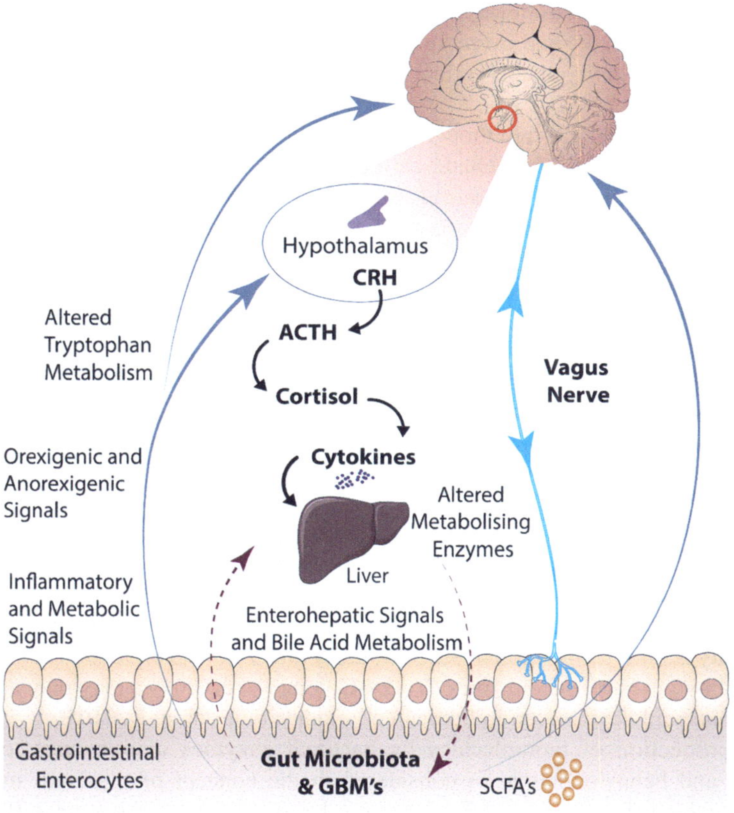

Fig. 2 Some significant signaling pathways of the gut-brain axis and external influencing factors. External factors such as stress and emotional changes are known to influence the brain and CNS function. Stress has the ability to influence the HPA axis and increase the amounts of circulating cortisol and cytokines. Cytokines may influence the hepatic metabolism of substances such as bile acids; the latter are also regulated and metabolized by the gut microbiota. The vagus nerve exerts a more direct bidirectional communication between the brain, gut, and microbiota. Gut microbiota via gut-brain modules regulates the metabolism of important molecules such as tryptophan and SCFAs, which have an impact on the CNS. Tryptophan is a precursor for serotonin and other bioactives, while SCFAs may alter HPA axis activity (see below). Other neuroactive metabolites and inflammatory signals produced by the gut microbiota may exert influence on the hypothalamus, resulting in the manifestation of certain eating behaviors (Gupta et al. 2020). These, along with orexigenic and anorexigenic biomolecules, contribute to eating habits. Factors such as diet, prebiotics, probiotics, and the environment have an impact on the gut microbiota. (Modified from Clarke et al. 2019). *CRH* corticotrophin-releasing hormone, *ACTH* adrenocorticotrophic hormone, *SCFAs* short-chain fatty acids, *GBMs* gut-brain modules

HPA axis. Also, the release of lipopolysaccharides and peptidoglycans can directly stimulate the HPA axis, and even the production of ClpB protein from *Escherichia coli* can have an upregulatory effect on adrenocorticotrophic hormone (ACTH) synthesis (Misiak et al. 2020).

It is here where SCFAs are thought to play a vital role as part of the microbiota-gut-brain axis. SCFAs are aliphatic carboxylic acids such as acetic, butyric, and propionic acids, which are produced in vivo by the fermentation of undigestible carbohydrates by species of bacteria within the gut (O'Riordan et al. 2022). Once absorbed, as demonstrated, they have the ability to reduce HPA axis activity and the overall cortisol response to acute psychosocial stress. This was the case in studies involving healthy males (Dalile et al. 2020). However, this study did not definitively show whether the site of action of the absorbed, colon-delivered SCFAs was within the CNS. SCFAs can limit inflammatory processes within the CNS by crossing the BBB and decreasing microglial activity, as well as downregulating the expression of genes that encode proteins integral to the HPA axis pathway (Misiak et al. 2020; van de Wouw et al. 2018).

Bile acids also participate in microbiota-gut-brain axis communication (see Fig. 2). These compounds are deconjugated by enzymes produced from the gut microbiota. During this process, there is a release of free taurine and glycine, and these, along with unconjugated bile acids, have the potential to function as protective neuroactive ligands (Spichak et al. 2021). Research has also shown that polyunsaturated fatty acids (PUFAs), such as docosahexaenoic acid and eicosapentaenoic acid produced by gut bacterial enzymes (Valles-Colomer et al. 2019), may play a role in driving bile acid degradation in human cell models (Cieślak et al. 2018).

The production of biomolecules by bacterial enzymes of relevance to brain function and behavior has been considered in the concept of gut-brain modules (GBMs), which are integral to microbiota-gut-brain axis communication. GBMs are specific pathways that are associated with brain-gut communication and involve particular microbes (Valles-Colomer et al. 2019; Spichak et al. 2021). Within each pathway, there is either the production or degradation of a specific neurochemical, and this process involves the action of specific enzymes. In order to fulfill the criteria for a GBM, the microbe must possess each enzyme within the specified pathway (Valles-Colomer et al. 2019; Spichak et al. 2021). In this manner, the gut microbiota may be involved in the regulation of certain neurotransmitters or their precursors and other molecules that can have an effect on the CNS. An example of this is the kynurenine pathway, in which metabolism along host pathways may be regulated by the gut microbiota, thus having implications for the availability of circulating tryptophan (Kennedy et al. 2017). There may be direct production of metabolites by the microbiota, or alternatively, the microbiota may influence the way in which, for example, the host processes precursors such as tryptophan. Some metabolites from GBM pathways, such as SCFAs, have the potential to stimulate serotonin production from host enterochromaffin cells by upregulating tryptophan 5-hydroxylase 1, one of the key enzymes involved in the synthesis of serotonin (Reigstad et al. 2015; Silva et al. 2020).

As mentioned earlier, GBMs are not only associated with the production of neurochemicals. Some are linked to degradation, for example, dopamine. GBM analysis may potentially identify the microbial pathways responsible for metabolizing molecules capable of traversing across both the intestinal epithelium and BBB, thus interacting with the CNS (Tian et al. 2021).

5 Updates to the Pathogenesis and Neurobiology of Depression: A Multifactorial Contribution

Clinical depression is a mental illness that features persistently depressed mood, low energy levels, and anhedonia, as well as other symptoms, such as sleep disturbances, loss of libido, appetite changes and cognitive deficits (Semple and Smyth 2009). Traditionally, it was thought that depression was the result of a deficit of catecholamines and monoamine transmitters within the CNS, for example, serotonin and noradrenaline (Schildkraut 1965). Many of the antidepressants that are used in the clinical management of depression were developed with the aim of increasing the levels of monoamines in the CNS (Du et al. 2020). Over the years, however, such a simplistic, pathological model for depression has been replaced by models that use a system and circuit-based approach (Jauhar et al. 2023).

As mentioned previously, stress has the potential to impact the gut-brain axis, and it is also considered a causative factor in many cases of depression. Psychosocial stress impacts the HPA axis, resulting in elevated levels of cortisol, as has been reported in many studies (Mayer et al. 2018). High levels of cortisol are associated with depression (Bertollo et al. 2020), and it is well appreciated that depression and fluctuations in mood are quite common in medical illnesses with high glucocorticoid levels, such as Cushing's disease (Davey 2007; de Kloet et al. 2005; Cruz-Pereira et al. 2020; Thau et al. 2022). Normally, the HPA axis operates with a negative feedback mechanism. Therefore, when activated and the levels of cortisol increase, there is subsequent downregulation of this pathway, which results in a return to homeostasis (Ramamoorthy and Cidlowski 2016; Thau et al. 2022). In illnesses such as Cushing's disease and depression, there is dysfunction of this negative feedback mechanism (de Kloet et al. 2005; Cruz-Pereira et al. 2020).

The amino acid tryptophan is a precursor to serotonin, as well as other bioactives involved in gut-brain axis signaling. Tryptophan changes may also be involved in the depression that accompanies various physical illnesses (Clarke et al. 2009; O'Mahony et al. 2015; Kennedy et al. 2017). For example, the tryptophan-kynurenine pathway appears to be activated in patients suffering from both inflammatory bowel disease and depression (Chen et al. 2021). It is important nonetheless to appreciate that an increased metabolism of tryptophan is not implicated exclusively in the depression associated with general medical conditions.

A more recently combined immunological and metabolic model has been suggested as a contributory factor for depression. This may have an impact on serotonin pathways, thus indirectly supporting earlier models of the pathogenesis of depression. Thus, the activation of the immune system has been associated with the induction of important enzymes such as indoleamine 2,3-dioxygenase (IDO1). This would be expected to increase the metabolism of tryptophan along the kynurenine pathway (Marx et al. 2021). A reduction in the availability of tryptophan could lead to a decrease in serotonin, with reduced plasma tryptophan being described in depression – indeed, several meta-analyses have confirmed this abnormality in unmedicated depressed patients (Jauhar et al. 2023). While acute tryptophan depletion does not cause a significant lowering of mood in healthy individuals (Ruhé et al. 2007), in those who are vulnerable to depression through a previous history of illness, tryptophan depletion results in a clinically significant lowering of mood (Ruhé et al. 2007; Jauhar et al. 2023).

Likewise, glucocorticoids released following a stress response may activate the enzyme tryptophan-2,3-dioxygenase (TDO) in the liver, another enzyme involved in tryptophan metabolism along the kynurenine pathway, again possibly resulting in diminished tryptophan availability (Kennedy et al. 2017). In terms of other metabolic pathways, the gut microbiota may play a role as well (see Sects. 4, 8, and 9).

It is important to appreciate that inflammatory molecules have the potential to influence neurobiological functions such as functional brain activity, perhaps by modifying neurotransmitter breakdown and synthesis (Marx et al. 2021). An example of this has been illustrated by the effects of lipopolysaccharide (LPS), which is a potent neuroinflammatory stimulator. In animal studies, LPS delivered in an intracerebellar fashion to mice resulted in the promotion of local pro-inflammatory molecules and apoptotic cell death of Purkinje cells, along with observed cerebellar dysfunction (Hong et al. 2020). Evidence from a human study suggests that LPS has the potential to activate the kynurenine pathway, resulting in tryptophan degradation (Millischer et al. 2021). This small study was not powered to account for sex differences and did not examine a correlation between tryptophan metabolism and clinical psychiatric features such as a depressed mood or illness behavior. Nevertheless, there is supporting clinical evidence that inflammation plays a role in depression. Certain intravenous cytokines, for example, interferon alpha, are known to cause significant depressive symptomatology (Raison et al. 2006). Additionally, there is evidence that anti-inflammatory agents such as nonsteroidal anti-inflammatory medications, certain cytokine inhibitors, as well as selected antibiotics have shown merit as adjuncts in the treatment of depression (Marx et al. 2021; Nettis et al. 2021).

Nonetheless, it should be noted that levels of other cytokines such as interleukin-6 (IL-6), interferon-gamma (IFN-γ), and other acute-phase proteins, as well as tumor necrosis factor (TNF), are significantly elevated in depressed patients, albeit at low levels (Cruz-Pereira et al. 2020). Additionally, the activation of the peripheral innate and adaptive immune system, especially in stressful situations, as well as the involvement of inflammasomes have been postulated to contribute toward depression. Inflammasomes are complexes of protein nature that originate from bone marrow cell lines in response to the body's infection with pathogenic microbes as well as psychological stress. They are considered to be innate immune sensors that are activated in response to the stimulation of pattern recognition receptors, e.g., Nod-like receptors (NLRs). Such stimulation leads to, for example, the activation of caspase-1, followed by the downstream activation of the IL-1 family of pro-inflammatory cytokines as part of the NLRP3 inflammasome. NLRP3 appears to be overexpressed in the circulating blood cells of patients with major depressive disorder (MDD). Subsequent treatment with antidepressants led to a reversal in this overexpression (Cruz-Pereira et al. 2020).

6 The Microbiome-Gut-Brain Axis and Depression

In a systematic review and meta-analysis, it was found that changes in the gut microbiota, especially a deficit in butyrate-producing species, as well as an increase in proinflammatory bacterial species, were associated with several psychiatric illnesses, such as depression, bipolar affective disorder, and schizophrenia (Nikolova et al. 2021). Apart from lower levels of butyrate-producing microbes being associated with these illnesses, higher rates of lactic-acid-producing bacterial species may also be contributory to these conditions (McGuinness et al. 2022).

Some studies have hinted at the prospect of alternative mechanisms playing a role in the neurobiology of depression. Decreased testosterone levels, such as those seen in clinical hypogonadism, have been linked to major depression (Gheorghe et al. 2022). Testosterone is thought to have an effect on neurotransmitters that are implicated in the pathogenesis of the condition (Gheorghe et al. 2022), e.g., serotonin (Celec et al. 2015) and γ-aminobutyric acid (GABA) (Sivcev et al. 2019). Certain species of bacteria within the gut microbiota, containing the enzyme 3-β-Hydroxysteroid dehydrogenase (3β-HSD), were found to a significant extent in testosterone-deficient patients with depression (Li et al. 2022). 3β-HSD has the ability to degrade testosterone, and isolates of *Mycobacterium neoaurum*, when administered to rodents, resulted in the expression of depression-like behaviors in the said animals (Li et al. 2022; Gheorghe et al. 2022).

Preclinical trials showed that fecal microbiota transplantation (FMT), an approach used to establish potential causality for disease-associated microbiota configurations (Gheorghe et al. 2021), from patients with MDD to microbiota-deficient rats induced depressive-like behaviors in the recipient rodents (Kelly et al. 2016). Some earlier clinical investigations have yielded results suggesting that patients suffering from MDD have altered microbial composition in comparison to nondepressed individuals (Bastiaanssen et al. 2019; Jiang et al. 2015). However, these studies were relatively small in terms of patient numbers, and there have been more recent larger-scale clinical investigations. For example, Radjabzadeh and colleagues found that genera such as *Eggerthella*, *Subdoligranulum*, *Coprococcus*, *Sellimonas*, *Lachnoclostridium*, and others were associated with depressive symptoms (Radjabzadeh et al. 2022). These findings are significant as the implicated taxa were involved in the synthesis of neurotransmitters such as GABA, serotonin, glutamate, and butyrate. Genera such as *Sellimonas* and *Hungatella* were also positively associated with depressive symptoms, and the former has been reported to be increased in inflammatory diseases. Despite this study correlating the taxa with "depressive symptoms" rather than a depressive disorder, it was nonetheless a relatively large study with significant results.

Another group of researchers took a further step to identify whether ethnicity played a role in the gut microbiota and depressive symptoms. They found that within a large and ethnically diverse population, there were strong associations between gut microbiota and depressive symptoms without a specific ethnic variation (Bosch et al. 2022). The families of taxa Christensenellaceae, Lachnospiraceae, and Ruminococcaceae were positively correlated with depressive symptoms. Such studies are important as, firstly, it is essential to ascertain whether demographic factors may have an impact on the gut microbiota, which in turn may have an influence on depression. Further, such findings (i.e., taxa associated with depressive symptoms) allow us potentially to identify targets for psychobiotic interventions (Bosch et al. 2022) (see Sect. 9).

Additional evidence that supports the role of the microbiome in depression includes studies involving antibiotic therapy. Depleting the microbiota with a single antibiotic appears to result in an increased risk of depression as well as anxiety (Lurie et al. 2015). However, one should be cautious when interpreting such results as there may be confounding factors related to antibiotic use. An example of this could be depression arising during the antibiotic treatment of *Clostridioides difficile* infection. The underlying infection may have a psychological impact due to physical illness (Royal College of Psychiatrists 2019), and therefore concomitant symptoms of depression and anxiety might be due to this rather than an effect of antibiotic treatment on the individual's microbiota.

7 Conventional Antidepressant Therapy

Having appreciated the multifactorial neurobiological basis of clinical depression, it should not be a surprise to realize the difficulty in knowing which antidepressant will be effective in treating a particular patient. There is no "one antidepressant treats all" model for managing depression. Indeed, on many occasions, treatment may have to follow a "trial and error" fashion. We may attribute this to the variation of pathological mechanisms between individuals with depression. Moreover, therapeutic options that target cortisol, immunological factors, metabolic pathways, inflammatory processes, and other mechanisms implicated in the pathogenesis of depression are not yet available for licensed treatment. As alluded to previously, potential candidates are still being developed and investigated. Results from a meta-analysis suggested that immunomodulatory agents such as some monoclonal antibodies may improve psychological symptoms of depression associated with an inflammatory medical illness (Wittenberg et al. 2019). This study, as well as others (Pitharouli et al. 2021), anticipates a future possibility of selecting clinically depressed patients with high C-reactive protein (CRP) levels for specific immune-modulating therapies, to alleviate their depressive symptomatology.

Nevertheless, currently in psychiatry, conventional antidepressant medications are the mainstay of treatment for moderate to severe depressive disorders (Cipriani et al. 2018; Taylor et al. 2012). The choice of antidepressants prescribed generally depends on several clinical factors, such as severity, new episodes, recurrent depressive episodes, treatment resistance, side effects, tolerability, medical and psychiatric comorbidities, and also, importantly, informed patient decisions (NICE Guidelines 2022).

Current pharmacological agents include selective serotonin reuptake inhibitors (SSRIs), which are generally considered first-line options (NICE Guidelines 2022; Taylor et al. 2012). Dual reuptake inhibitors of serotonin and noradrenaline, referred to as serotonin noradrenaline reuptake inhibitors (SNRIs), such as venlafaxine and duloxetine can also be considered, as well as some older options such as tricyclic antidepressants (TCAs) and monoamine oxidase inhibitors (MAOIs), the latter usually under specialist psychiatric care. TCAs can be associated with cardiovascular side effects, and toxicity due to overdose may be lethal. Generally, SSRIs are safer but may commonly cause side effects such as headaches and GI symptoms. Other side effects, such as sexual dysfunction, hyponatremia, and GI hemorrhage, may also be problematic (Anglin et al. 2014; Taylor et al. 2012).

Other antidepressants available for treatment include atypical antidepressants such as mirtazapine, bupropion, and agomelatine (Sheffler et al. 2022). Atypical agents have several pharmacodynamic targets, although most still impact monoamine neurotransmitters (Alfaifi and Althemery 2022).

8 Antidepressants and the Microbiota-Gut-Brain Axis

We have already mentioned that the microbiota-gut-brain axis is bidirectional in nature, and certain elements, such as the vagus nerve, facilitate this type of communication between the gut microbiota, the GI system, and the brain. Additionally, the brain may influence the gut microbiome via endocrine and neuronal pathways, as well as through higher-order behaviors, such as choice of diet and exercise (Clauss et al. 2021). In an analogous way, conventional antidepressants too may impact the microbiota-gut-brain axis and vice versa.

However, before we can discuss this, we need to appreciate the effects of antidepressants on the gut microbiota itself. Many medications and xenobiotics, in fact, have an impact on the gut microbiota (Clarke et al. 2019). Overall, studies have shown a strong association between antidepressant administration and changes in gut microbiota composition (Cussotto et al. 2019). The effects of these medications on the gut microbiota, however, vary depending on the antidepressant used. Some antidepressants have been associated with a reduction or increase in composition and relative abundance of certain taxa of organisms. Indeed, a number of antidepressants were initially used as antimicrobial agents (Butler et al. 2019a), but the association between the antimicrobial effects on the gut microbiota and the therapeutic benefits/side effects is unknown. A few antidepressants, on the other hand, may promote the growth of certain strains of organisms. These will be discussed below, and Table 1 summarizes the effects that some antidepressants and agents with antidepressant characteristics have on different microorganisms.

Some in vitro and preclinical animal studies have been conducted regarding TCAs. Amitriptyline was shown to exert antimicrobial activity in vitro against numerous strains of organisms, but most notably *Staphylococcus* spp., *Bacillus* spp., *Vibrio cholerae*, as well as fungal species such as *Cryptococcus* spp. and *Candida albicans*, at varying dosages (Cussotto et al. 2019; Mandal et al. 2010). In animal studies, amitriptyline also significantly protected mice from becoming infected with *Salmonella typhimurium* (Mandal et al. 2010). Imipramine has the ability to inhibit the growth of species such as *E. coli* and *Yersinia enterocolitica*, as well as parasites such as *Giardia lamblia*, along with clomipramine, which has the ability to exert cytotoxic effects against *Leishmania donovani* and *Leishmania major* protozoan parasites (Cussotto et al. 2019).

SSRIs have also been shown to exert a range of antibacterial and antifungal qualities in in vitro studies. Sertraline was found to inhibit the growth of *P. aeruginosa*, *S. aureus*, *E. coli*, as well as *Cryptococcus neoformans* and *Coccidioides immitis* (Cussotto et al. 2019). Both fluoxetine and escitalopram showed the ability to inhibit *E. coli*, while the former inhibits *L. rhamnosus* (Cussotto et al. 2018). In terms of in vivo animal studies, Cussotto and colleagues found that administering fluoxetine to rats resulted in the inhibition of the growth of *Succinivibrio* and *Prevotella* species (Cussotto et al. 2018).

In rats who displayed depressive-like behaviors, a higher relative abundance of Bacteroidetes but a lower relative abundance of Firmicutes occurred with fluoxetine and amitriptyline treatment (Zhang et al. 2021). Other results from animal studies have indicated that fluoxetine may increase bacterial richness (Cussotto et al. 2018).

Ketamine, with the exception of its S-enantiomer esketamine, is not a clinically licensed antidepressant (Royal College of Psychiatrists 2017). However, it is increasingly used in patients with difficult-to-treat depression (Naughton et al. 2014). Antimicrobial properties against *S. aureus*, *S. epidermidis*, *E. faecalis*, *S. pyogenes*, *P. aeruginosa*, and *E. coli* have been demonstrated within in vitro studies. However, very little is known about its effect on the gut microbiota in vivo (Cussotto et al. 2021a). Venlafaxine, an SNRI, did not appear to have an in vitro impact on bacterial strains, while phenelzine, an MAOI, showed antimicrobial activity against *E. rectale* and *F. prausnitzii* (Ait Chait et al. 2020).

As appreciated, there is a significant limitation in our knowledge about the interface between many antidepressants and the gut microbiota – in vitro, they have shown significant antimicrobial properties; however, in vivo, especially in humans, there is still a lack of studies reproducing these findings. Nevertheless, we can consider an example of an antidepressant whose effect on the gut microbiota has been assessed clinically – vortioxetine. In a study that correlated the abundance distribution differences in the gut microbiota, it was found that treatment with vortioxetine was associated with a significant increase in the proportion and amount

Table 1 Effects of some antidepressants and ketamine on different microorganisms

Antidepressant or agent with antidepressant characteristics	Effect on microorganisms	Predominantly targeted neurotransmitter system(s)	Some notable side effects
Amitriptyline	Antimicrobial effects on *Staphylococcus* spp., *Bacillus* spp., *Vibrio cholera*, *Cryptococcus* spp., *Candida albicans* (Cussotto et al. 2019; Mandal et al. 2010), and *Salmonella typhimurium* (Mandal et al. 2010) Possible antimicrobial effects against Firmicutes (Zhang et al. 2021) Possible promotion of Bacteroidetes (Zhang et al. 2021)	Serotonergic and noradrenergic (Frazer 2001; Taylor et al. 2012)	Sedation, cardiovascular side effects (e.g., hypotension, tachycardia, QT_c prolongation), visual changes, constipation, and urinary retention (Taylor et al. 2012)
Imipramine	Antimicrobial effects on *E. coli*, *Yersinia enterocolitica*, *Leishmania donovani*, and *Leishmania major* (Cussotto et al. 2019)	Serotonergic and noradrenergic (Frazer 2001; Taylor et al. 2012; Fayez and Gupta 2021)	As for amitriptyline but less sedative (Taylor et al. 2012; Fayez and Gupta 2021)
Clomipramine	Antimicrobial effects on *Leishmania donovani* and *Leishmania major* (Cussotto et al. 2019)	Serotonergic and noradrenergic (Taylor et al. 2012; Wilson and Tripp 2023)	As for amitriptyline but more toxic in overdose (Taylor et al. 2012)
Fluoxetine	Antimicrobial effects on *E .coli*, *L. rhamnosus*, *Succinivibrio*, and *Prevotella* (Cussotto et al. 2018)	Serotonergic (Taylor et al. 2012; Chu and Wadhwa 2022)	Insomnia, agitation, tremor, nausea, vomiting, abdominal pain, diarrhea, sexual dysfunction hyponatremia, and bleeding disorders (Taylor et al. 2012)
Sertraline	Antimicrobial effects on *P. aeruginosa*, *S. aureus*, *E. coli*, *Cryptococcus neoformans*, and *Coccidioides immitis* (Cussotto et al. 2019)	Serotonergic (Taylor et al. 2012; Chu and Wadhwa 2022)	As for fluoxetine
Escitalopram	Antimicrobial effects on *E.coli* (Cussotto et al. 2018)	Serotonergic (Taylor et al. 2012; Chu and Wadhwa 2022)	As for fluoxetine
Ketamine	Antimicrobial effects on *S. aureus*, *S. epidermidis*,	Glutaminergic and GABAergic (Allen	Dissociative symptoms, anxiety, nausea, increased blood

(continued)

Table 1 (continued)

Antidepressant or agent with antidepressant characteristics	Effect on microorganisms	Predominantly targeted neurotransmitter system(s)	Some notable side effects
	E. faecalis, *S. pyogenes*, *P. aeruginosa*, and *E. coli* (Cussotto et al. 2021a)	et al. 2015; Zanos and Gould 2018)	pressure, headache, and ulcerative cystitis (Taylor et al. 2012; Ceban et al. 2021)
Phenelzine (MAOI)	Antimicrobial effects on *E. rectale* and *F. prausnitzii* (Ait Chait et al. 2020)	Serotonergic, noradrenergic, dopaminergic, and GABAergic (Taylor et al. 2012; Sidhu and Marwaha 2023)	Postural hypotension, dizziness, and hypertensive crisis (e.g., tyramine induced) (Sub Laban and Saadabadi 2022; Taylor et al. 2012)
Vortioxetine	Possible antimicrobial effects on the Prevotellaceae family Possible promotion of *Bifidobacterium*, *Roseburia*, and *Faecalibacterium*	Serotonergic (D'Agostino et al. 2015)	Nausea, vomiting, and constipation (Connolly and Thase 2016)

of *Bifidobacterium*, *Roseburia*, and *Faecalibacterium* (Ye et al. 2021). The increase in *Roseburia* and *Faecalibacterium* was negatively correlated with scores on the Hamilton Rating Scale for Depression (HAM-D) (Carrozzino et al. 2020). In the same study, there was also a decrease in the *Prevotellaceae* family in patients treated with vortioxetine. Results from this study have led to the inference that some antidepressants may have the potential to promote and reduce certain species of organisms, thereby "reconstructing" the gut microbiota.

We have discussed how antidepressants have an effect on the gut microbiota. However, there is a small body of evidence that suggests that the gut microbiota may have an impact on antidepressants and possibly dictate their efficacy. The gut microbiota affects the metabolism and pharmacokinetics of some drugs (Zimmermann et al. 2019). Gut microorganisms play a role in the absorption and metabolism of certain medications (Cussotto et al. 2021a, b), a few of which we have alluded to earlier in the chapter. The gut microbiota may affect the availability and metabolism of antidepressants (Dong et al. 2022). It is thought that prolonged or high dosages of drugs may lead to the accumulation of medication within GI bacteria, thus compromising drug bioavailability. Additionally, there is the possibility for such an accumulation within the organisms to alter the metabolizing potential of the overall microbiota, thereby influencing the effectiveness and side effects of the medication (Klünemann et al. 2021). This is thought to be the case for the antidepressant duloxetine, which may undergo bioaccumulation within certain species of gut bacteria. Though this bioaccumulation may not be toxic to humans, there may be an altered metabolism of the antidepressant, resulting in less bioavailability and hence a lessened therapeutic effect. Regarding SSRIs, changes in the gut microbiota may reduce efficacy as a result of alterations in tryptophan metabolism within the serotonin synthesis pathway (Siopi et al. 2020).

In other scenarios, researchers have studied the effects of the gut microbiota associated with the depressive illness itself, having an impact on the antidepressants used in its treatment. While compositional alterations in the gut microbiota are characteristic of patients with MDD (Radjabzadeh et al. 2022), individuals with MDD who suffered from difficult-to-treat depression had significantly different gut microbiota in comparison to those who responded well to antidepressants (Fontana et al. 2020). In another clinical study, there was evidence to support the suggestion that changes in the gut microbiota may be relevant to antidepressant response, as well as mechanisms that impact the way by which antidepressants work (Dong et al. 2022). Dong and colleagues showed not only that there were significant differences between the gut microbiota of responders compared to nonresponders to antidepressants but also that there were baseline differences between the gut-microbiota-associated metabolic pathways of both groups. This leads us to believe that the mechanisms responsible for effective antidepressant therapy may have some dependence on the gut microbiota.

It should be noted that we have given examples where antidepressants have an impact on the gut microbiota. We have also examined the possibility of the microbiota having an influence on the overall efficacy of antidepressants. However, what currently eludes our knowledge are the downstream effects from both the impacted gut microbiota and antidepressants on the CNS, as well as the CNS on the gut microbiota, as depicted by the question marks in Fig. 3. More multistep, longitudinal, clinical studies involving the gut-brain axis are required.

As discussed above, we are aware that some antidepressants may have antimicrobial properties resulting in a reduction of bacterial species. Currently, we do not know whether such interactions are beneficial or adverse, or may even be associated with side effects. Nevertheless, regarding possible benefits, two points should be made. Firstly, certain organisms affected are pathogenic, such as some strains of *V. cholerae*. Secondly, some strains affected are “pro-inflammatory” (Ye et al. 2021) and may contribute to the inflammatory pathogenesis of depression. Antidepressant effects on other strains may have implications for physical health. Such strains may play a physiological role in, for example, preventing infections, metabolizing drugs, or even producing important nutrients. It should be pointed out that many of the studies investigating the antimicrobial effects of antidepressants were conducted in vitro or in animal models. Further human studies are definitely warranted; nonetheless, clinicians may still need to bear this interaction in mind, while following up-to-date guidelines for treating depression.

Another issue may arise when interpreting results from studies that depict that an antidepressant increases the proportion of a certain species. In fact, there is the possibility that the antidepressant may in fact be responsible for a reduction in some different species, resulting in the thriving of others; therefore, it may be that the antidepressant does not provide a direct “fertilizing” effect on a particular species.

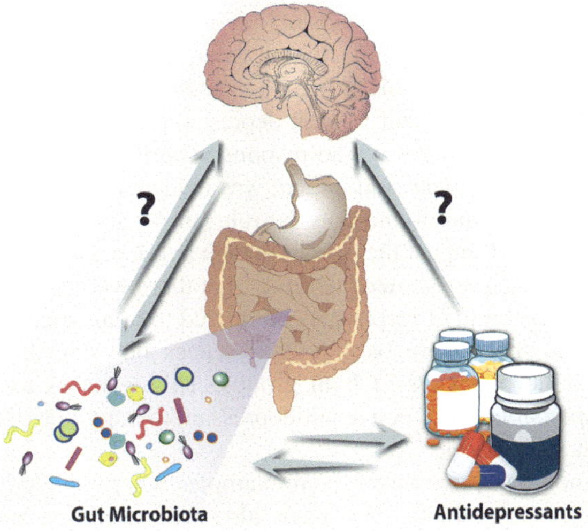

Fig. 3 Scope for research surrounding the CNS effects of antidepressants on gut microbiota and vice versa

Research has indicated that some strains of *Bifidobacteria* may be considered beneficial in modifying stress response and depression (Ye et al. 2021; Dinan et al. 2013). There is evidence to suggest that probiotic preparations containing *B. longum* may have the ability to attenuate HPA axis stress responses (Ye et al. 2021; Messaoudi et al. 2011). Results from studies have also shown that such probiotic strains, for example, *B. longum* 1714, may be associated with not only reduced stress but also improved memory (Allen et al. 2016; Wang et al. 2019a, b). Nonetheless, in a subsequent study, despite being associated with an overall improvement in sleep duration and quality, the same strain failed to show any significant efficacious association with improved cognitive domains such as working memory, visual memory, sustained attention, or perception (Moloney et al. 2021). Since it was found that this type of bacteria was promoted during treatment with vortioxetine, there may be an argument for us to focus more attention on the impact and role of psychobiotics, especially probiotics and prebiotics, as adjunct treatments to antidepressants. This will be covered in Sect. 9 and may ultimately help address an overreliance on pharmacological antidepressants as the main treatment modality for depression.

9 Prebiotics and Probiotics

When it comes to our overall health, one may be familiar with the terms "prebiotics" and "probiotics." More recently, these have been grouped under the category of "psychobiotics," which are exogenous factors that target the gut microbiota resulting in benefits to mental health (Dinan et al. 2021; Berding et al. 2021).

Many of us are aware that particular diets and food improve our mood, a concept that has been developed in the field of nutritional psychiatry (Parletta et al. 2017; Jacka et al. 2017). Whole-food diets, individual food, as well as supplements may play a role in maintaining and improving mood as well as preventing depression.

Jacka and colleagues were able to show that a modified Mediterranean-style diet has the potential to facilitate remission in populations suffering from moderate to severe depression (Jacka et al. 2017). The Mediterranean diet facilitated an improvement in depressive symptoms in young males diagnosed with moderate to severe depression, in contrast to supportive therapy such as befriending therapy (Bayes et al. 2022). Supplementation of a Mediterranean diet with PUFA-rich fish oils was also shown to have improved the mental health of those diagnosed with depression (Parletta et al. 2017). Additionally, certain nuts containing vitamins, PUFAs, and polyphenols are thought to be protective against mood disorders, as well as nutrients such as magnesium (Arab et al. 2019).

Currently, research is being conducted to ascertain which specific food and diet may produce beneficial effects on mood. These include whole-dietary approaches as well as individual food supplements and components. Prebiotics and probiotics are also being investigated in this context.

It is important to appreciate that prebiotics and probiotics purvey benefits to other aspects of health and are thus not exclusively "psychobiotics." Before we can begin to understand mechanisms, it is important to define the terms. A prebiotic is a substrate that is selectively utilized by host microorganisms, resulting in a health benefit, as per the International Scientific Association for Probiotics and Prebiotics (ISAPP) (ISAPP 2022).

Examples of prebiotics include some subsets of dietary fibers as well as some nonfiber substances such as flavanols (Davani-Davari et al. 2019). Probiotics, on the other hand, are "live microorganisms that when administered in adequate amounts, confer a health benefit on the host" (ISAPP 2022). The ISAPP stipulates that only characterized strains of microbes with a scientifically demonstrated effect on health should be called probiotics. Examples of these include *Lacticaseibacillus paracasei* subsp. *paracasei* (Poon et al. 2020) and *B. longum* (Haas et al. 2020). Possible health benefits associated with the former include contributing to the modulation of immune responses (Poon et al. 2020; Belkacem et al. 2017), antioxidant activity, and the scavenging of free radicals (Bengoa et al. 2021; Liu et al. 2011) and apoptotic activity against certain colorectal malignant cells (Bengoa et al. 2021; Song et al. 2021). In vitro, *Lacticaseibacillus paracasei* subsp. *paracasei* was shown to prevent *E. coli* invasion of epithelial cells in patients with inflammatory bowel

Table 2 Some prebiotics and their general properties and associated benefits

Prebiotics	Some general properties and associated benefits
Fructans and inulin	Laxative properties, vitamin and antioxidant production, contribute toward increased GI calcium absorption and bone health (Anderson et al. 2009; Slavin 2013; Whisner and Castillo 2017)
Galactooligosaccharides	Contributes toward increased GI calcium absorption and bone health (Whisner and Castillo 2017)
Xylitol	Antimicrobial, prevention of dental caries and would infections (Salli et al. 2019)
Bimuno-galactooligosaccharides (B-GOS)	Reduced waking cortisol response, possible anxiolytic and anti-depressant effects (Sarkar et al. 2016; Schmidt et al. 2015; Mannie et al. 2007)
Trans-galactooligosaccharides (T-GOS)	Possible improved anxiety and depressive symptoms in patients with irritable bowel syndrome (Peirce and Alviña 2019; Silk et al. 2009)

disease, promote antimicrobial activity, and possibly attenuate chemically induced intestinal injury (Poon et al. 2020; Llopis et al. 2005).

Some *B. longum* strains, e.g., BB536, are considered probiotic and have been reported to provide beneficial physiological effects such as anti-allergic actions, improvement in bowel transit times, as well as an overall benefit in the intestinal environment and stool characteristics (Sugahara et al. 2015). However, we do need to exercise some caution in evaluating these benefits as precise molecular mechanisms in humans remain unidentified. Such a caveat is also important from the mixed results from the studies discussed earlier in Sect. 8, pertaining to benefits to cognitive health. Also, as discussed earlier, strains of *B. longum*, such as 1714, appear to be associated with reduced stress and improved sleep (Moloney et al. 2021).

Prebiotics have a wide range of general health benefits. Firstly, they function as a carbon source for the growth of beneficial bacteria such as *Bifidobacteria* and *Lactobacilli*, resulting in their expansion within the GI tract (Carlson et al. 2018). The benefits of these taxa will be discussed in greater detail below. Additionally, prebiotics allow for an increase in the production of beneficial metabolites and aid in the increased absorption of calcium (Carlson et al. 2018).

Prebiotics such as fructans and galactooligosaccharides have been well studied in terms of their beneficial effects on bone health. A mechanism has been suggested whereby prebiotics may be fermented by strains of colonic bacteria, thereby producing SCFAs, which may in turn allow for the increased expression of calcium transport proteins such as calbindin and thus increased intestinal absorption of calcium (Whisner and Castillo 2017). This may contribute toward bone health and turnover, as alluded to in Table 2. Prebiotics may facilitate a reduction in the presence of pathogenic strains of gut microorganisms, improve immune system function, reduce the individual's risk of developing allergies, and assist in the modulation of gut barrier permeability (Carlson et al. 2018; Slavin 2013). They have the potential to reduce the prevalence and duration of GI symptoms such as diarrhea, as well as reduce the risk of colon cancers (Pandey et al. 2015; Anderson et al. 2009). By promoting the growth of *Bifidobacteria*, prebiotics such as inulin may indirectly allow for the production of vitamins and antioxidants as well as treat and prevent constipation while aiding in the bulking of stools (Anderson et al. 2009; Slavin 2013). Xylitol, another prebiotic, has been associated with a reduction in dental caries and decreased wound infection. Also, its prebiotic characteristics may lead to a reduction in pathogenic gram-negative bacteria as well as yeasts (Salli et al. 2019).

In infants, prebiotics have been demonstrated to be beneficial in several ways, such as preventing sepsis from necrotizing enterocolitis, as well as making defecation frequency and stool consistency in formula-fed babies similar to that of their breast-fed counterparts (Bertelsen et al. 2016).

When we review the effects of prebiotics on the gut microbiota, we note the potential for there to be stimulating effects on the growth of different taxa of organisms. We have already alluded to *Bifidobacteria* and *Lactobacilli*; however, prebiotics may promote other SCFA-producing species, such as *Faecalibacterium prausnitzii*, which is known to produce butyrate (Hadinia et al. 2022; Tsukuda et al. 2021; Holscher 2017). Such SCFAs in animal studies have displayed antidepressant and neuroprotective properties (Bambury et al. 2018). In general, SCFAs are protective against stress-induced disorders that encompass depressive-like behaviors. They have produced an overall reduction in HPA axis activity during acute psychological stress in healthy males (Dalile et al. 2020). Despite this, more clinical research is required into the benefit of SCFAs on mood.

However, the facilitation of SCFA production is not the only proposed mechanism by which prebiotics influence neurobiological mechanisms via the gut microbiota. Evidence exists to suggest that these substrates may be associated with BDNF elaboration, the importance of which to antidepressant action was discussed earlier. Prebiotics such as some fructooligosaccharides (FOS) and glucooligosaccharides (GOS) resulted in the increased expression of BDNF in the hippocampus of adult rats (Sarkar et al. 2016). This was also apparent with neonatal rats (Williams et al. 2016). The administration of FOS and GOS prebiotics also increased the expression of N-methyl-D-aspartate (NMDA) receptor subunits in rodents (Sarkar et al. 2016; Li and Tsien 2009). Additionally, both prebiotics, when administered in chronically stressed mice, were associated with alleviation in anxiety and depressive-like behaviors (Suganya and Koo 2020; Burokas et al. 2017).

Pertaining to human studies, Bimuno-GOS (B-GOS) resulted in a reduced waking cortisol response (Sarkar et al. 2016; Schmidt et al. 2015), an interesting finding as an exaggerated waking cortisol response may be associated with emotional disturbances and depression (Mannie et al. 2007). A 12-week course of supplementation with the prebiotic trans-GOS resulted in patients with irritable bowel syndrome (IBS) having improved anxiety and global assessment scores, obtained by validated rating scales, in comparison to those given placebo (Peirce and Alviña 2019; Silk et al. 2009). This evidence suggests that prebiotics may have the potential to improve anxiety and depression in humans.

Regarding the effects of prebiotics on CNS function, there may be other neurobiological mechanisms; however, research is still mostly limited to animal models. It was found that the administration of prebiotics to transgenic Alzheimer's disease mice resulted in improved cognitive performance, as well as decreased number of postmortem hippocampal amyloid plaques in comparison to untreated animals (Pluta et al. 2020). Prebiotics have also been reported to improve overall brain function and prevent dementias such as Parkinson's disease, as well as autistic spectrum disorders (Suganya and Koo 2020; Li et al. 2020; Siniscalco et al. 2018). The consumption of oligofructose-enriched inulin was associated with improved cognitive function in terms of episodic memory, and participants also reported improved mood and overall positive emotions after the intervention (Smith et al. 2015; Marx et al. 2020). There is much more scope for further research in human and clinical studies in terms of investigating the effects of prebiotics on the gut microbiota and the accompanying mental health benefits. Nonetheless, we should acknowledge the progress that has been made with preclinical and animal studies.

Probiotics, like prebiotics, also have an assortment of health benefits. Common probiotic strains consist of those that are part of the *Bifidobacterium* and *Lactobacillus* families (Sarkar et al. 2016). However, other strains have been mentioned earlier. Many probiotics will have similar benefits as those of prebiotics because several are functional live microorganisms that can potentially utilize prebiotic substrates. Probiotics may prevent antibiotic-associated diarrhea in both children and adults as well as necrotizing enterocolitis in preterm infants (Bertelsen et al. 2016). Their effects have been well studied and established in the prevention and reduction of allergies and atopic conditions in infants, improvement in mouth flora, and the prevention of caries, as well as improvement in the symptoms of lactose intolerance and autoimmune diseases (de Vrese and Schrezenmeir 2008).

Clinical research has shown that certain probiotics may lead to the remission of ulcerative colitis but not Crohn's disease (Markowiak and Śliżewska 2017). Though the thought of a possible role of probiotics in the elimination of malignant cells has been entertained and evidence from animal studies suggests a decreased risk of colorectal cancer, results from human trials have not proven this (Markowiak and Śliżewska 2017). It is important to appreciate that despite promising earlier preclinical research, some expected benefits of probiotics have failed to be translated clinically. This has been the case in cardiovascular disease (Tang et al. 2019) as well as in healthy individuals (Khalesi et al. 2019). Some experts in the field have even described the benefits of probiotics in healthy individuals to be akin to multivitamins (Khalesi et al. 2019), but evidence from research that has been conducted regarding the potential antidepressant benefit of specific probiotic strains and cocktails, as well as CNS function via the microbiota-gut-brain axis, does offer some important insights.

The potential antidepressant and other CNS benefits of probiotics have been well studied. Many studies have examined the effects of probiotics on mood exclusively, while others have taken a step further to investigate the associated microbiota changes, suggesting correlations with changes in the microbiota-gut-brain axis.

Evidence of this, as well as the therapeutic implications on depression, will be discussed toward the end of the chapter.

In animal studies, probiotics have displayed the ability to reduce or normalize depression-like behaviors in rodent models (Suganya and Koo 2020; Abildgaard et al. 2017). Specific strains, such as *B. breve* CCFM1025, upon administration to chronically stressed mice, have reversed depression and anxiety-like behaviors while influencing the gut microbiota of the animals (Tian et al. 2020). *Lactobacillus rhamnosus* JB-1 probiotic strain was reported to reduce depressive behaviors, as well as cause a blunted corticosterone response to stress and an altered expression of GABA receptors in key brain regions (Sarkar et al. 2016; Bravo et al. 2011). Unfortunately, there were no significant translated results when *Lactobacillus rhamnosus* JB-1 was investigated in human studies (Kelly et al. 2017).

When tested in rats, the effects of other probiotics, such as *Lactobacillus helveticus* NS8, were comparable to that of the positive control antidepressant citalopram and resulted in decreased levels of circulating corticosterone and adrenocorticotropic hormones, as well as increased hippocampal levels of serotonin, noradrenaline, and BDNF (Liang et al. 2015). In terms of beneficial cognitive effects, improvement in the brain function of mice was observed in animals administered with probiotics such as *L. plantarum* PS128 (Rezaei Asl et al. 2019).

Some human studies investigating the neurobiological benefits of probiotics have yielded promising results. The consumption of *Lactobacillus casei* strain Shirota (*Lc*S) in patients with chronic fatigue syndrome led to decreased anxiety, despite no significant difference in depressive symptoms (as indicated on the Beck Depression Scale) when compared to placebo (Rao et al. 2009). *Lc*S has also been associated with the prevention of somatic symptoms associated with stress and anxiousness around the time of a "naturalistic" stressful event in healthy individuals (Kato-Kataoka et al. 2016). In this study, subjects in the intervention group had a decreased cortisol response and increased stool serotonin concentrations in comparison to the placebo group. One should, however, be cautious when interpreting the results of these studies as they were relatively small. Rao and colleagues reported microbiota changes in their study. They also acknowledged several other studies depicting similar neurobiological benefits upon consuming LcS; however, the latter failed to show any significant effect on the gut microbiota. Also, we should appreciate that despite a significant overlap and comorbidity with each other, anxiety and depression are two differently classified mental health conditions in terms of both their psychopathology and the aspects of their management (Tiller 2013). Recent literature also suggests that specific anxiety disorders might have distinct microbiota-gut-brain axis signatures (Butler et al. 2023).

Other clinical studies have involved different probiotics. *B. breve* strain A-1 appeared to facilitate an improvement in depression and anxiety in patients with schizophrenia, along with an associated increase in the relative abundance of *Parabacteroides* within the gut microbiota of responders (Okubo et al. 2019). Apart from the small sample size, other caveats to consider here include the fact that the study was a single-arm, open-label trial, which could not exclude placebo factors.

Apart from single-strain probiotics, combinations have also been investigated for their benefits on mood. Combined *Lactobacillus helveticus* R0052 and *Bifidobacterium longum*, when given to healthy male and female volunteers, was associated with an overall decrease in self-reported negative mood and distress while at the same time being correlated with decreased urinary cortisol (Sarkar et al. 2016; Messaoudi et al. 2011). Further work involving multispecies probiotics has been undertaken and has shown a reduced cognitive reactivity to sadness when consumed (Peirce and Alviña 2019; Steenbergen et al. 2015). Probiotics may also be efficacious in improving symptoms of depression while causing minimal side effects (Wallace and Milev 2017; Huang et al. 2016) and also being acceptable and tolerable as adjunctive treatment in adults with MDD (Nikolova et al. 2023). They not only have the potential to interact with the gut microbiotia but may also have an impact on the metabolism of precursors to some neurochemicals, for example, tryptophan (Tian et al. 2022). Hence, consuming strains such as *Bifidobacterium breve* CCFM1025 may aid in the regulation of the serotonergic system, conveying antidepressant benefits.

Some probiotic strains were associated with an improvement in cognitive function in humans, especially when investigated in certain populations. In a cohort of stressed adults, *Lactobacillus plantarum* P8 was associated with an improvement in stress, accompanied by improved memory and cognition (Lew et al. 2019; Marx et al. 2020). In patients with MDD, *Lactobacillus Plantarum* 299v, in combination with SSRIs, facilitated a decrease in kynurenine concentrations as well as improved cognitive function (Rudzki et al. 2019; Marx et al. 2020). Researchers believe that lower kynurenine levels may help improve cognitive function in these patients (Rudzki et al. 2019).

Not all the findings from research regarding the mental health benefits of both prebiotics and probiotics have been favorable. Some studies, such as meta-analyses of randomized controlled trials (RCTs), have indicated that there have been no remarkable effects of prebiotics in relieving symptoms of depression and anxiety. In contrast, probiotics produced significant but small improvements in symptomatology (Liu et al. 2018, 2019). Additionally, other placebo-controlled studies failed to show significant neurobiological changes with prebiotic interventions despite improvement in depressive symptoms in patients with MDD (Kazemi et al. 2019; Purton et al. 2021). Also, as alluded to earlier, *Lactobacillus rhamnosus* JB-1 did not display significant anti-inflammatory effects, improvement in cognitive function, or modification of HPA response, in comparison to placebo (Kelly et al. 2017; Marx et al. 2020). Nonetheless, there is sufficient evidence to warrant further research as we contemplate the role of prebiotics and probiotics in adjunctive antidepressant therapy.

There is also the concept of "synbiotics," which was advanced about 25 years ago. According to ISAPP, in 2019, the definition was updated to "a mixture comprising live microorganisms and substrate(s) selectively utilized by host microorganisms that confers a health benefit on the host" (Swanson et al. 2020). Essentially, by operating in synergy, there should be the potential for live microorganisms and substrates to provide more benefits in terms of improving mental health and preventing and alleviating depression.

Consuming simple breakfast food, such as some yogurts (containing probiotics) and fruits together, can represent a form of synbiotics as both yogurt and fruit allow for the survival of probiotics through the GI tract (Fernandez and Marette 2017). Additionally, both provide selective substrates to enhance probiotic proliferation as well as the modulation of microbiota (Fernandez and Marette 2017). Overall, in trials, synbiotics have shown some promise regarding mental health benefits. In patients with hepatic encephalopathy, there was an improvement in cognitive function upon the administration of Synbiotic Forte™ (Suganya and Koo 2020; Vidot et al. 2019). Synbiotic Forte™ is a proprietary combination of *Lactobacillus paracasei subsp. paracasei*, *Lactobacillus plantarum*, *Leuconostoc mesenteroides*, and *Pediococcus pentosaceus*, together with four fibers – oat bran, pectin, resistant starch, and inulin. Importantly, synbiotics have been shown to improve depressive symptoms and increase serum BDNF levels in hemodialysis patients after 12 weeks of administration compared to placebo (Haghighat et al. 2021; Suganya and Koo 2020). A systematic review of interventional trials reported that synbiotic supplementation provided modest benefits in terms of reducing symptoms of depression in comparison to placebo (Alli et al. 2022).

Additionally, some evidence suggests the potential use of synbiotics as an adjunct treatment to antidepressants. Fluoxetine in combination with synbiotic capsules resulted in a significant reduction in HAM-D scores, especially toward the final 3 weeks of a 6-week trial in comparison to a fluoxetine and placebo combination (Ghorbani et al. 2018; Smith et al. 2021). This shows that synbiotics may have a possible adjuvant therapeutic role with fluoxetine (Smith et al. 2021). In addition to antidepressant effects, synbiotics also have the potential to play important roles in cognitive function as they may regulate neurotransmitters and proteins such as GABA, BDNF, and glutamate (Li et al. 2020).

10 Future Directions

When it comes to understanding the pathogenesis and neurobiological basis of depression, as well as investigating the mental health benefits of psychobiotics, much exploratory and basic research has been conducted over the previous decades. We have indeed come a long way, but much more work is yet to be done.

In terms of future directions, a strategy that may allow us to comprehend the precise role of the gut microbiota in depression is that of fecal microbiota transplantation (FMT). As mentioned previously, it has been demonstrated that fecal transplantation from depressed donor patients to microbiota-deficient recipient rats induced psychological changes and depressive-like behaviors in the rodents (Kelly et al. 2016). The concept of FMT is also being studied in patient populations, especially those diagnosed with irritable bowel syndrome (IBS). There appears to be significant improvement in symptoms in some studies; however, more of this type of research is required (El-Salhy et al. 2022a, b). Some phase 1 studies have shown that FMT delivered via enema can be well tolerated and safe in patients with MDD (Green et al. 2023b). This appears to be the way forward in terms of the choice of intraluminal GI delivery of FMTs. Green and colleagues, from their pilot study, have suggested methods for conducting RCTs, investigating future clinical efficacy and ongoing safety/tolerability of FMT enemas in patients with MDD (Green et al. 2023a).

Despite this clinical research, it is important to appreciate that FMT is currently only approved for the treatment of recurrent *Clostridioides difficile* infection (NICE 2022). We should also bear in mind that FMT is not side effect free and that there are reports of adverse events ranging from common abdominal discomfort and bloating to more serious issues such as infection and sepsis (Wang et al. 2019a, b). As per the US National Library of Medicine's "ClinicalTrials.gov" website, there are at least six active or planned trials involving FMT interventions being investigated in depressive disorders or illnesses in which depression is a comorbidity (US National Library of Medicine 2023). This is depicted in Table 3 below.

Of course, the potential therapeutic benefits of FMT need to be balanced by careful safety assessments. It will be essential to be vigilant and examine some potential pitfalls that may result in more harm than good. An example of this is whether depressive phenotypes have the ability to be transferred post-FMT as certain microbial strains not known as yet to us, associated with depression, may inadvertently be transferred during the process. Such considerations will help add to the already established guidelines for FMT donors (Cammarota et al. 2017). Results from some studies have already suggested the transmissibility of strains in individuals cohabiting and in close social contact for extended periods of time (Valles-Colomer et al. 2023). Examples include mothers and infants, as well as cohabiting individuals in close physical contact, who appear to share significant proportions of strains of their gut microbiota. Also, it was apparent that twins also experienced a decrease in their sharing of common strains after living approximately 3 decades apart (Valles-Colomer et al. 2023).

An aspect that we may need to pay attention to in the future is the cognitive component of depression, which currently may not be well served by conventional pharmacological antidepressants because most of these agents were developed or evaluated for their ability to improve cognitive deficits (Pan et al. 2018). We have briefly mentioned this earlier when highlighting the cognitive effects of some probiotics in rodent and human studies. However, this feature of depression is frequently neglected as a result of masking by the overlap of other depressive symptoms (Richardson and Adams 2018). As a matter of fact, a wide range of cognitive processes, including attention, memory, and executive function, are affected in conditions such as MDD (Woo et al. 2016). With novel candidates for adjunct antidepressant therapies, such as psychobiotics, it would be prudent to assess whether such entities have the potential to benefit cognitive function. We also know that cognitive therapies, such as mindfulness-based cognitive therapy (MBCT), have shown promise in preventing relapses in depression (Kuyken et al. 2016). Clinical studies have indeed shown that MBCT was significantly associated with an

Table 3 Registered FMT studies involving patients with depressive disorders or illnesses in which depression is a comorbidity (excerpt from https://clinicaltrials.gov/). Searched March 28, 2023

Study title	Date commenced (estimated)	Status	Condition	Interventions
The safety and efficacy of fecal microbiota transplantation in a population with major depressive disorder	March 4, 2021	Recruiting	• Depression • Treatment-resistant depression	• Biological: FMT oral capsules • Biological: placebo capsules
Safety and efficacy of fecal microbiota transplantation in a population with bipolar disorder	January 1, 2018	Active, not recruiting	• Bipolar depression	• Biological: allogenic FMT • Biological: autologous FMT
Fecal microbiota transplantation	December 1, 2019	Unknown	• Depression in schizophrenia	• Drug: experimental drug
Clinical study on fecal microbiota transplantation in the treatment of ulcerative colitis with depression	September 10, 2021	Recruiting	• Ulcerative colitis	• Drug: intestinal flora capsule • Procedure: placebo capsule
FMT treating constipation patients with depression and/or anxiety symptoms – clinical efficacy and potential mechanisms	July 30, 2017	Unknown	• Constipation – functional • Depressive symptoms • Anxiety symptoms • Gut-brain disorders	• Others: FMT
Neurocognitive effects of FMT in MDD patients with and without IBS	April 6, 2022	Recruiting	• Major depressive disorder • Irritable bowel syndrome	• Biological: fecal microbiota transplantation

increased gut microbiota diversity posttherapy for anxiety (Wang et al. 2022). Further work needs to be done to investigate similar effects of MBCT and other psychological therapies impacting the gut microbiota in patients with depression.

When it comes to the concept of antibiotics being investigated for their potential therapeutic role in mental illness via modulating the microbiota-gut-brain axis, results have been mixed. Some studies have depicted that early-life antibiotic exposure has been associated with an increased risk of psychiatric disorders, whereas the use of some antibiotics that were nonabsorbable was associated with improvement in symptoms of Autism (Karakan et al. 2021). As mentioned earlier, certain antibiotics may be useful in the treatment of depression, and an example that has been studied clinically and has yielded positive results is minocycline (Marx et al. 2021; Dean et al. 2012, 2017; Nettis et al. 2021). However, despite its amelioration of depressive symptoms in patients, some questions remain as to its mechanism of action – does it have an effect on the gut microbiota? Or does it act in a nonmicrobial fashion? In terms of future research pertaining to the impact of antibiotics on the microbiota-gut-brain axis for antidepressant effects, it would make sense to investigate the effects of nonabsorbable antibiotics, e.g., vancomycin, as indicated by Karakan and colleagues (Karakan et al. 2021). This would help to control the nonmicrobial effects, if any, of systemically absorbed antibiotics.

Owing to synbiotics appearing to have both features of prebiotics and probiotics, one may suggest that they could have better efficacy and are therefore worth investing more in terms of their research and development. The current status is that there are not enough studies investigating the efficacy of synbiotics independently on the gut microbiota as well as examining depressive symptoms. Also, there is a general deficit of studies comparing antidepressant and gut microbial effects between synbiotics, probiotics, and prebiotics. Additional RCTs and longitudinal studies are warranted in the future involving these three types of psychobiotics. In the development of synbiotics, other factors may need to be considered, such as differences in carbohydrate utilization patterns that vary with strains and species (Pandey et al. 2015).

Targeting specific colonic organisms for health benefit purposes would be the way to proceed in terms of future research; however, we should not forget that there may be merit in targeting individual fibers as prebiotics. The latter has the potential to mediate health benefits via the gut microbiota, and results from a cross-over trial have shown that individual purified fibers preferentially support specific microbiomes in a personalized manner (Lancaster et al. 2022). Though Lancaster and colleagues focused on cholesterol-lowering benefits, the underlying principle of a personalized response to fiber may be examined in terms of depression and mood improvement. More research examining this can be planned for the future.

In terms of probiotic research, future directions could include developing genetically modified microbes as next-generation probiotics, or what is referred to as "smart" bacteria (Sirisinha 2016). There may be genetically modified bacteria that have the ability to synthesize and secrete, for example, neurochemicals, as well as regulate mood and behavior (Sirisinha 2016). Already, some studies have shown that a relatively high proportion of the human circulating metabolome is associated with the gut microbiota (Diener et al. 2022). Nevertheless, some concerns exist with this type of biotechnology, especially when there could be the risk of bacterial overgrowth and unpredictable adverse effects. However, it is suggested that this may be countered with mechanisms such as "kill switch" genes (Sirisinha 2016). This would be possible with future advanced genetic engineering as we strive to develop therapeutic adjuncts for antidepressant therapy.

In conclusion, there is mounting evidence to suggest that the pathogenesis of depression, and its neurobiological basis, is multifactorial in origin. Recent evidence implicates the microbiota-gut-brain axis in depression. While novel antidepressant agents are being developed on an ongoing basis, some of these may have the potential to target the gut microbiota and vice versa. However, at the same time, there is the potential for the development of more effective psychobiotics such as specific whole-diet (Berding et al. 2023) prebiotics, probiotics, and synbiotics as adjunct antidepressant treatment options (Long-Smith et al. 2020). Future research will involve targeting specific microorganisms and studying the effects of specific prebiotic fibers in an individualized manner. Finally, more research will have to be undertaken to elucidate whether particular strains of gut organisms are linked to depression. Lastly, we ought not to forget the role that cognition plays in depression, and we should focus on the "top-down" effect that certain therapies, such as CBT and MBT, may have on the gut microbiota. This could potentially uncover a psychotherapy-induced remodeling of the gut microbiota and/or psychotherapy-responsive microbial taxa or guilds with implications for brain-gut-axis signaling pathways.

Acknowledgements The authors would like to thank Dr. Kenneth O'Riordan for his assistance in contributing to this chapter.

References

Abildgaard A, Elfving B, Hokland M, Wegener G, Lund S (2017) Probiotic treatment reduces depressive-like behaviour in rats independently of diet. Psychoneuroendocrinology 79:40–48. https://doi.org/10.1016/j.psyneuen.2017.02.014

Ait Chait Y, Mottawea W, Tompkins TA, Hammami R (2020) Unravelling the antimicrobial action of antidepressants on gut commensal microbes. Sci Rep 10(1). https://doi.org/10.1038/s41598-020-74934-9

Alcock J, Maley CC, Aktipis CA (2014) Is eating behavior manipulated by the gastrointestinal microbiota? Evolutionary pressures and potential mechanisms. Bioessays 36(10):940–949. https://doi.org/10.1002/bies.201400071

Alfaifi AA, Althemery AU (2022) Sociodemographic characteristics and health-related quality of life of individuals undergoing antidepressant therapy. Sci Rep 12(1). https://doi.org/10.1038/s41598-022-22164-6

Allen AP, Naughton M, Dowling J, Walsh A, Ismail F, Shorten G, Scott L, McLoughlin DM, Cryan JF, Dinan TG, Clarke G (2015) Serum BDNF as a peripheral biomarker of treatment-resistant depression and the rapid antidepressant response: a comparison of ketamine and ECT. J Affect Disord 186:306–311. https://doi.org/10.1016/j.jad.2015.06.033

Allen AP, Hutch W, Borre YE, Kennedy PJ, Temko A, Boylan G, Murphy E, Cryan JF, Dinan TG, Clarke G (2016) Bifidobacterium longum 1714 as a translational psychobiotic: modulation of stress, electrophysiology and neurocognition in healthy volunteers. Transl Psychiatry 6(11): e939. https://doi.org/10.1038/tp.2016.191

Alli SR, Gorbovskaya I, Liu JCW, Kolla NJ, Brown L, Müller DJ (2022) The gut microbiome in depression and potential benefit of prebiotics, probiotics and synbiotics: a systematic review of clinical trials and observational studies. Int J Mol Sci 23(9):4494. https://doi.org/10.3390/ijms23094494

Anderson JW, Baird P, Davis RH Jr, Ferreri S, Knudtson M, Koraym A, Waters V, Williams CL (2009) Health benefits of dietary fiber. Nutr Rev 67(4):188–205. https://doi.org/10.1111/j.1753-4887.2009.00189.x

Anglin R, Yuan Y, Moayyedi P, Tse F, Armstrong D, Leontiadis GI (2014) Risk of upper gastrointestinal bleeding with selective serotonin reuptake inhibitors with or without concurrent nonsteroidal anti-inflammatory use: a systematic review and meta-analysis. Am J Gastroenterol 109(6):811–819. https://doi.org/10.1038/ajg.2014.82

Arab A, Mehrabani S, Moradi S, Amani R (2019) The association between diet and mood: a systematic review of current literature. Psychiatry Res 271:428–437. https://doi.org/10.1016/j.psychres.2018.12.014

Bambury A, Sandhu K, Cryan JF, Dinan TG (2018) Finding the needle in the haystack: systematic identification of psychobiotics. Br J Pharmacol 175(24):4430–4438. https://doi.org/10.1111/bph.14127

Bastiaanssen TFS, Cowan CSM, Claesson MJ, Dinan TG, Cryan JF (2019) Making sense of . . . the microbiome in psychiatry. Int J Neuropsychopharmacol 22(1):37–52. https://doi.org/10.1093/ijnp/pyy067

Bayes J, Schloss J, Sibbritt D (2022) The effect of a Mediterranean diet on the symptoms of depression in young males (the "AMMEND" study): a randomized control trial. Am J Clin Nutr 116(2):nqac106. https://doi.org/10.1093/ajcn/nqac106

Belkacem N, Serafini N, Wheeler R, Derrien M, Boucinha L, Couesnon A, Cerf-Bensussan N, Gomperts Boneca I, Di Santo JP, Taha M-K, Bourdet-Sicard R (2017) Lactobacillus paracasei feeding improves immune control of influenza infection in mice. PloS One 12(9):e0184976. https://doi.org/10.1371/journal.pone.0184976

Bengoa AA, Dardis C, Garrote GL, Abraham AG (2021) Health-promoting properties of Lacticaseibacillus paracasei: a focus on kefir isolates and exopolysaccharide-producing strains. Foods 10(10):2239. https://doi.org/10.3390/foods10102239

Berding K, Vlckova K, Marx W, Schellekens H, Stanton C, Clarke G, Jacka F, Dinan TG, Cryan JF (2021) Diet and the microbiota–gut–brain axis: sowing the seeds of good mental health. Adv Nutr 12(4). https://doi.org/10.1093/advances/nmaa181

Berding K, Bastiaanssen TFS, Moloney GM, Boscaini S, Strain CR, Anesi A, Long-Smith C, Mattivi F, Stanton C, Clarke G, Dinan TG, Cryan JF (2023) Feed your microbes to deal with stress: a psychobiotic diet impacts microbial stability and perceived stress in a healthy adult population. Mol Psychiatry 28:601–610. https://doi.org/10.1038/s41380-022-01817-y

Bertelsen RJ, Jensen ET, Ringel-Kulka T (2016) Use of probiotics and prebiotics in infant feeding. Best Pract Res Clin Gastroenterol 30(1):39–48. https://doi.org/10.1016/j.bpg.2016.01.001

Bertollo AG, Grolli RE, Plissari ME, Gasparin VA, Quevedo J, Réus GZ, Bagatini MD, Ignácio ZM (2020) Stress and serum cortisol levels in major depressive disorder: a cross-sectional study. AIMS Neurosci 7(4):459–469. https://doi.org/10.3934/neuroscience.2020028

Björkholm C, Monteggia LM (2016) BDNF – a key transducer of antidepressant effects. Neuropharmacology 102:72–79. https://doi.org/10.1016/j.neuropharm.2015.10.034

Bordenstein SR, Theis KR (2015) Host biology in light of the microbiome: ten principles of holobionts and hologenomes. PLoS Biol 13(8):e1002226. https://doi.org/10.1371/journal.pbio.1002226

Bosch JA, Nieuwdorp M, Zwinderman AH, Deschasaux M, Radjabzadeh D, Kraaij R, Davids M, de Rooij SR, Lok A (2022) The gut microbiota and depressive symptoms across ethnic groups. Nat Commun 13(1):7129. https://doi.org/10.1038/s41467-022-34504-1

Bravo JA, Forsythe P, Chew MV, Escaravage E, Savignac HM, Dinan TG, Bienenstock J, Cryan JF (2011) Ingestion of Lactobacillus strain regulates emotional behavior and central GABA receptor expression in a mouse via the vagus nerve. Proc Natl Acad Sci 108(38):16050–16055. https://doi.org/10.1073/pnas.1102999108

Burokas A, Arboleya S, Moloney RD, Peterson VL, Murphy K, Clarke G, Stanton C, Dinan TG, Cryan JF (2017) Targeting the microbiota-gut-brain axis: prebiotics have anxiolytic and antidepressant-like effects and reverse the impact of chronic stress in mice. Biol Psychiatry 82(7):472–487. https://doi.org/10.1016/j.biopsych.2016.12.031

Butler MI, Sandhu K, Cryan JF, Dinan TG (2019a) From isoniazid to psychobiotics: the gut microbiome as a new antidepressant target. Br J Hosp Med 80(3):139–145. https://doi.org/10.12968/hmed.2019.80.3.139

Butler MI, Mörkl S, Sandhu KV, Cryan JF, Dinan TG (2019b) The gut microbiome and mental health: what should we tell our patients?: Le microbiote intestinal et la Santé Mentale : que Devrions-Nous dire à nos Patients? Can J Psychiatry 64(11):706743719874168. https://doi.org/10.1177/0706743719874168

Butler MI, Bastiaanssen TFS, Long-Smith C, Morkl S, Berding K, Ritz NL, Strain C, Patangia D, Patel S, Stanton C, O'Mahony SM, Cryan JF, Clarke G, Dinan TG (2023) The gut microbiome in social anxiety disorder: evidence of altered composition and function. Transl Psychiatry 13(1):1–12. https://doi.org/10.1038/s41398-023-02325-5

Cammarota G, Ianiro G, Tilg H, Rajilić-Stojanović M, Kump P, Satokari R, Sokol H, Arkkila P, Pintus C, Hart A, Segal J, Aloi M, Masucci L, Molinaro A, Scaldaferri F, Gasbarrini G, Lopez-Sanroman A, Link A, de Groot P, de Vos WM (2017) European consensus conference on faecal microbiota transplantation in clinical practice. Gut 66(4):569–580. https://doi.org/10.1136/gutjnl-2016-313017

Carbia C, Bastiaanssen TFS, Iannone LF, García-Cabrerizo R, Boscaini S, Berding K, Strain CR, Clarke G, Stanton C, Dinan TG, Cryan JF (2023) The microbiome-gut-brain axis regulates social cognition & craving in young binge drinkers. EBioMedicine. https://doi.org/10.1016/j.ebiom.2023.104442

Carlson JL, Erickson JM, Lloyd BB, Slavin JL (2018) Health effects and sources of prebiotic dietary fiber. Curr Dev Nutr 2(3). https://doi.org/10.1093/cdn/nzy005

Carrozzino D, Patierno C, Fava GA, Guidi J (2020) The Hamilton rating scales for depression: a critical review of clinimetric properties of different versions. Psychother Psychosom 89(3):133–150. https://doi.org/10.1159/000506879

Ceban F, Rosenblat JD, Kratiuk K, Lee Y, Rodrigues NB, Gill H, Subramaniapillai M, Nasri F, Lui LMW, Lipsitz O, Kumar A, Lee JG, Chau EH, Cao B, Lin K, Ho RC, Mansur RB, Swainson J, McIntyre RS (2021) Prevention and management of common adverse effects of ketamine and esketamine in patients with mood disorders. CNS Drugs 35(9):925–934. https://doi.org/10.1007/s40263-021-00846-5

Celec P, Ostatníková D, Hodosy J (2015) On the effects of testosterone on brain behavioral functions. Front Neurosci 9. https://doi.org/10.3389/fnins.2015.00012

Chen L-M, Bao C-H, Wu Y, Liang S-H, Wang D, Wu L-Y, Huang Y, Liu H-R, Wu H-G (2021) Tryptophan-kynurenine metabolism: a link between the gut and brain for depression in inflammatory bowel disease. J Neuroinflammation 18:135. https://doi.org/10.1186/s12974-021-02175-2

Chu A, Wadhwa R (2022) Selective serotonin reuptake inhibitors. StatPearls Publishing. https://www.ncbi.nlm.nih.gov/books/NBK554406/

Cieślak A, Trottier J, Verreault M, Milkiewicz P, Vohl M-C, Barbier O (2018) N-3 polyunsaturated fatty acids stimulate bile acid detoxification in human cell models. Can J Gastroenterol Hepatol. https://doi.org/10.1155/2018/6031074

Cipriani A, Furukawa TA, Salanti G, Chaimani A, Atkinson LZ, Ogawa Y, Leucht S, Ruhe HG, Turner EH, Higgins JPT, Egger M, Takeshima N, Hayasaka Y, Imai H, Shinohara K, Tajika A, Ioannidis JPA, Geddes JR (2018) Comparative efficacy and acceptability of 21 antidepressant drugs for the acute treatment of adults with major depressive disorder: a systematic review and network meta-analysis. Lancet 391(10128):1357–1366. https://doi.org/10.1016/s0140-6736(17)32802-7

Clarke G, Quigley EMM, Cryan JF, Dinan TG (2009) Irritable bowel syndrome: towards biomarker identification. Trends Mol Med 15(10):478–489. https://doi.org/10.1016/j.molmed.2009.08.001

Clarke G, Stilling RM, Kennedy PJ, Stanton C, Cryan JF, Dinan TG (2014) Minireview: gut microbiota: the neglected endocrine organ. Mol Endocrinol 28(8):1221–1238. https://doi.org/10.1210/me.2014-1108

Clarke G, Sandhu KV, Griffin BT, Dinan TG, Cryan JF, Hyland NP (2019) Gut reactions: breaking down xenobiotic–microbiome interactions. Pharmacol Rev 71(2):198–224. https://doi.org/10.1124/pr.118.015768

Clauss M, Gérard P, Mosca A, Leclerc M (2021) Interplay between exercise and gut microbiome in the context of human health and performance. Front Nutr 8. https://doi.org/10.3389/fnut.2021.637010

Collins SL, Patterson AD (2020) The gut microbiome: an orchestrator of xenobiotic metabolism. Acta Pharm Sin B 10(1):19–32. https://doi.org/10.1016/j.apsb.2019.12.001

Colucci-D'Amato L, Speranza L, Volpicelli F (2020) Neurotrophic factor BDNF, physiological functions and therapeutic potential in depression, neurodegeneration and brain cancer. Int J Mol Sci 21(20). https://doi.org/10.3390/ijms21207777

Connolly KR, Thase ME (2016) Vortioxetine: a new treatment for major depressive disorder. Expert Opin Pharmacother 17(3):421–431. https://doi.org/10.1517/14656566.2016.1133588

Cruz-Pereira JS, Rea K, Nolan YM, O'Leary OF, Dinan TG, Cryan JF (2020) Depression's unholy trinity: dysregulated stress, immunity, and the microbiome. Annu Rev Psychol 71(1):49–78. https://doi.org/10.1146/annurev-psych-122216-011613

Cryan JF, O'Riordan KJ, Cowan CSM, Sandhu KV, Bastiaanssen TFS, Boehme M, Codagnone MG, Cussotto S, Fulling C, Golubeva AV, Guzzetta KE, Jaggar M, Long-Smith CM, Lyte JM, Martin JA, Molinero-Perez A, Moloney G, Morelli E, Morillas E, O'Connor R (2019) The microbiota-gut-brain axis. Physiol Rev 99(4):1877–2013. https://doi.org/10.1152/physrev.00018.2018

Cussotto S, Strain CR, Fouhy F, Strain RG, Peterson VL, Clarke G, Stanton C, Dinan TG, Cryan JF (2018) Differential effects of psychotropic drugs on microbiome composition and gastrointestinal function. Psychopharmacology (Berl) 236(5):1671–1685. https://doi.org/10.1007/s00213-018-5006-5

Cussotto S, Clarke G, Dinan TG, Cryan JF (2019) Psychotropics and the microbiome: a chamber of secrets.... Psychopharmacology (Berl) 236(5):1411–1432. https://doi.org/10.1007/s00213-019-5185-8

Cussotto S, Clarke G, Dinan TG, Cryan JF (2021a) Psychotropic drugs and the microbiome. Modern Trends Psychiatry 32:113–133. https://doi.org/10.1159/000510423

Cussotto S, Walsh J, Golubeva AV, Zhdanov AV, Strain CR, Fouhy F, Stanton C, Dinan TG, Hyland NP, Clarke G, Cryan JF, Griffin BT (2021b) The gut microbiome influences the bioavailability of olanzapine in rats. EBioMedicine 66:103307. https://doi.org/10.1016/j.ebiom.2021.103307

D'Agostino A, English CD, Rey JA (2015) Vortioxetine (brintellix): a new serotonergic antidepressant. P T 40(1):36–40. https://www.ncbi.nlm.nih.gov/pmc/articles/PMC4296590/

Dalile B, Vervliet B, Bergonzelli G, Verbeke K, Van Oudenhove L (2020) Colon-delivered short-chain fatty acids attenuate the cortisol response to psychosocial stress in healthy men: a randomized, placebo-controlled trial. Neuropsychopharmacology 45(13):2257–2266. https://doi.org/10.1038/s41386-020-0732-x

Davani-Davari D, Negahdaripour M, Karimzadeh I, Seifan M, Mohkam M, Masoumi S, Berenjian A, Ghasemi Y (2019) Prebiotics: definition, types, sources, mechanisms, and clinical applications. Foods 8(3):92. https://www.ncbi.nlm.nih.gov/pmc/articles/PMC6463098/

Davey P (ed) (2007) Medicine at a glance, 2nd edn. Blackwell Publishing, p 290

de Kloet ER, Joëls M, Holsboer F (2005) Stress and the brain: from adaptation to disease. Nat Rev Neurosci 6(6):463–475. https://doi.org/10.1038/nrn1683

de Vrese M, Schrezenmeir J (2008) Probiotics, prebiotics, and synbiotics. Food Biotechnol 111:1–66. https://doi.org/10.1007/10_2008_097

Dean OM, Data-Franco J, Giorlando F, Berk M (2012) Minocycline. CNS Drugs 26(5):391–401. https://doi.org/10.2165/11632000-000000000-00000

Dean OM, Kanchanatawan B, Ashton M, Mohebbi M, Ng CH, Maes M, Berk L, Sughondhabirom A, Tangwongchai S, Singh AB, McKenzie H, Smith DJ, Malhi GS, Dowling N, Berk M (2017) Adjunctive minocycline treatment for major depressive disorder: a proof of concept trial. Aust N Z J Psychiatry 51(8):829–840. https://doi.org/10.1177/0004867417709357

Diener C, Dai CL, Wilmanski T, Baloni P, Smith B, Rappaport N, Hood L, Magis AT, Gibbons SM (2022) Genome–microbiome interplay provides insight into the determinants of the human blood metabolome. Nat Metab 4(11):1560–1572. https://doi.org/10.1038/s42255-022-00670-1

Dinan TG, Cryan JF (2012) Regulation of the stress response by the gut microbiota: implications for psychoneuroendocrinology. Psychoneuroendocrinology 37(9):1369–1378. https://doi.org/10.1016/j.psyneuen.2012.03.007

Dinan TG, Stanton C, Cryan JF (2013) Psychobiotics: a novel class of psychotropic. Biol Psychiatry 74(10):720–726. https://doi.org/10.1016/j.biopsych.2013.05.001

Dinan TG, Butler MI, Cryan JF (2021) Psychobiotics: evolution of novel antidepressants. Modern Trends Psychiatry 32:134–143. https://doi.org/10.1159/000510424

Dong Z, Shen X, Hao Y, Li J, Xu H, Yin L, Kuang W (2022) Gut microbiome: a potential indicator for predicting treatment outcomes in major depressive disorder. Front Neurosci 16. https://doi.org/10.3389/fnins.2022.813075

Du Y, Gao X, Peng L, Jin-Fang G (2020) Crosstalk between the microbiota-gut-brain axis and depression. Heliyon 6(6):e04097. https://doi.org/10.1016/j.heliyon.2020.e04097

El-Salhy M, Mazzawi T, Hausken T, Hatlebakk JG (2022a) The fecal microbiota transplantation response differs between patients with severe and moderate irritable bowel symptoms. Scand J Gastroenterol 57(9):1036–1045. https://doi.org/10.1080/00365521.2022.2064725

El-Salhy M, Winkel R, Casen C, Hausken T, Gilja OH, Hatlebakk JG (2022b) Efficacy of fecal microbiota transplantation for patients with irritable bowel syndrome at three years after transplantation. Gastroenterology S0016-5085(22):006254. https://doi.org/10.1053/j.gastro.2022.06.020

Fayez R, Gupta V (2021) Imipramine. StatPearls Publishing. https://pubmed.ncbi.nlm.nih.gov/32491588/

Fernandez MA, Marette A (2017) Potential health benefits of combining yogurt and fruits based on their probiotic and prebiotic properties. Adv Nutr 8(1):155S164S. https://doi.org/10.3945/an.115.011114

Fontana A, Manchia M, Panebianco C, Paribello P, Arzedi C, Cossu E, Garzilli M, Montis MA, Mura A, Pisanu C, Congiu D, Copetti M, Pinna F, Carpiniello B, Squassina A, Pazienza V (2020) Exploring the role of gut microbiota in major depressive disorder and in treatment resistance to antidepressants. Biomedicine 8(9):311. https://doi.org/10.3390/biomedicines8090311

Frazer A (2001) Serotonergic and noradrenergic reuptake inhibitors: prediction of clinical effects from in vitro potencies. J Clin Psychiatry 62(Suppl 12):16–23. https://pubmed.ncbi.nlm.nih.gov/11430614/

Fülling C, Dinan TG, Cryan JF (2019) Gut microbe to brain Signaling: what happens in Vagus. . . . Neuron 101(6):998–1002. https://doi.org/10.1016/j.neuron.2019.02.008

Gareau MG, Wine E, Rodrigues DM, Cho JH, Whary MT, Philpott DJ, MacQueen G, Sherman PM (2010) Bacterial infection causes stress-induced memory dysfunction in mice. Gut 60(3):307–317. https://doi.org/10.1136/gut.2009.202515

Generoso JS, Giridharan VV, Lee J, Macedo D, Barichello T (2020) The role of the microbiota-gut-brain axis in neuropsychiatric disorders. Braz J Psychiatry 43(3):293–305. https://doi.org/10.1590/1516-4446-2020-0987

Gheorghe CE, Ritz NL, Martin JA, Wardill HR, Cryan JF, Clarke G (2021) Investigating causality with fecal microbiota transplantation in rodents: applications, recommendations and pitfalls. Gut Microbes 13(1). https://doi.org/10.1080/19490976.2021.1941711

Gheorghe CE, Cryan JF, Clarke G (2022) Debugging the gut-brain axis in depression. Cell Host Microbe 30(3):281–283. https://doi.org/10.1016/j.chom.2022.02.007

Ghorbani Z, Nazari S, Etesam F, Nourimajd S, Ahmadpanah M, Razeghi Jahromi S (2018) The effect of synbiotic as an adjuvant therapy to fluoxetine in moderate depression: a randomized multicenter trial. Arch Neurosci 5(2). https://doi.org/10.5812/archneurosci.60507

Green JE, McGuinness AJ, Berk M, Castle D, Athan E, Hair C et al (2023a) Safety and feasibility of faecal microbiota transplant for major depressive disorder: study protocol for a pilot randomised controlled trial. Pilot Feasibility Stud 9(1). https://doi.org/10.1186/s40814-023-01235-z

Green JE, Berk M, Mohebbi M, Loughman A, McGuinness AJ, Castle D, Chatterton ML, Perez J, Strandwitz P, Athan E, Hair C, Nierenberg AA, Cryan JF, Jacka F (2023b) Feasibility, acceptability, and safety of faecal microbiota transplantation in the treatment of major depressive disorder: a pilot randomized controlled trial. Can J Psychiatry 070674372211505. https://doi.org/10.1177/07067437221150508

Gupta A, Osadchiy V, Mayer EA (2020) Brain–gut–microbiome interactions in obesity and food addiction. Nat Rev Gastroenterol Hepatol 17(11):655–672. https://doi.org/10.1038/s41575-020-0341-5

Haas GS, Wang W, Saffar M, Mooney-Leber SM, Brummelte S (2020) Probiotic treatment (Bifidobacterium longum subsp. longum 35624™) affects stress responsivity in male rats after chronic corticosterone exposure. Behav Brain Res 393:112718. https://doi.org/10.1016/j.bbr.2020.112718

Hadinia N, Edalatian Dovom MR, Yavarmanesh M (2022) The effect of fermentation conditions (temperature, salt concentration, and pH) with lactobacillus strains for producing short chain fatty acids. LWT 165:113709. https://doi.org/10.1016/j.lwt.2022.113709

Haghighat N, Rajabi S, Mohammadshahi M (2021) Effect of synbiotic and probiotic supplementation on serum brain-derived neurotrophic factor level, depression and anxiety symptoms in hemodialysis patients: a randomized, double-blinded, clinical trial. Nutr Neurosci 24(6): 490–499. https://doi.org/10.1080/1028415x.2019.1646975

Hansson GC (2019) Mucus and mucins in diseases of the intestinal and respiratory tracts. J Intern Med 285(5):479–490. https://doi.org/10.1111/joim.12910

Heijtz RD, Wang S, Anuar F, Qian Y, Bjorkholm B, Samuelsson A, Hibberd ML, Forssberg H, Pettersson S (2011) Normal gut microbiota modulates brain development and behavior. Proc Natl Acad Sci 108(7):3047–3052. https://doi.org/10.1073/pnas.1010529108

Holscher HD (2017) Dietary fiber and prebiotics and the gastrointestinal microbiota. Gut Microbes 8(2):172–184. https://doi.org/10.1080/19490976.2017.1290756

Hong J, Yoon D, Nam Y, Seo D, Kim J-H, Kim MS, Lee TY, Kim KS, Ko P-W, Lee H-W, Suk K, Kim SR (2020) Lipopolysaccharide administration for a mouse model of cerebellar ataxia with neuroinflammation. Sci Rep 10(1):13337. https://doi.org/10.1038/s41598-020-70390-7

Hooper LV, Littman DR, Macpherson AJ (2012) Interactions between the microbiota and the immune system. Science 336(6086):1268–1273. https://doi.org/10.1126/science.1223490

Huang R, Wang K, Hu J (2016) Effect of probiotics on depression: a systematic review and meta-analysis of randomized controlled trials. Nutrients 8(8):483. https://doi.org/10.3390/nu8080483

ISAPP (2022) Prebiotics. ISAPP. https://isappscience.org/for-scientists/resources/prebiotics/

Jacka FN, O'Neil A, Opie R, Itsiopoulos C, Cotton S, Mohebbi M, Castle D, Dash S, Mihalopoulos C, Chatterton ML, Brazionis L, Dean OM, Hodge AM, Berk M (2017) A randomised controlled trial of dietary improvement for adults with major depression (the "SMILES" trial). BMC Med 15(1):23. https://doi.org/10.1186/s12916-017-0791-y

Järbrink-Sehgal E, Andreasson A (2020) The gut microbiota and mental health in adults. Curr Opin Neurobiol 62:102–114. https://doi.org/10.1016/j.conb.2020.01.016

Jauhar S, Cowen PJ, Browning M (2023) Fifty years on: serotonin and depression. J Psychopharmacol 37(3):237–241. https://doi.org/10.1177/02698811231161813

Jiang H, Ling Z, Zhang Y, Mao H, Ma Z, Yin Y, Wang W, Tang W, Tan Z, Shi J, Li L, Ruan B (2015) Altered fecal microbiota composition in patients with major depressive disorder. Brain Behav Immun 48:186–194. https://doi.org/10.1016/j.bbi.2015.03.016

Karakan T, Ozkul C, Küpeli Akkol E, Bilici S, Sobarzo-Sánchez E, Capasso R (2021) Gut-brain-microbiota axis: antibiotics and functional gastrointestinal disorders. Nutrients 13(2):389. https://doi.org/10.3390/nu13020389

Kato-Kataoka A, Nishida K, Takada M, Suda K, Kawai M, Shimizu K, Kushiro A, Hoshi R, Watanabe O, Igarashi T, Miyazaki K, Kuwano Y, Rokutan K (2016) Fermented milk containing Lactobacillus casei strain Shirota prevents the onset of physical symptoms in medical students under academic examination stress. Benefic Microbes 7(2):153–156. https://doi.org/10.3920/bm2015.0100

Kazemi A, Noorbala AA, Azam K, Eskandari MH, Djafarian K (2019) Effect of probiotic and prebiotic vs placebo on psychological outcomes in patients with major depressive disorder: a randomized clinical trial. Clin Nutr 38(2):522–528. https://doi.org/10.1016/j.clnu.2018.04.010

Kelly JR, Kennedy PJ, Cryan JF, Dinan TG, Clarke G, Hyland NP (2015) Breaking down the barriers: the gut microbiome, intestinal permeability and stress-related psychiatric disorders. Front Cell Neurosci 9:392. https://doi.org/10.3389/fncel.2015.00392

Kelly JR, Borre Y, O'Brien C, Patterson E, El Aidy S, Deane J, Kennedy PJ, Beers S, Scott K, Moloney G, Hoban AE, Scott L, Fitzgerald P, Ross P, Stanton C, Clarke G, Cryan JF, Dinan TG (2016) Transferring the blues: depression-associated gut microbiota induces neurobehavioural changes in the rat. J Psychiatr Res 82:109–118. https://doi.org/10.1016/j.jpsychires.2016.07.019

Kelly JR, Allen AP, Temko A, Hutch W, Kennedy PJ, Farid N, Murphy E, Boylan G, Bienenstock J, Cryan JF, Clarke G, Dinan TG (2017) Lost in translation? The potential psychobiotic Lactobacillus rhamnosus (JB-1) fails to modulate stress or cognitive performance in healthy male subjects. Brain Behav Immun 61:50–59. https://doi.org/10.1016/j.bbi.2016.11.018

Kennedy PJ, Cryan JF, Dinan TG, Clarke G (2017) Kynurenine pathway metabolism and the microbiota-gut-brain axis. Neuropharmacology 112:399–412. https://doi.org/10.1016/j.neuropharm.2016.07.002

Khalesi S, Bellissimo N, Vandelanotte C, Williams S, Stanley D, Irwin C (2019) A review of probiotic supplementation in healthy adults: helpful or hype? Eur J Clin Nutr 73(1):24–37. https://doi.org/10.1038/s41430-018-0135-9

Kim C-S, Cha L, Sim M, Jung S, Chun WY, Baik HW, Shin D-M (2020) Probiotic supplementation improves cognitive function and mood with changes in gut microbiota in community-dwelling elderly: a randomized, double-blind, placebo-controlled, multicenter trial. J Gerontol Ser A 76(1). https://doi.org/10.1093/gerona/glaa090

Klünemann M, Andrejev S, Blasche S, Mateus A, Phapale P, Devendran S, Vappiani J, Simon B, Scott TA, Kafkia E, Konstantinidis D, Zirngibl K, Mastrorilli E, Banzhaf M, Mackmull M-T, Hövelmann F, Nesme L, Brochado AR, Maier L, Bock T (2021) Bioaccumulation of therapeutic drugs by human gut bacteria. Nature 597(7877):533–538. https://doi.org/10.1038/s41586-021-03891-8

Kuyken W, Warren FC, Taylor RS, Whalley B, Crane C, Bondolfi G, Hayes R, Huijbers M, Ma H, Schweizer S, Segal Z, Speckens A, Teasdale JD, Van Heeringen K, Williams M, Byford S, Byng R, Dalgleish T (2016) Efficacy of mindfulness-based cognitive therapy in prevention of depressive relapse. JAMA Psychiatry 73(6):565. https://doi.org/10.1001/jamapsychiatry.2016.0076

Lancaster SM, Lee-McMullen B, Abbott CW, Quijada JV, Hornburg D, Park H, Perelman D, Peterson DJ, Tang M, Robinson A, Ahadi S, Contrepois K, Hung C-J, Ashland M, McLaughlin T, Boonyanit A, Horning A, Sonnenburg JL, Snyder MP (2022) Global, distinctive, and personal changes in molecular and microbial profiles by specific fibers in humans. Cell Host Microbe 30(6):848–862.e7. https://doi.org/10.1016/j.chom.2022.03.036

Lew L-C, Hor Y-Y, Yusoff NAA, Choi S-B, Yusoff MSB, Roslan NS, Ahmad A, Mohammad JAM, Abdullah MFIL, Zakaria N, Wahid N, Sun Z, Kwok L-Y, Zhang H, Liong M-T (2019) Probiotic Lactobacillus plantarum P8 alleviated stress and anxiety while enhancing memory and

cognition in stressed adults: a randomised, double-blind, placebo-controlled study. Clin Nutr 38(5):2053–2064. https://doi.org/10.1016/j.clnu.2018.09.010
Li F, Tsien JZ (2009) Memory and the NMDA receptors. N Engl J Med 361(3):302–303. https://doi.org/10.1056/nejmcibr0902052
Li W, Guo J, Shen Y, Huang L, Leng B, Fan D, Shui L, Chen C (2020) Probiotics, prebiotics, and synbiotics for the treatment of dementia. Medicine 99(5):e18608. https://doi.org/10.1097/md.0000000000018608
Li D, Liu R, Wang M, Peng R, Fu S, Fu A, Le J, Yao Q, Yuan T, Chi H, Mu X, Sun T, Liu H, Yan P, Wang S, Cheng S, Deng Z, Liu Z, Wang G, Li Y (2022) 3β-hydroxysteroid dehydrogenase expressed by gut microbes degrades testosterone and is linked to depression in males. Cell Host Microbe 30(3):329–339.e5. https://doi.org/10.1016/j.chom.2022.01.001
Liang S, Wang T, Hu X, Luo J, Li W, Wu X, Duan Y, Jin F (2015) Administration of Lactobacillus helveticus NS8 improves behavioral, cognitive, and biochemical aberrations caused by chronic restraint stress. Neuroscience 310:561–577. https://doi.org/10.1016/j.neuroscience.2015.09.033
Liu C-F, Tseng K-C, Chiang S-S, Lee B-H, Hsu W-H, Pan T-M (2011) Immunomodulatory and antioxidant potential of Lactobacillus exopolysaccharides. J Sci Food Agric 91(12). https://doi.org/10.1002/jsfa.4456
Liu B, He Y, Wang M, Liu J, Ju Y, Zhang Y, Liu T, Li L, Li Q (2018) Efficacy of probiotics on anxiety-A meta-analysis of randomized controlled trials. Depress Anxiety 35(10):935–945. https://doi.org/10.1002/da.22811
Liu RT, Walsh RFL, Sheehan AE (2019) Prebiotics and probiotics for depression and anxiety: a systematic review and meta-analysis of controlled clinical trials. Neurosci Biobehav Rev 102: 13–23. https://doi.org/10.1016/j.neubiorev.2019.03.023
Llopis M, Antolín M, Guarner F, Salas A, Malagelada J-R (2005) Mucosal colonisation with Lactobacillus casei mitigates barrier injury induced by exposure to trinitronbenzene sulphonic acid. Gut 54(7):955–959. https://doi.org/10.1136/gut.2004.056101
Long-Smith C, O'Riordan KJ, Clarke G, Stanton C, Dinan TG, Cryan JF (2020) Microbiota-gut-brain axis: new therapeutic opportunities. Annu Rev Pharmacol Toxicol 60(1):477–502. https://doi.org/10.1146/annurev-pharmtox-010919-023628
Luczynski P, McVey Neufeld K-A, Oriach CS, Clarke G, Dinan TG, Cryan JF (2016) Growing up in a bubble: using germ-free animals to assess the influence of the gut microbiota on brain and behavior. Int J Neuropsychopharmacol 19(8):pyw020. https://doi.org/10.1093/ijnp/pyw020
Lurie I, Yang Y-X, Haynes K, Mamtani R, Boursi B (2015) Antibiotic exposure and the risk for depression, anxiety, or psychosis: a nested case-control study. J Clin Psychiatry 76(11): 1522–1528. https://doi.org/10.4088/JCP.15m09961
Lurie-Weinberger MN, Gophna U (2015) Archaea in and on the human body: health implications and future directions. PLoS Pathog 11(6):e1004833. https://doi.org/10.1371/journal.ppat.1004833
Mandal A, Sinha C, Kumar Jena A, Ghosh S, Samanta A (2010) An investigation on in vitro and in vivo antimicrobial properties of the antidepressant: amitriptyline hydrochloride. Braz J Microbiol 41(3):635–645. https://doi.org/10.1590/S1517-83822010000300014
Mannie ZN, Harmer CJ, Cowen PJ (2007) Increased waking salivary cortisol levels in young people at familial risk of depression. Am J Psychiatry 164(4):617–621. https://doi.org/10.1176/ajp.2007.164.4.617
Markowiak P, Śliżewska K (2017) Effects of probiotics, prebiotics, and synbiotics on human health. Nutrients 9(9):1021. https://doi.org/10.3390/nu9091021
Marx W, Scholey A, Firth J, D'Cunha NM, Lane M, Hockey M, Ashton MM, Cryan JF, O'Neil A, Naumovski N, Berk M, Dean OM, Jacka F (2020) Prebiotics, probiotics, fermented foods and cognitive outcomes: a meta-analysis of randomized controlled trials. Neurosci Biobehav Rev 118:472–484. https://doi.org/10.1016/j.neubiorev.2020.07.036
Marx W, Lane M, Hockey M, Aslam H, Berk M, Walder K, Borsini A, Firth J, Pariante CM, Berding K, Cryan JF, Clarke G, Craig JM, Su K-P, Mischoulon D, Gomez-Pinilla F, Foster JA,

Cani PD, Thuret S, Staudacher HM (2021) Diet and depression: exploring the biological mechanisms of action. Mol Psychiatry 26:1–17. https://doi.org/10.1038/s41380-020-00925-x
Matijašić M, Meštrović T, Čipčić Paljetak H, Perić M, Barešić A, Verbanac D (2020) Gut microbiota beyond bacteria – Mycobiome, Virome, Archaeome, and eukaryotic parasites in IBD. Int J Mol Sci 21(8):2668. https://doi.org/10.3390/ijms21082668
Mayer SE, Lopez-Duran NL, Sen S, Abelson JL (2018) Chronic stress, hair cortisol and depression: a prospective and longitudinal study of medical internship. Psychoneuroendocrinology 92:57–65. https://doi.org/10.1016/j.psyneuen.2018.03.020
McGuinness AJ, Davis JA, Dawson SL, Loughman A, Collier F, O'Hely M, Simpson CA, Green J, Marx W, Hair C, Guest G, Mohebbi M, Berk M, Stupart D, Watters D, Jacka FN (2022) A systematic review of gut microbiota composition in observational studies of major depressive disorder, bipolar disorder and schizophrenia. Mol Psychiatry 27(4):1920–1935. https://doi.org/10.1038/s41380-022-01456-3
Messaoudi M, Violle N, Bisson J-F, Desor D, Javelot H, Rougeot C (2011) Beneficial psychological effects of a probiotic formulation (Lactobacillus helveticusR0052 and Bifidobacterium longumR0175) in healthy human volunteers. Gut Microbes 2(4):256–261. https://doi.org/10.4161/gmic.2.4.16108
Millischer V, Heinzl M, Faka A, Resl M, Trepci A, Klammer C, Egger M, Dieplinger B, Clodi M, Schwieler L (2021) Intravenous administration of LPS activates the kynurenine pathway in healthy male human subjects: a prospective placebo-controlled cross-over trial. J Neuroinflammation 18(1). https://doi.org/10.1186/s12974-021-02196-x
Misiak B, Łoniewski I, Marlicz W, Frydecka D, Szulc A, Rudzki L, Samochowiec J (2020) The HPA axis dysregulation in severe mental illness: can we shift the blame to gut microbiota? Prog Neuropsychopharmacol Biol Psychiatry 102:109951. https://doi.org/10.1016/j.pnpbp.2020.109951
Moloney RD, Johnson AC, O'Mahony SM, Dinan TG, Greenwood-Van Meerveld B, Cryan JF (2015) Stress and the microbiota-gut-brain axis in visceral pain: relevance to irritable bowel syndrome. CNS Neurosci Ther 22(2):102–117. https://doi.org/10.1111/cns.12490
Moloney GM, Long-Smith CM, Murphy A, Dorland D, Hojabri SF, Ramirez LO, Marin DC, Bastiaanssen TFS, Cusack A-M, Berding K, Fouhy F, Allen AP, Stanton C, Clarke G, Dinan TG, Cryan JF (2021) Improvements in sleep indices during exam stress due to consumption of a Bifidobacterium longum. Brain Behav Immun Health 10:100174. https://doi.org/10.1016/j.bbih.2020.100174
Morey JN, Boggero IA, Scott AB, Segerstrom SC (2015) Current directions in stress and human immune function. Curr Opin Psychol 5(1):13–17. https://doi.org/10.1016/j.copsyc.2015.03.007
Naughton M, Clarke G, O'Leary OF, Cryan JF, Dinan TG (2014) A review of ketamine in affective disorders: current evidence of clinical efficacy, limitations of use and pre-clinical evidence on proposed mechanisms of action. J Affect Disord 156:24–35. https://doi.org/10.1016/j.jad.2013.11.014
Nettis MA, Lombardo G, Hastings C, Zajkowska Z, Mariani N, Nikkheslat N, Worrell C, Enache D, McLaughlin A, Kose M, Sforzini L, Bogdanova A, Cleare A, Young AH, Pariante CM, Mondelli V (2021) Augmentation therapy with minocycline in treatment-resistant depression patients with low-grade peripheral inflammation: results from a double-blind randomised clinical trial. Neuropsychopharmacology 46(5):939–948. https://doi.org/10.1038/s41386-020-00948-6
NICE (2022) Recommendations | Depression in adults: treatment and management | Guidance https://www.nice.org.uk/guidance/ng222/chapter/Recommendations#choice-of-treatments
Nikolova VL, Smith MRB, Hall LJ, Cleare AJ, Stone JM, Young AH (2021) Perturbations in gut microbiota composition in psychiatric disorders. JAMA Psychiatry 78(12). https://doi.org/10.1001/jamapsychiatry.2021.2573
Nikolova VL, Cleare AJ, Young AH, Stone JM (2023) Acceptability, tolerability, and estimates of putative treatment effects of probiotics as adjunctive treatment in patients with depression. JAMA Psychiatry:e231817. https://doi.org/10.1001/jamapsychiatry.2023.1817

Nunez N, Réot L, Menu E (2021) Neonatal immune system ontogeny: the role of maternal microbiota and associated factors. How might the non-human primate model enlighten the path? Vaccines 9(6):584. https://doi.org/10.3390/vaccines9060584

O'Mahony SM, Clarke G, Borre YE, Dinan TG, Cryan JF (2015) Serotonin, tryptophan metabolism and the brain-gut-microbiome axis. Behav Brain Res 277:32–48. https://doi.org/10.1016/j.bbr.2014.07.027

O'Riordan KJ, Collins MK, Moloney GM, Knox EG, Aburto MR, Fülling C, Morley SJ, Clarke G, Schellekens H, Cryan JF (2022) Short chain fatty acids: microbial metabolites for gut-brain axis signalling. Mol Cell Endocrinol 546:111572. https://doi.org/10.1016/j.mce.2022.111572

Okubo R, Koga M, Katsumata N, Odamaki T, Matsuyama S, Oka M, Narita H, Hashimoto N, Kusumi I, Xiao J, Matsuoka YJ (2019) Effect of bifidobacterium breve A-1 on anxiety and depressive symptoms in schizophrenia: a proof-of-concept study. J Affect Disord 245:377–385. https://doi.org/10.1016/j.jad.2018.11.011

Oleskin AV, Shenderov BA (2019) Probiotics and psychobiotics: the role of microbial neurochemicals. Probiotics Antimicrob Proteins 11. https://doi.org/10.1007/s12602-019-09583-0

Pan W, Stone KP, Hsuchou H, Manda VK, Zhang Y, Kastin AJ (2011) Cytokine signaling modulates blood-brain barrier function. Curr Pharm Des 17(33):3729–3740. https://doi.org/10.2174/138161211798220918

Pan Z, Park C, Brietzke E, Zuckerman H, Rong C, Mansur RB, Fus D, Subramaniapillai M, Lee Y, McIntyre RS (2018) Cognitive impairment in major depressive disorder. CNS Spectr 24(1): 22–29. https://doi.org/10.1017/s1092852918001207

Pandey KR, Naik SR, Vakil BV (2015) Probiotics, prebiotics and synbiotics – a review. J Food Sci Technol 52(12):7577–7587. https://doi.org/10.1007/s13197-015-1921-1

Parletta N, Zarnowiecki D, Cho J, Wilson A, Bogomolova S, Villani A, Itsiopoulos C, Niyonsenga T, Blunden S, Meyer B, Segal L, Baune BT, O'Dea K (2017) A Mediterranean-style dietary intervention supplemented with fish oil improves diet quality and mental health in people with depression: a randomized controlled trial (HELFIMED). Nutr Neurosci 22(7): 474–487. https://doi.org/10.1080/1028415x.2017.1411320

Peirce JM, Alviña K (2019) The role of inflammation and the gut microbiome in depression and anxiety. J Neurosci Res 97(10):1223–1241. https://doi.org/10.1002/jnr.24476

Pham VT, Dold S, Rehman A, Bird JK, Steinert RE (2021) Vitamins, the gut microbiome and gastrointestinal health in humans. Nutr Res 95. https://doi.org/10.1016/j.nutres.2021.09.001

Phillips-Farfán B, Gómez-Chávez F, Medina-Torres EA, Vargas-Villavicencio JA, Carvajal-Aguilera K, Camacho L (2021) Microbiota signals during the neonatal period forge life-long immune responses. Int J Mol Sci 22(15):8162. https://doi.org/10.3390/ijms22158162

Pitharouli MC, Hagenaars SP, Glanville KP, Coleman JRI, Hotopf M, Lewis CM, Pariante CM (2021) Elevated C-reactive protein in patients with depression, independent of genetic, health, and psychosocial factors: results from the UK Biobank. Am J Psychiatry 178(6):522–529. https://doi.org/10.1176/appi.ajp.2020.20060947

Pluta R, Ułamek-Kozioł M, Januszewski S, Czuczwar SJ (2020) Gut microbiota and pro/prebiotics in Alzheimer's disease. Aging 12(6):5539–5550. https://doi.org/10.18632/aging.102930

Poon T, Juana J, Noori D, Jeansen S, Pierucci-Lagha A, Musa-Veloso K (2020) Effects of a fermented dairy drink containing Lacticaseibacillus paracasei subsp. paracasei CNCM I-1518 (Lactobacillus casei CNCM I-1518) and the standard yogurt cultures on the incidence, duration, and severity of common infectious diseases: a systematic review and meta-analysis of randomized controlled trials. Nutrients 12(11):3443. https://doi.org/10.3390/nu12113443

Purton T, Staskova L, Lane MM, Dawson SL, West M, Firth J, Clarke G, Cryan JF, Berk M, O'Neil A, Dean O, Hadi A, Honan C, Marx W (2021) Prebiotic and probiotic supplementation and the tryptophan-kynurenine pathway: a systematic review and meta analysis. Neurosci Biobehav Rev 123:1–13. https://doi.org/10.1016/j.neubiorev.2020.12.026

Radjabzadeh D, Bosch JA, Uitterlinden AG, Zwinderman AH, Ikram MA, van Meurs JBJ, Luik AI, Nieuwdorp M, Lok A, van Duijn CM, Kraaij R, Amin N (2022) Gut microbiome-wide association study of depressive symptoms. Nat Commun 13(1):7128. https://doi.org/10.1038/s41467-022-34502-3

Raison CL, Capuron L, Miller AH (2006) Cytokines sing the blues: inflammation and the pathogenesis of depression. Trends Immunol 27(1):24–31. https://doi.org/10.1016/j.it.2005.11.006

Ramamoorthy S, Cidlowski JA (2016) Corticosteroids: mechanisms of action in health and disease. Rheum Dis Clin North Am 42(1):15–31. https://doi.org/10.1016/j.rdc.2015.08.002

Rao AV, Bested AC, Beaulne TM, Katzman MA, Iorio C, Berardi JM, Logan AC (2009) A randomized, double-blind, placebo-controlled pilot study of a probiotic in emotional symptoms of chronic fatigue syndrome. Gut Pathogens 1(1):6. https://doi.org/10.1186/1757-4749-1-6

Reigstad CS, Salmonson CE, Rainey JF, Szurszewski JH, Linden DR, Sonnenburg JL, Farrugia G, Kashyap PC (2015) Gut microbes promote colonic serotonin production through an effect of short-chain fatty acids on enterochromaffin cells. FASEB J 29(4):1395–1403. https://doi.org/10.1096/fj.14-259598

Rezaei Asl Z, Sepehri G, Salami M (2019) Probiotic treatment improves the impaired spatial cognitive performance and restores synaptic plasticity in an animal model of Alzheimer's disease. Behav Brain Res 376:112183. https://doi.org/10.1016/j.bbr.2019.112183

Richardson L, Adams S (2018) Cognitive deficits in patients with depression. J Nurse Pract 14(6):437–443.e3. https://doi.org/10.1016/j.nurpra.2018.03.006

Royal College of Psychiatrists (2017) Statement on ketamine to treat depression. https://www.rcpsych.ac.uk/docs/default-source/about-us/who-we-are/ectcommittee-ketamine-to-treat-depression-statement-feb17.pdf?sfvrsn=db4e7a95_2

Royal College of Psychiatrists (2019) Physical illness. RC Psychiatry Royal College of Psychiatrists. https://www.rcpsych.ac.uk/mental-health/problems-disorders/physical-illness

Rudzki L, Ostrowska L, Pawlak D, Małus A, Pawlak K, Waszkiewicz N, Szulc A (2019) Probiotic lactobacillus plantarum 299v decreases kynurenine concentration and improves cognitive functions in patients with major depression: a double-blind, randomized, placebo controlled study. Psychoneuroendocrinology 100:213–222. https://doi.org/10.1016/j.psyneuen.2018.10.010

Ruhé HG, Mason NS, Schene AH (2007) Mood is indirectly related to serotonin, norepinephrine and dopamine levels in humans: a meta-analysis of monoamine depletion studies. Mol Psychiatry 12(4):331–359. https://doi.org/10.1038/sj.mp.4001949

Salli K, Lehtinen MJ, Tiihonen K, Ouwehand AC (2019) Xylitol's health benefits beyond dental health: a comprehensive review. Nutrients 11(8):1813. https://doi.org/10.3390/nu11081813

Sarkar A, Lehto SM, Harty S, Dinan TG, Cryan JF, Burnet PWJ (2016) Psychobiotics and the manipulation of bacteria–gut–brain signals. Trends Neurosci 39(11):763–781. https://doi.org/10.1016/j.tins.2016.09.002

Schildkraut JJ (1965) The catecholamine hypothesis of affective disorders: a review of supporting evidence. Am J Psychiatry 122(5):509–522. https://doi.org/10.1176/ajp.122.5.509

Schmidt K, Cowen PJ, Harmer CJ, Tzortzis G, Errington S, Burnet PWJ (2015) Prebiotic intake reduces the waking cortisol response and alters emotional bias in healthy volunteers. Psychopharmacology (Berl) 232(10):1793–1801. https://doi.org/10.1007/s00213-014-3810-0

Semple D, Smyth R (2009) Oxford handbook of psychiatry, Oxford University Press, pp 266, 788, 790

Sheffler Z, Patel P, Abdijadid S (2022) Antidepressants. StatPearls Publishing

Sidhu G, Marwaha R (2023) Phenelzine. StatPearls Publishing. https://www.ncbi.nlm.nih.gov/books/NBK554508/

Silk DBA, Davis A, Vulevic J, Tzortzis G, Gibson GR (2009) Clinical trial: the effects of a trans-galactooligosaccharide prebiotic on faecal microbiota and symptoms in irritable bowel syndrome. Aliment Pharmacol Ther 29(5):508–518. https://doi.org/10.1111/j.1365-2036.2008.03911.x

Silva YP, Bernardi A, Frozza RL (2020) The role of short-chain fatty acids from gut microbiota in gut-brain communication. Front Endocrinol 11(25). https://doi.org/10.3389/fendo.2020.00025

Simpson CA, Diaz-Arteche C, Eliby D, Schwartz OS, Simmons JG, Cowan CSM (2021) The gut microbiota in anxiety and depression – a systematic review. Clin Psychol Rev 83:101943. https://doi.org/10.1016/j.cpr.2020.101943

Siniscalco D, Schultz S, Brigida A, Antonucci N (2018) Inflammation and neuro-immune dysregulations in autism spectrum disorders. Pharmaceuticals 11(2):56. https://doi.org/10.3390/ph11020056

Siopi E, Chevalier G, Katsimpardi L, Saha S, Bigot M, Moigneu C, Eberl G, Lledo P-M (2020) Changes in gut microbiota by chronic stress impair the efficacy of fluoxetine. Cell Rep 30(11): 3682–3690.e6. https://doi.org/10.1016/j.celrep.2020.02.099

Sirisinha S (2016) The potential impact of gut on your health: current status and future challenges. Asian Pac J Allergy Immunol 34(4):249–264. https://doi.org/10.12932/ap0803

Sivcev S, Slavikova B, Rupert M, Ivetic M, Nekardova M, Kudova E, Zemkova H (2019) Synthetic testosterone derivatives modulate rat P2X2 and P2X4 receptor channel gating. J Neurochem 150(1):28–43. https://doi.org/10.1111/jnc.14718

Slavin J (2013) Fiber and prebiotics: mechanisms and health benefits. Nutrients 5(4):1417–1435. https://doi.org/10.3390/nu5041417

Smith A, Sutherland D, Hewlett P (2015) An investigation of the acute effects of oligofructose-enriched inulin on subjective wellbeing, mood and cognitive performance. Nutrients 7(11): 8887–8896. https://doi.org/10.3390/nu7115441

Smith KS, Greene MW, Babu JR, Frugé AD (2021) Psychobiotics as treatment for anxiety, depression, and related symptoms: a systematic review. Nutr Neurosci 24(12):963–977. https://doi.org/10.1080/1028415x.2019.1701220

Song W, Hu P, Guo S, Hu J, Song C, Wang T, Gao Z, Yue T (2021) Oxidative stress and endoplasmic reticulum stress contribute to *L. paracasei* subsp. *paracasei* M5L exopolysaccharide-induced apoptosis in HT-29 cells. Food Sci Nutr 9(3):1676–1687. https://doi.org/10.1002/fsn3.2142

Spichak S, Bastiaanssen TFS, Berding K, Vlckova K, Clarke G, Dinan TG, Cryan JF (2021) Mining microbes for mental health: determining the role of microbial metabolic pathways in human brain health and disease. Neurosci Biobehav Rev 125. https://doi.org/10.1016/j.neubiorev.2021.02.044

Steenbergen L, Sellaro R, van Hemert S, Bosch JA, Colzato LS (2015) A randomized controlled trial to test the effect of multispecies probiotics on cognitive reactivity to sad mood. Brain Behav Immun 48:258–264. https://doi.org/10.1016/j.bbi.2015.04.003

Sub Laban T, Saadabadi A (2022) Monoamine oxidase inhibitors (MAOI). StatPearls Publishing

Sugahara H, Odamaki T, Fukuda S, Kato T, Xiao J, Abe F, Kikuchi J, Ohno H (2015) Probiotic bifidobacterium longum alters gut luminal metabolism through modification of the gut microbial community. Sci Rep 5. https://doi.org/10.1038/srep13548

Suganya K, Koo B-S (2020) Gut–brain axis: role of gut microbiota on neurological disorders and how probiotics/prebiotics beneficially modulate microbial and immune pathways to improve brain functions. Int J Mol Sci 21(20):7551. https://doi.org/10.3390/ijms21207551

Sun C-Y, Zheng Z-L, Chen C-W, Lu B-W, Liu D (2022) Targeting gut microbiota with natural polysaccharides: effective interventions against high-fat diet-induced metabolic diseases. Front Microbiol 13. https://doi.org/10.3389/fmicb.2022.859206

Swanson KS, Gibson GR, Hutkins R, Reimer RA, Reid G, Verbeke K, Scott KP, Holscher HD, Azad MB, Delzenne NM, Sanders ME (2020) The international scientific Association for Probiotics and Prebiotics (ISAPP) consensus statement on the definition and scope of synbiotics. Nat Rev Gastroenterol Hepatol 17(11):687–701. https://doi.org/10.1038/s41575-020-0344-2

Tang WHW, Bäckhed F, Landmesser U, Hazen SL (2019) Intestinal microbiota in cardiovascular health and disease. J Am Coll Cardiol 73(16):2089–2105. https://doi.org/10.1016/j.jacc.2019.03.024

Taylor DM, Paton C, Kapur S (2012) The Maudsley prescribing guidelines in psychiatry. Wiley, pp 202–239

Thau L, Gandhi J, Sharma S (2022) Physiology, cortisol. StatPearls Publishing. https://pubmed.ncbi.nlm.nih.gov/30855827/

Tian P, O'Riordan KJ, Lee Y, Wang G, Zhao J, Zhang H, Cryan JF, Chen W (2020) Towards a psychobiotic therapy for depression: bifidobacterium breve CCFM1025 reverses chronic stress-induced depressive symptoms and gut microbial abnormalities in mice. Neurobiol Stress 12: 100216. https://doi.org/10.1016/j.ynstr.2020.100216

Tian P, Bastiaanssen TFS, Song L, Jiang B, Zhang X, Zhao J, Zhang H, Chen W, Cryan JF, Wang G (2021) Unraveling the microbial mechanisms underlying the psychobiotic potential of a bifidobacterium breve strain. Mol Nutr Food Res 65(8):e2000704. https://doi.org/10.1002/mnfr.202000704

Tian P, Chen Y, Zhu H, Wang L, Qian X, Zou R, Zhao J, Zhang H, Qian L, Wang Q, Wang G, Chen W (2022) Bifidobacterium breve CCFM1025 attenuates major depression disorder via regulating gut microbiome and tryptophan metabolism: a randomized clinical trial. Brain Behav Immun 100:233–241. https://doi.org/10.1016/j.bbi.2021.11.023

Tiller J (2013) Depression and anxiety. Med J Aust 199(6):S28–S31. https://doi.org/10.5694/mja12.10628

Tsukuda N, Yahagi K, Hara T, Watanabe Y, Matsumoto H, Mori H, Higashi K, Tsuji H, Matsumoto S, Kurokawa K, Matsuki T (2021) Key bacterial taxa and metabolic pathways affecting gut short-chain fatty acid profiles in early life. ISME J 15(9):2574–2590. https://doi.org/10.1038/s41396-021-00937-7

U.S. National Library of Medicine (2023) Home – ClinicalTrials.gov. NIH U.S. National Library of Medicine. https://clinicaltrials.gov/

Valles-Colomer M, Falony G, Darzi Y, Tigchelaar EF, Wang J, Tito RY, Schiweck C, Kurilshikov A, Joossens M, Wijmenga C, Claes S, Van Oudenhove L, Zhernakova A, Vieira-Silva S, Raes J (2019) The neuroactive potential of the human gut microbiota in quality of life and depression. Nat Microbiol 4(4):623–632. https://doi.org/10.1038/s41564-018-0337-x

Valles-Colomer M, Blanco-Míguez A, Manghi P, Asnicar F, Dubois L, Golzato D, Armanini F, Cumbo F, Huang KD, Manara S, Masetti G, Pinto F, Piperni E, Punčochář M, Ricci L, Zolfo M, Farrant O, Goncalves A, Selma-Royo M, Binetti AG (2023) The person-to-person transmission landscape of the gut and oral microbiomes. Nature 614:125–135. https://doi.org/10.1038/s41586-022-05620-1

van de Wouw M, Boehme M, Lyte JM, Wiley N, Strain C, O'Sullivan O, Clarke G, Stanton C, Dinan TG, Cryan JF (2018) Short-chain fatty acids: microbial metabolites that alleviate stress-induced brain-gut axis alterations. J Physiol 596(20):4923–4944. https://doi.org/10.1113/jp276431

Vidot H, Cvejic E, Finegan LJ, Shores EA, Bowen DG, Strasser SI, McCaughan GW, Carey S, Allman-Farinelli M, Shackel NA (2019) Supplementation with synbiotics and/or branched chain amino acids in hepatic encephalopathy: a pilot randomised placebo-controlled clinical study. Nutrients 11(8):1810. https://doi.org/10.3390/nu11081810

Wallace CJK, Milev R (2017) The effects of probiotics on depressive symptoms in humans: a systematic review. Ann Gen Psychiatry 16(1). https://doi.org/10.1186/s12991-017-0138-2

Wang H, Braun C, Murphy EF, Enck P (2019a) Bifidobacterium longum 1714™ strain modulates brain activity of healthy volunteers during social stress. Am J Gastroenterol 114(7):1152–1162. https://doi.org/10.14309/ajg.0000000000000203

Wang J-W, Kuo C-H, Kuo F-C, Wang Y-K, Hsu W-H, Yu F-J, Hu H-M, Hsu P-I, Wang J-Y, Wu D-C (2019b) Fecal microbiota transplantation: review and update. J Formos Med Assoc 118(1): S23–S31. https://doi.org/10.1016/j.jfma.2018.08.011

Wang Z, Liu S, Xu X, Xiao Y, Yang M, Zhao X, Jin C, Hu F, Yang S, Tang B, Song C, Wang T (2022) Gut microbiota associated with effectiveness and responsiveness to mindfulness-based cognitive therapy in improving trait anxiety. Front Cell Infect Microbiol 12:719829. https://doi.org/10.3389/fcimb.2022.719829

Whisner CM, Castillo LF (2017) Prebiotics, bone and mineral metabolism. Calcif Tissue Int 102(4): 443–479. https://doi.org/10.1007/s00223-017-0339-3

Wiertsema SP, van Bergenhenegouwen J, Garssen J, Knippels LMJ (2021) The interplay between the gut microbiome and the immune system in the context of infectious diseases throughout life and the role of nutrition in optimizing treatment strategies. Nutrients 13(3):886. https://doi.org/10.3390/nu13030886

Williams S, Chen L, Savignac HM, Tzortzis G, Anthony DC, Burnet PW (2016) Neonatal prebiotic (BGOS) supplementation increases the levels of synaptophysin, GluN2A-subunits and BDNF proteins in the adult rat hippocampus. Synapse 70(3):121–124. https://doi.org/10.1002/syn.21880

Wilson M, Tripp J (2023) Clomipramine. StatPearls Publishing. https://www.ncbi.nlm.nih.gov/books/NBK541006/#article-42905.s5

Wittenberg GM, Stylianou A, Zhang Y, Sun Y, Gupta A, Jagannatha PS, Wang D, Hsu B, Curran ME, Khan S, Chen G, Bullmore ET, Drevets WC (2019) Effects of immunomodulatory drugs on depressive symptoms: a mega-analysis of randomized, placebo-controlled clinical trials in inflammatory disorders. Mol Psychiatry 25(6). https://doi.org/10.1038/s41380-019-0471-8

Woo YS, Rosenblat JD, Kakar R, Bahk W-M, McIntyre RS (2016) Cognitive deficits as a mediator of poor occupational function in remitted major depressive disorder patients. Clin Psychopharmacol Neurosci 14(1):1–16. https://doi.org/10.9758/cpn.2016.14.1.1

Wu W-L, Adame MD, Liou C-W, Barlow JT, Lai T-T, Sharon G, Schretter CE, Needham BD, Wang MI, Tang W, Ousey J, Lin Y-Y, Yao T-H, Abdel-Haq R, Beadle K, Gradinaru V, Ismagilov RF, Mazmanian SK (2021a) Microbiota regulate social behaviour via stress response neurons in the brain. Nature 595(7867):409–414. https://doi.org/10.1038/s41586-021-03669-y

Wu G, Zhao N, Zhang C, Lam YY, Zhao L (2021b) Guild-based analysis for understanding gut microbiome in human health and diseases. Genome Med 13(1). https://doi.org/10.1186/s13073-021-00840-y

Ye X, Wang D, Zhu H, Wang D, Li J, Tang Y, Wu J (2021) Gut microbiota changes in patients with major depressive disorder treated with vortioxetine. Front Psych 12. https://doi.org/10.3389/fpsyt.2021.641491

Zanos P, Gould TD (2018) Mechanisms of ketamine action as an antidepressant. Mol Psychiatry 23(4):801–811. https://doi.org/10.1038/mp.2017.255

Zhang W, Qu W, Wang H, Yan H (2021) Antidepressants fluoxetine and amitriptyline induce alterations in intestinal microbiota and gut microbiome function in rats exposed to chronic unpredictable mild stress. Transl Psychiatry 11(1). https://doi.org/10.1038/s41398-021-01254-5

Zheng P, Zeng B, Zhou C, Liu M, Fang Z, Xu X, Zeng L, Chen J, Fan S, Du X, Zhang X, Yang D, Yang Y, Meng H, Li W, Melgiri ND, Licinio J, Wei H, Xie P (2016) Gut microbiome remodeling induces depressive-like behaviors through a pathway mediated by the host's metabolism. Mol Psychiatry 21(6):786–796. https://doi.org/10.1038/mp.2016.44

Zimmermann M, Zimmermann-Kogadeeva M, Wegmann R, Goodman AL (2019) Mapping human microbiome drug metabolism by gut bacteria and their genes. Nature 570(7762):462–467. https://doi.org/10.1038/s41586-019-1291-3

Anti-inflammatory Drugs in the Treatment of Depression

Jonathan T. Cavanagh

Contents

Abstract The last two decades have seen a flourishing of research into the immunobiology of psychiatric phenotypes, in particular major depressive disorder. Both preclinical and clinical data have highlighted pathways and possible mechanisms that might link changes in immunobiology, most especially inflammation, to clinically relevant behaviour. From a therapeutics perspective, a major impetus has been the action of Biologics, often monoclonal antibodies, that target specific cytokines acting as "molecular scalpels" helping to uncover the actions of those proteins. These interventions have been associated with improvements in mood and

J. T. Cavanagh (✉)
Centre for Immunobiology, School of Infection and Immunity, College of Medical, Veterinary and Life Sciences, University of Glasgow, Glasgow, UK
e-mail: Jonathan.Cavanagh@glasgow.ac.uk

Curr Topics Behav Neurosci (2024) 66: 217–232
https://doi.org/10.1007/7854_2023_459
Published Online: 20 December 2023

related symptoms. There are now enough studies and participants to permit meta-analytic analyses of the actions of these and other anti-inflammatory agents.

In this chapter, the focus is on the evidence for the role of inflammation biology in depression and the meta-analytic data from trials. The putative mechanisms that might underpin the antidepressant effect of anti-inflammatory drugs are also explored. Lastly, I describe the more stubborn difficulties around heterogeneity, deep phenotyping and stratification as well as improved animal models and greater understanding of the biology that might be addressed by future studies.

Keywords Biologics · Inflammation · Major depression

1 Immune-Mediated Inflammation in the Neurobiology of Depression

1.1 General Background

The challenges of understanding existing drug mechanisms of action and for new drug development in depression include the clinical heterogeneity of the condition, poorly predictive animal models, and a lack of clear molecular targets. An immunobiological explanation of at least part of the variance in the condition is a potential solution to some of these challenges.

Over the last 20 years or so, the role of inflammation biology in depression has attracted considerable contemporary scientific interest. The definition of inflammation is the *immune system's response to harmful stimuli*, such as pathogens, damaged cells, toxic compounds, etc. There are compelling data implicating these mechanisms in at least a subset of people suffering from major depressive disorder (MDD). Preclinical evidence implicates the effects of proinflammatory cytokines on monoaminergic neurotransmission, neurotrophic factors, and measures of synaptic plasticity. Clinical neuroimaging studies reveal the potential consequences of these changes on neural connectivity and clinical trial data highlight the beneficial effects of immune modulation on depression with and without comorbid immune-mediated inflammatory disease (IMID).

Evidence from therapeutic data includes (a) de novo occurrence of depression in individuals treated with therapeutic cytokines, e.g. the type 1 interferon, IFN-α and the continued remitting/relapsing pattern of depression despite the cessation of the IFN-α treatment (Machado et al. 2017; Chiu et al. 2017). (b) Increased levels of proinflammatory cytokines and chemokines in the peripheral blood and CSF of depressed individuals in contrast to healthy controls. This has been shown in meta-analytic data of cytokines such as IL-6, IL-1β and TNF-α and the acute phase protein CRP (Eyre et al. 2016). Chemokines, including increased CCL2, CXCL4 and CXCL7 but also lower levels of CCL4 have been related in inflammatory disorders and associated depression (Leighton et al. 2018). (c) The antidepressant effects of anti-inflammatory therapies, especially biologics which is particularly relevant for those with evidence of inflammation (Drevets et al. 2022). (d) A strong association noted between resistance to conventional antidepressant therapy and altered immune status (Chamberlain et al. 2019).

1.2 Genetic Data

There is further support for the role of immunobiological mechanisms from the recent crop of GWAS, genetic and epigenetic meta-analyses. In general, GWAS of MDD have identified 178 genetic risk loci and proposed more than 200 candidate genes including immune-relevant loci (Levey et al. 2021). Data from one of the largest GWAS in MDD identified cytokine and immune response pathways (Wray et al. 2018). A Mendelian randomization study reported a functional SNP in the IL-6 receptor promoter region which predicted levels of IL-6, CRP and depressive symptom severity in a population cohort (Khandaker et al. 2018).

Kappelmann et al. (2021) used large-scale prior GWAS data to analyse the effects of CRP, IL-6 and body mass index (BMI) on depressive symptoms and identified BMI as putatively causal for several depressive symptoms and upregulated genes for IL-6 for suicidal ideation.

1.3 Epigenetics

Genome-wide methylation studies have highlighted differentially methylated regions in immune-inflammation-related genes. For example, in post-mortem brain tissue taken from patients with depression who died by suicide, highlighted the psoriasis susceptibility gene (PSORS1C3) and TAPBP located in the MHC encoding a glycoprotein that mediates interaction between MHC class1 molecules and transporter associated with antigen processing (Murphy 2017).

Most recently, Lynall et al. (2022) used three independent epigenetic datasets representing multiple organ systems and immune cell subsets. Trans-diagnostic and cis-diagnostic risk variants (for schizophrenia and depression) were found to be enriched at epigenetically active sites in brain tissues and in lymphoid cells, especially stimulated $CD4^{+}$ T cells, but not myeloid cells.

1.4 Cell Biology in Depression

1.4.1 Cell Populations

As described above, the evidence for increased inflammatory cytokine and chemokine levels in depression enjoys the support of meta-analytic data, but the cellular source of this remains rather opaque and the literature is conflicted as to the roles of different immune cell populations. To date several studies have explored the immune cell sub-type populations in depression.

1.4.2 Peripheral

In terms of the peripheral cellular immunophenotype in depression, Foley et al. (2023) meta-analysed 27 studies ($n = 2{,}277$) and found changes in both myeloid and lymphoid populations including increased mean absolute counts of the total white blood cell count. The following cells were involved: granulocytes, neutrophils, monocytes, CD4+ T cells, NK cells, CD19+ B cells and activated T cells, thus characterized by widespread alterations in circulating myeloid and lymphoid cells, consistent with dysfunction in both innate and adaptive immunity.

This meta-analysis also highlighted the relatively low number of studies looking at cell subtypes such as lymphocytes and granulocytes. However, it remains unclear as to whether and how altered immune cells can causally influence the risk of depression.

1.4.3 Transcriptional Differences

Whole-transcriptome studies offer another variant of the genome-wide search for disease-related mechanisms by measuring mRNA expression levels of each gene in a relevant tissue. One of the largest studies examined a total sample of 922 people (463 with MDD and 459 healthy controls) and sequenced RNA from whole blood (Mostafavi et al. 2014). A relatively small number of genes were found to be associated with MDD (29) at a relaxed false discovery rate (FDR) threshold of 0.25. With the more customary and restrictive FDR threshold of 0.05, no significant genes were found. They also showed modest enrichment for the IFN α/β pathway, which included three significant genes at FDR < 0.25. However, examining a specific cell subtype (PBMCs) rather than whole blood, Cole et al. (2021) found no evidence for differential gene expression in major depressive disorder compared to healthy controls. This is an active field of research and there is, as yet, no settled view as to the cell-specific transcriptional signature in depression. It's worth reflecting on the potential role of other cell types in this context.

The other major group of non-PBMC leukocytes, neutrophils, produce a large variety of chemokines and cytokines upon stimulation and can differentially switch phenotypes, displaying distinct subpopulations in different microenvironments (Rosales 2018).

Similarly, *endothelial cells* are ubiquitous throughout brain and body and Blank et al. (2016) demonstrated a specific role in relation to aspects of depression-relevant behaviour in mice by showing that downstream signalling of brain endothelial cells induces fatigue and cognitive impairment via impaired neurotransmission in the hippocampus. Nevertheless, considering findings presented in a recent GWAS of

MDD, it is important to consider that peripheral tissue biomarkers may be less relevant than brain tissue biomarkers. Wray et al. (2018) integrated their GWAS data with functional genomic data, comparing their findings with bulk tissue RNAseq from genotype tissue expression (GTEx). Here only brain tissue showed enrichment, with the areas showing the most significant enrichment being cortical. This was in contradistinction to other tissue types, including whole blood.

However, the relative inaccessibility of the brain to biomarker testing makes this an impractical source of biomarker measurement or anti-inflammatory response index.

1.4.4 Cells Within the CNS

As discussed above, PBMCs are a key source of peripheral cytokines and pre-clinical models have suggested some PBMC subsets can enter the brain and contribute to the onset of sickness behaviour in the context of stress. Monocytes recruited to the brain have been linked to behavioural changes associated with anxiety and with direct effects on neuronal dendritic spine remodelling linked to cognitive deficits (McKim et al. 2018; Menard et al. 2017; Garré et al. 2017). Similarly, CD4 + T cells have been linked to stress-related behavioural changes via mitochondrial fission leading to xanthine upregulation (Fan et al. 2019). There is no confirmation of these phenomena in humans yet, and again there is no obvious practical use in the context of drug treatment response assessment.

It remains unclear how cellular "machinery" can contribute to the biological environment that either promotes or maintains depressive phenotypes and this is how anti-inflammatory interventions might operate at this level. In this next section, we will explore one possible mechanism.

1.5 Cellular Mechanisms: The Inflammasome

Inflammasomes are central to the innate immune response via recognition of pathogen and damage-associated molecular patterns (PAMPS and DAMPS), homeostasis-altering molecular processes (HAMPs) and effector-triggered immunity (ETI). Inflammatory caspases act upstream or downstream of inflammasome assembly and are the principal enzymes controlling the activities of these organelles. There are several variants including NLRP1, CARD8, NLRP6, NLRC4/NAIP, AIM2, pyrin, caspases-4/-5/-11 and NLRP3 (Barnett et al. 2023). In this section, I will focus on the Nod-like receptor protein 3 (NLPR3) inflammasome as this is regarded as a key contributor to neuroinflammation. Specifically, the NLRP3 inflammasome complex regulates adaptive and innate immunity and promotes the maturation and secretion of IL-1β and IL-18 during the immune response to damage-associated molecular patterns (DAMPs) by activating caspase-1 (Miller and Raison 2016).

There is evidence that levels of NLRP3, casp-1 and IL1β are increased in the serum of patients with depression and in rodent models. Mouse behaviours consistent with depression are improved by inhibition of the NLRP3.

Most recently, Li et al. (2022) have shown that NLRP3 inflammasome, specifically in microglia, activates astrocytes to produce neurotoxins in a mouse model of depression-like behaviour.

DAMPs associated with depression include high mobility group box-1 (HMGB-1) and extracellular ATP. Another is circulating cell-free mitochondrial DNA. In depression there is evidence of mitochondrial dysfunction and lack of autophagy with loss of mitochondrial quality control leading to excessive DAMPs release which can additionally promote NLRP3 inflammasome production (Song et al. 2020). Furthermore, the NLRP3 inflammasome being activated also appears to cause glucocorticoid resistance which suggests it could mitigate anti-inflammatory effects of endogenous corticosteroids. The P2X7 receptor is a potential target for novel antidepressant therapy (see below).

1.6 Adiposity and Aging

While in depression there is consistent evidence for raised IL-6 and CRP, these effects reduce when BMI is controlled for. There is a metabolic aspect of the function of IL-6, which promotes lipolysis and IL-6 levels reflect the storage of energy in the form of deposited fat. The depression symptom profile is also important here – atypical features have been associated with elevated sIL-6R, indicative of upregulated trans-signalling, while melancholic features were not. Atypical features also correlate with obesity and higher BMI. In the depression GWAS of over 1.2 million individuals, in addition to immune relevant genes such as *TRAF3* linked to type1 interferon response, significant associations were reported for *CCDC71* linked to cell lipid metabolism and fat cell differentiation and *FADS1* linked to fatty acid regulation (Levey et al. 2021).

Linked to disrupted fat metabolism is accelerated biological ageing seen in MDD. Immune cell senescence has a well-documented effect on both epigenome and transcriptome (Benayoun et al. 2019). MDD has also been linked to the senescence-associated secretory phenotype (SASP), a dynamic secretory molecular pathway indicative of cellular senescence. MDD exhibited greater molecular senescence in young and middle-aged adults by examining the impact of MDD on the senescence-associated secretory phenotype (SASP), a dynamic secretory molecular pathway indicative of cellular senescence (Diniz et al. 2019). More severe episodes of depression present with higher SASP indices and a significant interaction between current MDD episode and obesity, thus comorbid current MDD in addition to being overweight had the highest SASP index.

This complex interaction between depression, immune factors and obesity/fat metabolism illustrates the importance of careful phenotyping and patient stratification in studies that are assessing or trialling anti-inflammatory interventions.

2 Evidence for the Antidepressant Effects of Anti-inflammatory Medication

2.1 *Drug Types*

A considerable variety of drugs with primarily anti-inflammatory action have been noted for their antidepressant effects. The data are available from three main sources: trials of immunomodulatory therapies for other diseases in which depressive symptoms were measured as a secondary factor; randomized controlled trials of immunomodulatory therapies where change in depressive symptoms is the primary outcome; meta-analyses of change in depressive symptoms from clinical trials of immunomodulation in IMIDs. The targets of these drugs are also varied and include specific (proinflammatory) cytokines and/or relevant receptors, for example, TNF and IL-6 and the systemic inflammatory response via prostaglandins (non-steroidal anti-inflammatory drugs).

There are also studies of drugs where the anti-inflammatory effect is secondary to the primary mode of action such as statins, glucocorticoids and specific drugs such as the antibiotic minocycline and anti-diabetic pioglitazone. The bulk of evidence for the beneficial effects of anti-inflammatories on depressive symptoms is drawn from meta-analytic studies of cytokine antagonism or NSAIDS.

2.2 *Meta-Analytic Data of Drug Effects*

Secondary data analysis is a key issue when thinking about the literature on antidepressant effects of anti-inflammatory drugs. While a number of RCTs that have depression as the primary outcome have been registered and have begun recruitment, these remain in the minority in terms of the data available.

There are several recurring difficulties in this literature including the heterogeneity of MDD itself, patient population diversity, and variance in outcome measures. Specific to studies where depression is comorbid or a secondary measure, those patients with a primary IMID (immune-mediated inflammatory disorder) who also have a significant score on MDD are in the minority, so it is more difficult to detect an antidepressant effect. Parsing mood from pain and fatigue can be challenging.

2.2.1 Meta-Analyses Studies with Depression as a Secondary Measure

Köhler et al. (2014) reported a meta-analysis on trials of NSAIDS (10 studies) and TNF or IL-12/23 inhibition (4 studies) representing a total of >6,000 participants. There were significant pooled effects on depressive symptoms with positive standardized mean difference for NSAIDS (SMD 0.27) and for cytokine inhibition (SMD 0.38).

Kappelmann et al. (2018) found a significant effect for cytokine inhibition (SMD 0.4) and especially for TNF inhibition (SMD 0.33) in a meta-analysis of RCTs that included a total of >2,000 participants. Köhler-Forsberg et al. (2019) reported on a broader group of interventions and found the following pooled positive effects for glucocorticoids (SMD 0.9); minocycline (SMD 0.87); cytokine inhibitors (SMD 0.56); NSAIDS (SMD 0.4) and statins (SMD 0.26).

Wittenberg et al. (2020) carried out a meta-analysis on data from 18 RCTS conducted by two pharmaceutical companies (total patient population of >10,700). The large sample size afforded the opportunity of stratifying based on high and low depressive symptoms at baseline, a key issue in ascertaining the effectiveness of anti-inflammatory medication. Similarly, it was possible to adjust for potential confounds such as pain. The study reported that anti-IL-6 (SMD 0.8) and anti-IL-12/23 (SMD 0.48) treatments demonstrated the greatest benefit in depressive symptom improvement.

In a recent systematic review of all existing meta-analyses by Simon et al. (2023), no clear recommendations for treatment were made due to heterogeneity, diversity of patient populations, treatment regimen, and outcomes, and limited scientific quality. These are all key problem issues for the field. The authors did, however, note that meta-analytic evidence was strongest for celecoxib and weakest for omega 3 fatty acids.

In terms of specifically addressing the question of depression as the primary disease and effectiveness of anti-inflammatories in reducing depressive symptoms, a pioneering trial using a cytokine antagonist explicitly for depression by Raison et al. (2013) found that TNF antagonism did not have generalized efficacy in treatment-resistant depression but did improve depressive symptoms in patients with high baseline inflammatory biomarkers. It is perhaps unsurprising that antidepressant efficacy of an anti-inflammatory may only be evident in the context of ongoing inflammation. Hellmann-Regen (2022) published a large randomized clinical trial with minocycline at a dose of 200 mg/d added to antidepressant treatment as usual for 6 weeks, minocycline was well tolerated but not superior to placebo in reducing depressive symptoms in patients with treatment-resistant depression, and stratification of patients by peripheral immune markers did not influence the effect of minocycline in this study.

The protocols of several planned RCTs are now published, for example, the protocol for the efficacy of celecoxib as an adjuvant treatment for immune-metabolic depression (Zwiep et al. 2022).

In the summary above we see that meta-analyses provide evidence of improvement in depressive symptoms with NSAIDS (e.g. celecoxib); cytokine inhibition (e.g. IL-6, TNF, IL-12/23); glucocorticoids; statins and minocycline. The variety of interventions that have a putative antidepressant effect might be an indicator of the biological diversity that underpins the mechanisms that lead to similar phenotypes described under the heading of major depressive disorder. This illustrates the pressing need to understand these mechanisms in their diversity.

3 The (Possible) Antidepressant Mechanism of Anti-inflammatory Drugs

It is important to remember that as yet no drug has been developed de novo for an immune target in depression as the primary disease. The immune-inflammatory mechanisms that might underpin the resolution of depression are the subject of active research, but below is a summary of what are thought to be the key mechanisms that link inflammation to depression and thus are candidates for the antidepressant mode of action of anti-inflammatory interventions.

Synaptic transmission has been a key target for modern antidepressant therapy such as SSRIs, and inflammation is known to have an impact on this.

3.1 Glutamate

Inflammation influences both glutamate release and reuptake mechanisms. Inflammatory cytokines can amplify and alter the magnitude and propagation of intracellular Ca^{2+} influx to transfer immune signals and mediate changes in neural activity/signalling and synaptic plasticity (Haroon and Miller 2017, 2017). Thus, systemic inflammation may impact on gliotransmission through increased glutamate release. Taking the specific example of the cytokine TNF, this stimulates TNFR1 receptors, expressed on hippocampal neurons, to increase extra-synaptic glutamate levels by inducing glutamate exocytosis and inhibiting glutamate reuptake. Thus excess TNF-α signalling leads to glutamate-mediated excitotoxicity resulting in chaotic and incoherent neuronal signalling activity in the short term, and synaptic toxicity by the suppression of intracellular survival mechanisms in the long term. Of therapeutic relevance, experimental rodent models have shown that anti-TNF treatment, with etanercept, lowers brain glutamate levels (Clark and Vissel 2016). However, the link between the effects of anti-cytokine therapy on glutamate and improved depressive symptoms in humans remains to be made convincingly.

3.2 Serotonin

Serotonin is a key neurotransmitter modulating multiple processes intrinsic to depression, including emotional regulation, reward, memory and cognition (Massart et al. 2012). As with other monoamine transporters, the serotonin transporter (5-HTT) is central to the homeostasis of neurotransmission in the active reuptake of serotonin from the synapse. This protein is linked to both depression itself and to its treatment via commonly prescribed SSRIs (Savitz and Drevets 2013; Couch et al. 2013). In preclinical studies, raised proinflammatory cytokines have been correlated with increased 5-HTT expression. Anti-TNF therapy has been shown to increase the scope of MAPK-dependent 5-HTT uptake along with maximum uptake velocity and mRNA expression of 5-HTT (Malynn et al. 2013; Krishnadas et al. 2016).

One recent human study using [123I]-beta-CIT SPECT brain imaging and assessing circulating cytokine concentrations found no major association between higher levels of TNF and dopamine transporter (DAT) availability. However, increased TNF positively correlated with 5-HTT brainstem availability suggesting that the main effect of raised inflammation was on 5-HTT availability rather than DAT. There is further clinical evidence of this relationship with decreased brainstem 5-HTT availability following treatment with anti-TNF biologic therapy (Krishnadas et al. 2016).

Again, as with glutamate, the link remains to be made between antagonism of specific inflammatory molecules, the impact of this on monoamine neurotransmission, and improved depression.

3.3 IDO and the Kynurenine Pathway

Serotonin is derived from the essential amino acid tryptophan (TRY). The bulk of TRY metabolism (~95%) is through the Kynurenine pathway; pro-inflammatory cytokines and cortisol can upregulate indoleamine 2,3-dioxygenase (IDO), a key enzyme in this pathway.

Upregulation of IDO in the context of inflammation shunts tryptophan away from serotonin synthesis, reducing the bioavailability of serotonin, and toward the production of neuroactive kynurenine metabolites. One such metabolite is quinolinic acid (QA), an N-Methyl-D-aspartate (NMDA) agonist. Increases in QA have been correlated with higher extra-synaptic glutamate within the brain and this in turn leads to neuronal and oligodendrocyte cell damage/death through excitotoxicity (Ogbechi et al. 2020; Huang et al. 2020; Marszalek-Grabska et al. 2021). Most studies on these metabolites have been conducted on murine samples, but one human study showed a higher serum concentration of QA in suicidal patients compared to healthy controls (Steiner et al. 2011).

In some, but not all, studies (see summary of meta-analyses above) minocycline has been shown to improve depressive symptoms. This drug is a brain-penetrant tetracycline antibiotic known to inhibit proinflammatory cytokine release and to promote anti-apoptotic and anti-inflammatory action. As a result of this inhibition of cytokine release, minocycline can indirectly reduce activation of the kynurenine pathway with subsequent reduction in NMDA agonists like QA.

However, as with the other anti-inflammatory interventions described above, a convincing link between the putative mechanisms explored here and sustained improvement in depressive symptoms remains to be made.

3.4 Measuring Mechanism in Clinical Studies

Experimental medicine studies involving clinical populations are aimed at uncovering mechanisms in humans. While these have been prominent in other clinical areas, they remain relatively rare in psychiatry. Neuroimaging technologies combined with biomarker measurements offer a route into understanding the mechanisms that underlie the antidepressant effects of anti-inflammatory treatments.

Magnetic resonance spectroscopy (MRS) allows brain measurement of various neurochemicals. Of relevance to this area is the improved resolution of high-field-strength MRS for neurotransmitters such as GABA and glutamate. For example, Haroon et al. (2016) reported higher glutamate concentrations in MDD patients with evidence of peripheral inflammation (raised CRP).

Magnetic resonance imaging (MRI) methods such as functional MRI constitute the majority of neuroimaging studies in this area. A meta-analysis of studies that were both correlative (fMRI metrics with blood biomarkers of inflammation) and experimental (imaging before and after inflammatory stimuli) revealed consistent evidence that systemic inflammation is related to changes in connectivity within limbic, prefrontal and temporal cortices and related subcortical nuclei (Felger et al. 2016).

Molecular imaging such as positron emission tomography (PET) allows ligand imaging of inflammation-relevant targets such as TSPO expressed on mitochondrial membranes (e.g. [^{11}C] PK11195 and [^{11}C] PBR28. Typically, this has been used to determine microglial activity, but TSPO is not specific to microglia, being expressed also on other glial cells and on neurons themselves.

4 Future

The innate heterogeneity of the construct, major depressive disorder, is a constant refrain in the literature. This fundamental issue continues to bedevil attempts to understand mechanisms, develop new targets and thus new interventions.

Flint (2023) raises this most eloquently in the discussion around genetic findings in Depression. The phenotyping is variable at best and minimal at worst. Flint argues that there is a large genetic component unique to MDD that remains inaccessible to minimal phenotyping strategies and the majority of genetic risk loci identified with minimal phenotyping are unlikely to be MDD risk loci. Inflammatory biomarkers are one possible way of reducing the heterogeneity by stratifying patient cohorts according to readily available measures.

Having said that, understanding the "bio" in the biomarkers is a pressing issue – especially given that, for the foreseeable future, most biomarkers are in the peripheral blood – very distinctive from the brain. At present, stratification according to inflammatory biomarkers remains a realistic prospect but with the caveat that the link between putative mechanisms and "biomarkers" remains unclear.

A key question is what mechanisms drive the improvement in depressive symptoms with anti-inflammatory drugs. To address this, new ways of modelling the disorder preclinically are desperately needed. More realistic and ecologically valid in vivo models are essential alongside ex vivo and in silico modelling.

Linked to this is the need for a greater understanding of the immunology driving the inflammatory responses seen in depression. At present there is evidence of both innate and adaptive immune responses but little information as to the instigators of these.

In parallel, experimental medicine methods using clinical populations are developing using anti-inflammatory agents as tools to dissect neural mechanisms in humans. This has been helped significantly by the rapid development of high-field-strength neuroimaging technologies.

5 Conclusions

There are indeed compelling data supportive of a role for inflammation biology in the pathophysiology of major depressive disorder. This is based on extensive preclinical data and increasing clinical data. The latter is based in major on the evidence from clinical trials – with depression as a secondary outcome still the most common.

The variety of anti-inflammatory drugs with antidepressant effects is striking and speaks to the multiple possible mechanisms that might underpin this neurobiology. These different mechanisms of action are variously tractable in terms of therapeutic targeting, and these are being further explored in existing animal models.

There is a clear need for better-designed and more carefully controlled RCTs. This will involve a more careful consideration of how we can reduce causes of heterogeneity, including the linking of particular symptom clusters to inflammatory signals. This brings into focus the need to develop better more holistic animal behaviour read-outs that will translate to symptoms in humans more convincingly than they do presently.

There is a critical need for studies that will elucidate the relevant mechanistic pathways that explain the interaction between inflammatory changes and neural responses. Part of the difficulty is modelling chronic, low-level inflammation within the preclinical setting. Another issue is relating the neural changes to behaviour. The advances within clinical neuroimaging are making the possibility of mechanistically sensitive studies in humans more realistic. But the need to explore the molecular biology will require ever more sophisticated preclinical models if we are to deliver on new molecular targets that are tractable therapeutically. Only then can the prospect of a specific immunomodulator for depression be a realistic prospect.

This field is a relatively new departure for drug development in psychiatry and will need more significant academic–industry collaboration. The testing of libraries of tool compounds is one place to start as well as considering the realities of how drugs are re-purposed for depression when they are designed for quite different diseases, but where the symptoms of co-morbid depression are clear and contribute to excess morbidity.

References

Benayoun BA, Pollina EA, Singh PP, Mahmoudi S, Harel I, Casey KM et al (2019) Remodeling of epigenome and transcriptome landscapes with aging in mice reveals widespread induction of inflammatory responses. Genome Res 29:697–670

Blank T, Detje CN, Spieß A, Hagemeyer N, Brendecke SM, Wolfart J et al (2016) Brain endothelial- and epithelial-specific interferon receptor chain 1 drives virus-induced sickness behavior and cognitive impairment. Immunity 44:901–912

Barnett KC, Li S, Liang K, Ting JP (2023) A 360° view of the inflammasome: mechanisms of activation, cell death, and diseases. Cell 186(11):2288–2312

Chamberlain S, Cavanagh J, De Boer P, Mondelli V, Jones D, Drevets W, Bullmore E (2019) Treatment-resistant depression and peripheral C-reactive protein. Br J Psychiatry 214(1):11–19. https://doi.org/10.1192/bjp.2018.66

Chiu W, Su Y, Su K (2017) Chen P recurrence of depressive disorder after interferon-induced depression. Transl Psychiatry 7:e 1026

Clark IA, Vissel B (2016) Excess cerebral TNF causing glutamate excitotoxicity rationalizes treatment of neurodegenerative diseases and neurogenic pain by anti-TNF agents. J Neuroinflammation 13(1):1–6

Cole JJ, McColl A, Shaw R, Lynall M-E, Cowen PJ, de Boer P, Drevets WC, Harrison N, Pariante C, Pointon L, NIMA Consortium, Goodyear C, Bullmore E, Cavanagh J (2021) No evidence for differential gene expression in major depressive disorder PBMCs, but robust evidence of elevated biological ageing. Transl Psychiatry 11:404. https://doi.org/10.1038/s41398-021-01506-4

Couch Y, Anthony DC, Dolgov O, Revischin A, Festoff B, Santos AI, Steinbusch HW, Strekalova T (2013) Microglial activation, increased TNF and SERT expression in the prefrontal cortex define stress-altered behaviour in mice susceptible to anhedonia. Brain Behav Immun 29:136–146

Diniz BS, Reynolds Iii CF, Sibille E, Bot M, Penninx B (2019) Major depression and enhanced molecular senescence abnormalities in young and middle-aged adults. Transl Psychiatry 9:198

Drevets WC, Wittenberg GM, Bullmore ET, Manji HK (2022) Immune targets for therapeutic development in depression: towards precision medicine. Nat Rev Drug Discov 21(3):224–244

Eyre HA, Air T, Pradhan A, Johnston J, Lavretsky H, Stuart MJ, Baune BT (2016) A meta-analysis of chemokines in major depression. Prog Neuropsychopharmacol Biol Psychiatry 68:1–8

Fan KQ, Li YY, Wang HL, Mao XT, Guo JX, Wang F et al (2019) Stress-induced metabolic disorder in peripheral CD4+ T cells leads to anxiety-like behavior. Cell 179:864–879

Felger JC et al (2016) Inflammation is associated with decreased functional connectivity within corticostriatal reward circuitry in depression. Mol Psychiatry 21:1358–1365

Flint J (2023) The genetic basis of major depressive disorder. Mol Psychiatry. https://doi.org/10.1038/s41380-023-01957-9

Foley ÉM, Parkinson JT, Mitchell RE, Turner L, Khandaker GM (2023) Peripheral blood cellular immunophenotype in depression: a systematic review and meta-analysis. Mol Psychiatry 28(3):1004–1019. https://doi.org/10.1038/s41380-022-01919-7

Garré JM, Silva HM, Lafaille JJ, Yang G (2017) CX3CR1(+) monocytes modulate learning and learning-dependent dendritic spine remodeling via TNF-alpha. Nat Med 23:714–722

Haroon E et al (2016) Conceptual convergence: increased inflammation is associated with increased basal ganglia glutamate in patients with major depression. Mol Psychiatry 21:1351–1357

Haroon E, Miller AH (2017) Inflammation effects on brain glutamate in depression: mechanistic considerations and treatment implications. Inflammation-associated depression: evidence, mechanisms and implications, pp 173–198

Haroon E, Miller AH, Sanacora G (2017) Inflammation, glutamate, and glia: a trio of trouble in mood disorders. Neuropsychopharmacology 42(1):193–215

Hellmann-Regen J, Clemens V, Grözinger M, Kornhuber J, Reif A, Prvulovic D, Goya-Maldonado-R, Wiltfang J, Gruber O, Schüle C, Padberg F, Ising M, Uhr M, Friede T, Huber C, Manook A, Baghai TC, Rupprecht R, Heuser I (2022) JAMA Netw Open 5(9):e2230367. https://doi.org/10.1001/jamanetworkopen.2022.30367

Huang YS, Ogbechi J, Clanchy FI, Williams RO, Stone TW (2020) IDO and kynurenine metabolites in peripheral and CNS disorders. Front Immunol 11:388

Kappelmann N, Lewis G, Dantzer R, Jones PB, Khandaker GM (2018) Antidepressant activity of anti-cytokine treatment: a systematic review and meta-analysis of clinical trials of chronic inflammatory conditions. Mol Psychiatry 23(2):335–343

Kappelmann N et al (2021) Dissecting the association between inflammation, metabolic dysregulation, and specific depressive symptoms: a genetic correlation and 2-sample Mendelian randomization study. JAMA Psychiatry 78:161–170

Khandaker GM, Zammit S, Burgess s, Lewis G, Jones PB (2018) Association between a functional interleukin 6 receptor genetic variant and risk of depression and psychosis in a population-based birth cohort. Brain Behav Immun. 69:264–272. https://doi.org/10.1016/j.bbi.2017.11.020. Epub 2017 Dec 2.

Köhler O et al (2014) Effect of anti-inflammatory treatment on depression, depressive symptoms, and adverse effects: a systematic review and meta-analysis of randomized clinical trials. JAMA Psychiatry 71:1381–1391

Köhler-Forsberg O, Lydholm CN, Hjorthøj C, Nordentoft M, Mors O, Benros ME (2019) Efficacy of anti-inflammatory treatment on major depressive disorder or depressive symptoms: meta-analysis of clinical trials. Acta Psychiatr Scand 139(5):404–419

Krishnadas R, Nicol A, Sassarini J, Puri N, Burden AD, Leman J, Combet E, Pimlott S, Hadley D, McInnes IB, Cavanagh J (2016) Circulating tumour necrosis factor is highly correlated with brainstem serotonin transporter availability in humans. Brain Behav Immun 51:29–38

Leighton SP, Nerurkar L, Krishnadas R, Johnman C, Graham GJ, Cavanagh J (2018) Chemokines in depression in health and in inflammatory illness: a systematic review and meta-analysis. Mol Psychiatry 23:48–58

Levey DF, Stein MB, Wendt FR, Pathak GA, Zhou H, Aslan M, Quaden R, Harrington KM, Nuñez YZ, Overstreet C, Radhakrishnan K (2021) Bi-ancestral depression GWAS in the million veteran program and meta-analysis in> 1.2 million individuals highlight new therapeutic directions. Nat Neurosci 24(7):954–963

Li S, Fang Y, Zhang Y, Song M, Zhang X, Ding X, Yao H, Chen M, Sun Y, Ding J, Wang Q, Lu M, Wu G, Hu G (2022) Microglial NLRP3 inflammasome activates neurotoxic astrocytes in depression-like mice. Cell Rep 41(4):111532

Lynall M-E et al (2022) Genetic variants associated with cross-disorder and disorder-specific risk for psychiatric disorders are enriched at epigenetically active sites in peripheral lymphoid cells. Nat Commun 13:6102

Machado MO et al (2017) Biological mechanisms of depression following treatment with interferon for chronic hepatitis C: a critical systematic review. J Affect Disord 209:235–245

Malynn S, Campos-Torres A, Moynagh P, Haase J (2013) The pro-inflammatory cytokine TNF-α regulates the activity and expression of the serotonin transporter (SERT) in astrocytes. Neurochem Res 38(4):694–704

Marszalek-Grabska M, Walczak K, Gawel K, Wicha-Komsta K, Wnorowska S, Wnorowski A, Turski WA (2021) Kynurenine emerges from the shadows–current knowledge on its fate and function. Pharmacol Ther 225:107845

Massart R, Mongeau R, Lanfumey L (2012) Beyond the monoaminergic hypothesis: neuroplasticity and epigenetic changes in a transgenic mouse model of depression. Philos Trans R Soc B Biol Sci 367(1601):2485–2494

McKim DB, Weber MD, Niraula A, Sawicki CM, Liu X, Jarrett BL et al (2018) Microglial recruitment of IL-1beta-producing monocytes to brain endothelium causes stress-induced anxiety. Mol Psychiatry 23:1421–1431

Menard C, Pfau ML, Hodes GE, Kana V, Wang VX, Bouchard S et al (2017) Social stress induces neurovascular pathology promoting depression. Nat Neurosci 20:1752–1760. https://doi.org/10.1038/s41593-017-0010-

Miller AH, Raison CL (2016) The role of inflammation in depression: from evolutionary imperative to modern treatment target. Nat Rev Immunol 16(1):22–34

Mostafavi S, Battle A, Zhu X, Potash JB, Weissman MM, Shi J et al (2014) Type I interferon signaling genes in recurrent major depression: increased expression detected by whole-blood RNA sequencing. Mol Psychiatry 19:1267–1274

Murphy T et al (2017) Methylomic profiling of cortex samples from completed suicide cases implicates a role for PSORS1C3 in major depression and suicide. Transl Psychiatry 7:e989

Ogbechi J, Clanchy FI, Huang YS, Topping LM, Stone TW, Williams RO (2020) IDO activation, inflammation and musculoskeletal disease. Exp Gerontol 131:110820

Raison CL, Rutherford RE, Woolwine BJ, Shuo C, Schettler P, Drake DF, Haroon E, Miller AH (2013 Jan) A randomized controlled trial of the tumor necrosis factor antagonist infliximab for treatment-resistant depression: the role of baseline inflammatory biomarkers. JAMA Psychiatry 70(1):31–41

Rosales C (2018) Neutrophil: a cell with many roles in inflammation or several cell types? Front Physiol 9:113–113

Savitz JB, Drevets WC (2013) Neuroreceptor imaging in depression. Neurobiol Dis 52:49–65

Simon MS, Arteaga-Henríquez G, Fouad Algendy A, Siepmann T, Illigens BMW (2023) Anti-inflammatory treatment efficacy in major depressive disorder: a systematic review of meta-analyses. Neuropsychiatr Dis Treat 19:1–25. https://doi.org/10.2147/NDT.S385117. eCollection 2023

Song Y, Zhou Y, Zhou X (2020) The role of mitophagy in innate immune responses triggered by mitochondrial stress. Cell Commun Signal 18(1):1–4

Steiner J, Walter M, Gos T, Guillemin GJ, Bernstein HG, Sarnyai Z, Mawrin C, Brisch R, Bielau H, Zu Schwabedissen LM, Bogerts B (2011) Severe depression is associated with increased microglial quinolinic acid in subregions of the anterior cingulate gyrus: evidence for an immune-modulated glutamatergic neurotransmission? J Neuroinflammation 8(1):1–9

Wittenberg GM, Greene J, Vértes PE, Drevets WC, Bullmore ET (2020) Major depressive disorder is associated with differential expression of innate immune and neutrophil-related gene networks in peripheral blood: a quantitative review of whole-genome transcriptional data from case-control studies. Biol Psychiatry 88(8):625–637

Wray NR, Ripke S, Mattheisen M, Trzaskowski M, Byrne EM, Abdellaoui A, Adams MJ, Agerbo E, Air TM, Andlauer TM, Bacanu SA (2018) Genome-wide association analyses identify 44 risk variants and refine the genetic architecture of major depression. Nat Genet 50(5):668–681

Zwiep JC, Bet PM, Rhebergen D, Nurmohamed MT, Vinkers CH, Penninx BWJH, Milaneschi Y, Lamers F (2022) Brain Behav Immun Health 27:100585

Mechanisms of Action of TMS in the Treatment of Depression

Jonathan Downar, Shan H. Siddiqi, Anish Mitra, Nolan Williams, and Conor Liston

Contents

J. Downar (✉)
Department of Psychiatry, Faculty of Medicine, Institute of Medical Science, University of Toronto, Toronto, ON, Canada

Temerty Centre for Therapeutic Brain Intervention, Centre for Addiction and Mental Health, Toronto, ON, Canada
e-mail: jonathan.downar@utoronto.ca

Curr Topics Behav Neurosci (2024) 66: 233–278
https://doi.org/10.1007/7854_2024_483
Published Online: 7 June 2024

Abstract Transcranial magnetic stimulation (TMS) is entering increasingly widespread use in treating depression. The most common stimulation target, in the dorsolateral prefrontal cortex (DLPFC), emerged from early neuroimaging studies in depression. Recently, more rigorous casual methods have revealed whole-brain target networks and anti-networks based on the effects of focal brain lesions and focal brain stimulation on depression symptoms. Symptom improvement during therapeutic DLPFC-TMS appears to involve directional changes in signaling between the DLPFC, subgenual and dorsal anterior cingulate cortex, and salience-network regions. However, different networks may be involved in the therapeutic mechanisms for other TMS targets in depression, such as dorsomedial prefrontal cortex or orbitofrontal cortex. The durability of therapeutic effects for TMS involves synaptic neuroplasticity, and specifically may depend upon dopamine acting at the D1 receptor family, as well as NMDA-receptor-dependent synaptic plasticity mechanisms. Although TMS protocols are classically considered 'excitatory' or 'inhibitory', the actual effects in individuals appear quite variable, and might be better understood at the level of populations of synapses rather than individual synapses. Synaptic meta-plasticity may provide a built-in protective mechanism to avoid runaway facilitation or inhibition during treatment, and may account for the relatively small number of patients who worsen rather than improve with TMS. From an ethological perspective, the antidepressant effects of TMS may involve promoting a whole-brain attractor state associated with foraging/hunting behaviors, centered on the rostrolateral periaqueductal gray and salience network, and suppressing an attractor state associated with passive threat defense, centered on the ventrolateral periaqueductal gray and default-mode network.

Keywords ACC · Anterior cingulate · Attractor · Causal network mapping · Default mode · Depression · DLPFC · DMPFC · Dopamine · fMRI · Foraging · Functional connectivity · Glutamate · LTD · LTP · Meta-plasticity · NMDA · Non-reward · Orbitofrontal · PAG · Passive threat defense · Periaqueductal gray · Pro-kurtosis · rTMS · Salience · Subgenual · Synapse · TMS · Transcranial magnetic stimulation

---

S. H. Siddiqi
Center for Brain Circuit Therapeutics, Brigham & Women's Hospital, Boston, MA, USA

Department of Psychiatry, Brigham & Women's Hospital, Harvard Medical School, Boston, MA, USA

A. Mitra and N. Williams
Department of Psychiatry and Behavioral Sciences, Stanford University, Stanford, CA, USA

C. Liston
Department of Psychiatry, Brain and Mind Research Institute, Weill Cornell Medicine, New York, NY, USA

1 Introduction

In 2002, a 36-year-old woman in Toronto underwent neurosurgical implantation of deep brain stimulation (DBS) electrodes bilaterally in the subthalamic nucleus, for medically refractory Parkinson's disease. After implantation, as expected, stimulation of the left STN yielded prompt reductions in the tremor of the contralateral hand and improved motor functioning. However, with stimulation of the right STN, the effects were quite unexpected. Within 30 s, the patient descended into an intense, deep dysphoric state: "panicky, tearful, diaphoretic. . .the experience was tolerable only for minutes at a time". Describing her subjective experience, "[I] did not care about myself or anything else. . .someone could have come in to shoot me and I could not have cared less" (Stefurak et al. 2003).

As it turned out, the patient had a previous history of depression through much of her life, and although she had been in remission for the 2 years prior to surgery, she described the immediate effects of the stimulation as "similar in some respects to my depression but a thousand times worse". The neurosurgical team was able to use functional MRI to compare the effects of stimulation in each hemisphere. As intended, left STN stimulation produced deactivation in the region of the supplementary motor area (SMA), with rapid effects on motor functioning. However, stimulation of the contacts in the right STN did not affect the SMA, but instead deactivated regions further anterior along the medial wall, including the dorsomedial prefrontal cortex (DMPFC) and medial frontopolar cortex (mFPC). Disruption of activity in these off-target projections from the STN, even for short periods of time, was evidently sufficient to allow an immediate re-emergence of the patient's negative mood state, despite 2 prior years of remission.

Some 20 years later, this time in San Francisco, another 36-year-old woman underwent DBS implantation – not for Parkinson's Disease, but for severe, longstanding, treatment-refractory major depression (Scangos et al. 2021a). The process involved a 10-day preliminary phase in which the patient underwent intracranial electrode mapping of a diverse set of cortical and subcortical brain regions, to identify specific patterns of activity and stimulation that relieved her symptoms of depression. Using this data, the neurosurgical team was able to program and implant a personalized, closed-loop DBS system that would detect patterns of activity previously associated with her depression symptoms, and immediately disrupt these patterns with a 6 s burst of stimulation. Her resultant experience was effectively a reciprocal of the previous case; in her words, "you're experiencing that whole negative, depressive, whatever the triggering thing is, and then it's like suddenly the ultra-rational side of you comes on and those emotions can be separated" (Belluck 2021). The instrument of her recovery was, in effect, analogous to an implantable cardiac defibrillator, only in this case disrupting an arrhythmia of the brain, rather than an arrhythmia of the heart.

These two remarkable DBS cases provide the backdrop for our consideration of the mechanisms of a non-invasive brain stimulation technique, transcranial magnetic stimulation (TMS), in the treatment of depression. There are several key lessons to be drawn:

First, although depression is often conceptualized as a neurochemical syndrome with slow dynamics, emerging and resolving over weeks and months, these cases illustrate a very different picture: a syndrome at whose core lies a specific mode of brain activity, akin to an arrhythmia or pathological attractor state, that may emerge within seconds and may be disrupted within seconds.

Second, this mode of brain activity plays out within a specific set of neural pathways, distinct from those mediating other functions such as motor or sensory processing. Identifying and targeting these pathways, perhaps in a personalized fashion, can yield robust and very rapid improvement of depression symptoms – even in cases where the illness is severe and has persisted for decades.

Third, this mode of activity may re-establish itself rather quickly, again in a manner akin to a stubborn cardiac arrhythmia. Sustained recovery thus requires either a permanently implanted stimulator, or else, in the case of a non-invasive technique such as TMS, the ability to induce durable, neuroplastic changes in the target pathways themselves, so that they become less likely to re-enter the self-perpetuating attractor state of activity associated with depression.

Given these observations, our consideration of the therapeutic mechanisms of TMS in depression needs to cover several levels of explanation. First, it is essential to understand which of the brain's networks and circuits are not merely *correlationally* but instead *causally* linked to depression. Identifying and targeting these circuits successfully is what allows TMS to exert therapeutic effects on depression specifically, as opposed to other established indications such as Parkinson's disease (Dong et al. 2022) or tinnitus (Liang et al. 2020). Second, it is essential to understand what particular changes TMS needs to exert upon the activity of these target circuits, in order to disrupt the mode of activity associated with the depressed state. Third, it is essential to understand what mechanisms of neuroplasticity are recruited by therapeutic TMS protocols, in order for the treatment effects to endure for weeks or months beyond the stimulation sessions themselves. As a corollary, this understanding may allow us to develop pharmacological adjuvants to enhance the potency and durability of TMS-induced neuroplasticity, and thus of the treatment itself.

Finally, it may be helpful to understand the evolutionary and ethological role of the depressed mode itself, in contributing to the survival of the individual. Understanding the adaptive role of depression offers a clearer perspective on what external factors (beyond the brain stimulation alone) may promote durable functional recovery, versus rapid relapse, during the post-treatment period.

With this multi-layered account of the antidepressant mechanisms of TMS in place, we will be in the best position to optimize (and perhaps personalize) the treatment parameters for any given patient. We will also gain a more holistic perspective on how TMS treatment may be incorporated into the broader goal of helping the patient emerge from a state of learned helplessness to sustained wellness – reclaiming access to the full range of motivations, thoughts, and activities that make a life worth living.

2 Neuroanatomical Targets for TMS in Depression

2.1 Correlational Mapping of Target Brain Regions for Depression

TMS has the interesting distinction of being one of the few treatments in psychiatry whose development was informed directly by neuroimaging findings. Positron emission tomography (PET) studies using 18-fluorodeoxglucouse (18-FDG) in the late 1980s identified a set of regions that were hypoactive in patients with major depression compared to healthy controls. Even in some of the earliest of these studies (Fig. 1a), one can identify a symmetrical, whole-brain network of hypoactive regions including the left and right dorsolateral prefrontal cortex (DLPFC), the dorsomedial prefrontal cortex (DMPFC), and left and right inferior parietal lobule (IPL) (Baxter et al. 1989). However, the consensus hypothesis of the time (originating with Folstein's study of post-stroke depression (Folstein et al. 1977)) was that the left hemisphere should be hypoactive and the right hemisphere should be hyperactive in patients with depression. As a consequence, the earliest imaging studies often

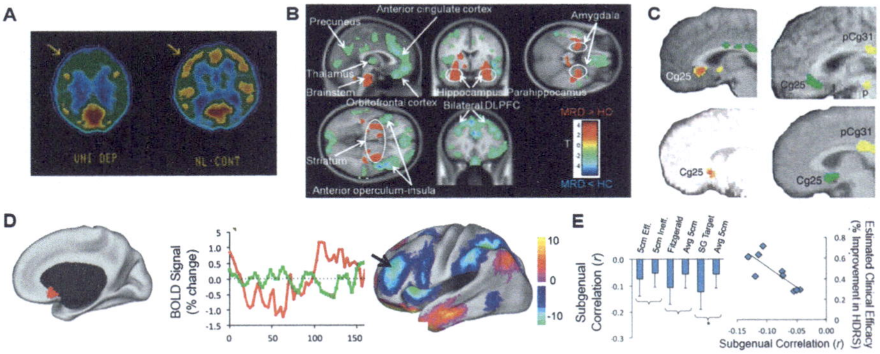

Fig. 1 Correlation-based mapping of target brain regions for depression. (**a**) Early PET studies in depression, such as that of Baxter et al. (1989), identified a network of regions with reduced metabolic activity in depressed patients versus healthy controls. Of these, the left DLPFC (arrow) was subsequently favored as a target for high-frequency TMS in depression; however, the right DLPFC, DMPFC, and lateral parietal regions showed similarly reduced metabolism in depression, both in this early study and in subsequent studies using both TMS and PET, such as that of Li et al. (2010) (**b**). (**c**) Another line of PET studies (reviewed in Ressler and Mayberg 2007) found hyperactivity in the sgACC during negative mood states (upper left); successful treatment with the SSRI fluoxetine (upper right), or cognitive behavioral therapy (lower left), or even placebo (lower right) normalized this hyperactivity among the treatment responders. (**d**) Fox et al. (2012) noted that in resting-state functional MRI scans, the ongoing activity of the sgACC (red plot) was anticorrelated to the activity of a network of other brain areas that included the DLPFC (green plot and blue regions; note their similarity to panel **a**); examining various DLPFC targets used in clinical trials of TMS for depression, those targets with higher anticorrelation to the sgACC yielded better overall treatment outcomes in each trial (**e**)

focused on the finding of hypoactivity in the left DLPFC, while declining to comment on the equal degree of hypoactivity evident in the right DLPFC, DMPFC, and IPL (Fig. 1a, b). Likewise, most of the earliest TMS studies in depression focused on delivering high-frequency stimulation to the left DLPFC (Pascual-Leone et al. 1996a; George et al. 1996), although low-frequency stimulation to the right DLPFC entered use soon after (Feinsod et al. 1998; Klein et al. 1999), and low-frequency right IPL stimulation has also been investigated (Schutter et al. 2009).

In parallel with this work, in the late 1990s and early 2000s another line of PET studies on depression, led by Mayberg and colleagues (reviewed in (Ressler and Mayberg 2007)), identified the subgenual anterior cingulate cortex (sgACC) as a region that was consistently hyperactive in depression. sgACC activity also consistently normalized downward among patients successfully treated for depression, regardless of whether the remission was achieved through antidepressant medication, cognitive behavioral therapy, ECT, or even placebo response (Ressler and Mayberg 2007; Mayberg et al. 2002) (Fig. 1c). These observations led to pioneering studies showing successful treatment of severe and highly refractory depression via deep brain stimulation (DBS) electrodes implanted in the sgACC to provide locally inhibitory stimulation (Mayberg et al. 2005). While the sgACC was considered too deep a target to be viable for the TMS equipment of the time, it was nonetheless noted that PET studies of patients undergoing DBS for depression also showed changes in metabolic activity of more superficial targets, including the DLPFC bilaterally (Lozano et al. 2008). This work provided additional evidence on the whole-brain, network-level neuroanatomical mechanisms of the antidepressant effect for DBS and TMS in depression (Seminowicz et al. 2004).

However, the DLPFC as a region covers a large area of the lateral convexity, thus requiring some approach for choosing the exact target for TMS coil placement. In the absence of well-developed neuronavigation techniques, early TMS studies targeting DLPFC employed scalp-based heuristics, such as the "5 cm rule" by which the DLPFC target is defined at a position 5 cm anteriorly from the upper extremity motor hotspot over primary motor cortex (George et al. 1995). Scalp heuristics are still in widespread use today, and early studies variously employed a 5 cm rule, 5.5 cm rule, 6 cm rule, or the F3 site of the 10-20 EEG electrode montage (reviewed in (Fox et al. 2012)). Later studies did indeed employ MRI-guided neuronavigation, and were thus obliged to choose an anatomical target based on putative Brodmann areas (Fitzgerald et al. 2009), sulcal landmarks (Mylius et al. 2013), or MNI coordinates (Blumberger et al. 2018).

This profusion of different targets within the broad region of the DLPFC raised the question of which specific location might be optimal, either at the group level or the individual level. An influential line of work by Fox and colleagues argued that a link could be drawn between effective targets for DBS and effective targets for TMS, across a wide variety of neurological and psychiatric disorders (Fox et al. 2014). Namely, by examining maps of whole-brain resting-state functional connectivity on blood oxygenation level-dependent (BOLD) fMRI, and 'seeding' the map at the site of an effective DBS target, one could then identify a network of regions whose activity was either correlated or anticorrelated to the activity of the DBS-based seed. Effective TMS targets seemed to be located on either the network or the anti-network of the DBS target. For example, if inhibitory DBS to the sgACC proved effective for depression, then the most effective sites for TMS stimulation should be at the precise DLPFC coordinate where the resting-state signal was maximally anticorrelated to the signal in the sgACC seed region (Fig. 1d).

At the group level, Fox and colleagues tested this hypothesis by reviewing data from a series of TMS trials in depression conducted by various groups using various scalp- or imaging-based heuristics. They noted that trials whose heuristics happened to land the coil closer to the DLPFC coordinate of maximum anti-sgACC connectivity (based on the 1,000-patient Human Functional Connectome project data) enjoyed better patient outcomes. Indeed, a strong correlation appeared to be present between the overall percent clinical improvement in each trial and the degree of sgACC anticorrelation for its specific DLPFC target (Fig. 1e).

These findings implied that optimal outcomes would be achieved at the group level by targeting a specific set of DLPFC target coordinates with maximal anticorrelation. Indeed, this precise set of coordinates was subsequently adopted in several large randomized trials of TMS in depression, including the THREE-D study (Blumberger et al. 2018), among others (Blumberger et al. 2021, 2022). In addition, a scalp heuristic was developed to target this particular DLPFC site, for use in situations where neuroimaging was available (Mir-Moghtadaei et al. 2015); this 'modified BeamF3 heuristic' has since been widely adopted in clinical use and training courses worldwide (F3 Measurement System n.d.).

However, on a note of caution, it is worth pointing out that much of the work leading to present TMS practice has been based primarily on i) lines of evidence that use correlational methods (PET, fMRI) rather than causal methods, and ii) an assumption that the sgACC (however defined) plays a crucial role in depression, despite the emergence of this target only from correlational PET imaging, as well as the failure of this target to demonstrate efficacy against depression in at least one major DBS trial (Holtzheimer et al. 2017). To remedy these issues, ideally one would wish to establish a whole-brain network for depression using no a priori assumptions about which brain regions formed its core, as well as purely causal (lesion, stimulation) methods rather than correlational methods. Fortunately, much progress has been made in this regard in recent years. Let us now turn to examine this progress.

2.2 *Causal Mapping of Whole-Brain Networks for Depression*

When the DLPFC target emerged from the PET studies reviewed above, this raised the possibility that neuroimaging correlates might be used to identify TMS targets more broadly (Cash et al. 2021). However, targeting a treatment based on correlations is fraught with peril (Siddiqi et al. 2022). The correlate may be causal in nature, but may also be spurious or epiphenomenal, in which case it would have no effect. Perhaps more concerningly, a correlative finding may be 'reverse causal' – if the correlate is compensating for an underlying cause, intervening on that correlate could be counterproductive (Siddiqi et al. 2022). For instance, infection is correlated with elevated white blood cell counts, but it could be potentially fatal to lower white blood cell counts in a patient with an active infection.

Counterproductive results have sometimes been observed when TMS is targeted based on correlates. This is perhaps best exemplified by three studies on fear extinction. First, in 41 patients with arachnophobia and 42 healthy controls, Notzon et al. applied real or sham intermittent theta burst stimulation (TBS) to the posterior DLPFC along with a virtual reality condition intended to induce fear. Contrary to the hypothesis, stimulation led to increased autonomic arousal relative to sham in the arachnophobia group (Notzon et al. 2015). In another study of 28 healthy controls, Teferi et al. applied continuous TBS or sham to the anterior DLPFC. Continuous TBS is believed to induce effects opposite to intermittent TBS and the anterior DLPFC is functionally anticorrelated to the posterior DLPFC – because of this double negative, the Teferi study may be expected to have a similar network-level effect to the Notzon study. Indeed, TMS led to increased fear-potentiated startle (Teferi et al. 2023). Finally, in a clinical trial 125 patients with posttraumatic stress disorder, Isserles et al. randomized patients to receive real or sham stimulation to the medial prefrontal cortex along with a brief exposure protocol. TMS was found to be inferior to sham, suggesting that it interfered with fear extinction and may have exacerbated symptoms (Isserles et al. 2021).

By contrast, causality can be inferred using natural experiments in which different brain regions are incidentally stimulated or lesioned (Siddiqi et al. 2022). Causal knowledge can only be obtained by comparing to a counterfactual – what would have happened if the stimulation or lesion had not occurred – an alternate reality which is unobservable by definition (Williamson 2019). Since a true counterfactual can never be observed, scientists often use natural experiments to estimate the effect of the counterfactual. For instance, since TMS is usually applied using scalp landmarks, this yields incidental variability in the actual stimulation site. Thus, if stimulation of some locations leads to better antidepressant response than others, these locations may have different causal roles in depression (Siddiqi et al. 2022).

The effect of stimulation location was first investigated by Herbsman et al. who found that more anterolateral stimulation sites appear to be more effective (Herbsman et al. 2009). A subsequent meta-analysis by Fox et al. found similar results across studies that used different targets. However, this result was not driven solely by anterolateral location – resting-state functional connectivity analysis found that the most effective sites showed stronger negative connectivity to the sgACC (Fox et al. 2012). A subsequent prospective study confirmed that the most effective TMS sites were defined by stronger negative connectivity to the subgenual cingulate, independently of anterolateral location (Weigand et al. 2018).

These studies were based on the assumption that scalp-based targeting leads to near-random variability in stimulation location. If the randomness assumption holds true, then all confounders are addressed – patients are alike in every relevant way other than the stimulation site, thus enabling strong causal inference. However, it is rare for the randomness assumption to be completely true. Even well-designed randomized controlled trials can be confounded by selective dropout bias. For scalp-based TMS targeting, variability in stimulation location is related to the patient's head size, which is related to height, which in turn is related to gender. On average, the '5 cm rule' would yield more anterolateral targets in female patients. If more anterolateral targets are truly better, this may explain why TMS appears to be more effective for women (Sackeim et al. 2020; Kryatova et al. 2024). Alternatively, it is also plausible that TMS is more effective in women for some other reason (Kryatova et al. 2024), and the effect of the stimulation site is just an epiphenomenon of the height difference between women and men.

For this reason, causal inference from a single retrospective modality is often inherently limited (Siddiqi et al. 2022). It is almost impossible to control for all confounders in a natural experiment. TMS studies are confounded by gender differences in scalp dimensions. DBS studies may be confounded by adjustment of stimulation location based on intraoperative testing, which has unclear significance for long-term outcomes. Lesion studies may be confounded by the fact that different risk factors can predispose to stroke in different vascular territories, while penetrating head trauma is more likely to occur near the frontal bone. This challenge may seem insurmountable when studying each data type in isolation, but when combining data types, one modality can mitigate the confounders from another modality. If a finding converges across multiple modalities that are not subject to the same confounders, this strengthens causal inference (Siddiqi et al. 2022).

This principle was recently used to map causal circuitry for depression across TMS, DBS, and lesions (Siddiqi et al. 2021a) (Fig. 2a). While these three modalities do not generally affect the same brain regions, their effects can be compared using the human connectome, a large-scale brain imaging database that can be used as a wiring diagram of the human brain. Even if lesions or stimulation sites do not directly intersect each other, they often intersect the same circuit (Siddiqi et al. 2022).

Using this convergent causal mapping approach, a common depression circuit was found across 151 TMS sites, 101 DBS sites, and 461 lesions (Siddiqi et al. 2021a). This circuit included many regions that had long been known to be involved in depression, including positive connectivity to the DLPFC and negative connectivity to the subgenual cingulate (Fig. 2b, c). High-frequency TMS sites were more likely to relieve depression if they were positively connected to the circuit, while the opposite was true for lesions and DBS. This is consistent with the longstanding notion that high-frequency TMS is believed to be excitatory, while DBS and lesions are believed to be inhibitory.

Subsequent studies confirmed these findings, applied them to other modalities, and extended them to other symptoms and disorders. First, in 283 patients with multiple sclerosis, lesions to the depression circuit were more likely to induce depression. This effect was specific to depression relative to 11 other symptoms measured by the Quality of Life in Neurological Disorders scale (Siddiqi et al. 2023).

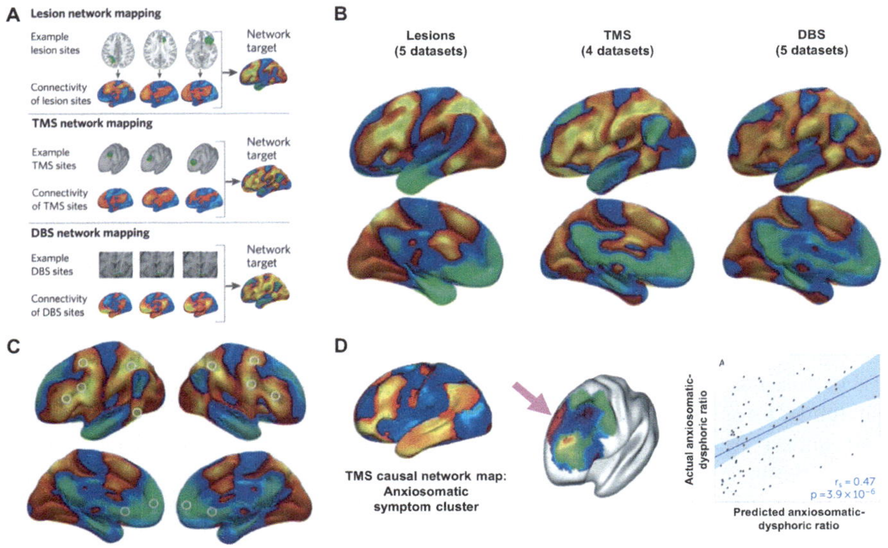

Fig. 2 Identification of symptom-specific TMS targets by causal network mapping. (**a**) Lesions, TMS stimulation sites, and DBS stimulation sites associated with improvement or worsening of depression in individual patients can be used as region-of-interest seeds to generate whole-brain network maps of functional connectivity, based on large datasets such as the Human Connectome Project. This enables the creation of whole-brain maps of regions that are causally linked (rather than merely correlated) to depression severity across individuals (modified from Siddiqi et al. 2023). (**b**) Causal network maps for depression generated from patients with focal brain lesions (left), patients undergoing TMS for depression (center), and patients undergoing DBS implantation (right) show good agreement, despite the independent datasets and the different modalities used to generate the maps (modified from Siddiqi et al. 2021a, b). (**c**) Combining the lesion, TMS, and DBS maps yields a whole-brain map of specific targets for excitatory (orange) or inhibitory (blue) TMS in depression, supported by causal rather than merely correlational evidence, and without requiring any a priori assumptions about the importance of any particular seed region (such as the sgACC) to depression. Note that this target map contains not one but two predicted optimal foci in the DLPFC, one anterior and one posterior; lateral parietal targets are also predicted to be effective, as is inhibitory neuromodulation of the frontopolar cortex (a mechanism posited for ECT and DBS). (**d**) The causal network mapping approach appears to be effective across many other neurological and psychiatric disorders aside from depression. For example, the same approach was recently applied to generate target sites for TMS in anxiety, independent of depression severity (Siddiqi et al. 2024), and the optimal target in DMPFC (center, right green arrow) was recently targeted with good effect in a DMPFC-TMS case series in generalized anxiety (Croarkin et al. 2024)

Second, in 386 patients receiving electroconvulsive therapy for MDD, volumetric change in this circuit was correlated with antidepressant efficacy, independent of nonspecific volumetric changes attributed to the ECT procedure itself (Argyelan et al. 2023). Finally, multiple studies have applied the causal mapping approach to other disorders, revealing convergence across lesions and stimulation sites for motor symptoms of Parkinson's disease (PD) (Siddiqi et al. 2021a), cognitive symptoms in

PD (Reich et al. 2022), Tourette syndrome (Ganos et al. 2022), epilepsy (Schaper et al. 2023), vertigo (Schaper et al. 2023; Li et al. 2023), essential tremor (Joutsa et al. 2018), and addiction (Joutsa et al. 2022).

Of note, most of these examples have used lesion studies to optimize existing stimulation sites. Brain stimulation is already known to be effective for depression, PD, Tourette syndrome, epilepsy, essential tremor, and nicotine use disorder. This begs the question of whether this approach can also be used to identify a novel target. Siddiqi et al. recently addressed this question using convergent causal mapping to identify a brain circuit target for anxiety (Siddiqi et al. 2024) (Fig. 2d). TMS, DBS, and lesions converged on a common substrate for anxiety that largely overlapped with the default mode network. Importantly, the anxiety circuit was negatively connected to the depression circuit (Siddiqi et al. 2024), suggesting the existence of distinct targets in the DLPFC for anxiety versus depression. The anxiety target was relatively posterior and medial, closer to the original "5 cm" or "5.5 cm" stimulation site and also overlapping much of the dorsomedial prefrontal cortex (Siddiqi et al. 2020). The depression target was relatively anterior and lateral, closer to the BeamF3 heuristic (Siddiqi et al. 2020). This may explain why clinical outcomes have not markedly improved as stimulation sites have progressively moved to more anterolateral locations over the last decade – these sites may be better for some symptoms, but worse for others. A recent clinical trial incidentally supported this distinction by randomizing patients to the 5.5 cm or BeamF3 site. Both targets were equally effective overall, but the 5.5 cm target was more effective for anxiety while controlling for depression (Trapp et al. 2023). Of note, another recent case series in GAD targeted the most prominent peak of the anxiety causal network map (Fig. 2d, center) in DMPFC, achieving remission in 50% of patients and response in 69% of patients after 30 sessions of treatment (Croarkin et al. 2024).

These emerging findings suggest that the efficacy of TMS for different symptoms involves stimulation of different circuits. Because these circuits were derived using causal data, they can be used to infer a mechanistic relationship between the stimulation site and the clinical outcome. However, this mechanistic understanding remains incomplete without understanding the direction of information flow from the TMS site to the underlying network. Many approaches have emerged to study the direction of information flow, such as interleaved TMS-fMRI experiments and recordings of TMS-induced transcranial or intracranial electrophysiological changes, which can provide strong spatial and temporal data on the causal interactions between different brain regions (Bergmann and Hartwigsen 2021; Tik et al. 2023a; Oathes et al. 2021). Directional effects may also be estimated indirectly using time-lagged resting-state fMRI correlations, which can be compared with clinical outcomes before and after a clinical course of TMS (Mitra et al. 2023). Next, we discuss how this approach may provide a neural signature of TMS response.

3 Functional Connectivity and Therapeutic Effects of TMS in Depression

3.1 Characterizing Changes in Functional Connectivity Associated with Response to DLPFC-TMS

The neural mechanisms underlying both the MDD and the treatment effects of DLPFC-TMS remain uncertain. However, recent work has identified candidate mechanisms. A wide array of human neuroimaging studies have largely agreed that the neural mechanisms of MDD are not restricted to any single lesion in the brain, but instead, likely involve aberrant or imbalanced communication across brain-wide networks (Fox et al. 2012; Siddiqi et al. 2020; Avissar et al. 2017; Dinga et al. 2019; Drevets 2000; Drevets et al. 1997; Drysdale et al. 2017; Hamilton et al. 2013; Mayberg 1997; Sylvester et al. 2012; Devinsky et al. 1995; Padmanabhan et al. 2019; Kirkby et al. 2018).

Although multiple hypotheses remain regarding the mechanisms of depression, several studies have converged on the idea that imbalanced signaling between the anterior cingulate cortex (ACC) and other brain areas involved in emotional processing may underlie MDD in many cases (Fox et al. 2012; Avissar et al. 2017; Baeken et al. 2017; Williams et al. 2018; Wu et al. 1992, 2016; George et al. 1994). Thus, one of the theories of how repetitive transcranial magnetic stimulation (rTMS) protocols function is that stimulating the left dorsolateral prefrontal cortex (DLPFC) (Fox et al. 2012; Avissar et al. 2017; Williams et al. 2018; Cole et al. 2020, 2022a; Jing et al. 2020; Kito et al. 2008; Tik et al. 2017; Dowdle et al. 2018; George et al. 1997, 2010) inhibits aberrant neural activity in the ACC. The primary evidence for this mechanism has been the observation that ongoing, spontaneous neural activity in the DLPFC is generally anticorrelated with activity in the ACC. Yet direct evidence for an inhibitory effect from rTMS delivered to the DLPFC has been difficult to obtain.

Historically, the search for neurobiological biomarkers and mechanisms of MDD has been hampered by both conceptual and experimental barriers. Most human neuroimaging studies of MDD rely on a technique known as resting state fMRI (rs-fMRI), in which experimenters measure spontaneous patterns of neural activity in the absence of any explicit task. The objective of rs-fMRI studies is to investigate differences in baseline neural communication at a large scale between typical individuals and an experimental group, such as individuals diagnosed with MDD. Although several resting-state fMRI (rs-fMRI) studies of human brain function have reported functional differences between patients with MDD and healthy controls, these differences are rarely subject to targeted network manipulations. Consequently, it has been difficult to assign causal relationships between mechanistic biomarkers and MDD. Second, although MDD is believed to arise from disordered communication in the brain, the vast majority of rs-fMRI studies examine either activity levels within individual brain regions or interactions between brain regions using zero-lag correlation (called "functional connectivity") (Avissar et al. 2017;

Drevets 2000; Drysdale et al. 2017; Hamilton et al. 2013; Sylvester et al. 2012; Jing et al. 2020; Kito et al. 2008; Zhou et al. 2017). A limitation of conventional functional connectivity is that there is no assignment of signaling directionality. Thus, a circumstance in which brain region A signals to brain region B can appear identical to a circumstance in which the direction of communication is reversed, even though the neurobiological implications of signaling direction may be profound.

In recent work, Mitra et al. developed an analytic strategy to discover that rs-fMRI contains a robust, highly reproducible temporal structure characterized by stereotyped directional signaling patterns across the human brain (Mitra and Raichle 2016; Mitra et al. 2015a; Mitra et al. 2014; Raut et al. 2020; Raut et al. 2019; Bolt et al. 2022). Directed patterns in rs-fMRI correspond with infra-slow (<0.01 Hz) human electrocorticography, mouse optical calcium imaging, and mouse electrophysiology (Mitra et al. 2016, 2018), establishing that propagated activity patterns in the blood-flow-based rs-fMRI signal agree with underlying neurophysiological signals. Additionally, there is evidence that the directional flow of rs-fMRI activity is exquisitely sensitive to brain state, more so than the zero-lag correlations that define conventional functional connectivity (Mitra et al. 2015b, 2016). Taken together, these developments raised the intriguing possibility that directional signaling patterns in rs-fMRI may be a sensitive marker of aberrant network activity in neuropsychiatric disorders such as MDD.

On this basis, Mitra and colleagues directly investigated the relationship between brain-wide directed signaling patterns and MDD by leveraging rs-fMRI data acquired during clinical trials for Stanford Neuromodulation Therapy, an FDA-approved TMS protocol. In brief, Stanford Neuromodulation Therapy (SNT), previously published in two open-label studies as well as a randomized controlled trial (Williams et al. 2018; Cole et al. 2020, 2022a), involves daily administration of ten TMS sessions each day for 5 days using a spaced intermittent theta burst pulse design applied to individualized left DLPFC targets (Cole et al. 2020, 2022a). While the SNT depression protocol has proven effective in the acute clinical treatment of treatment-resistant MDD (TRD) (Williams et al. 2018; Cole et al. 2020, 2022a), the biological basis of the treatment effect has been unclear.

To shed light on potential biological mechanisms of rTMS treatment effect, Mitra and colleagues analyzed both conventional functional connectivity and directed signaling patterns in rs-fMRI data acquired immediately before and after SNT treatment in two clinical trials: first, a double-blind randomized clinical trial in which patients with TRD received either active or sham SNT, and second, an open-label clinical trial in which patients with TRD received active SNT. In addition, they also analyzed brain-wide directed signaling patterns in typical control subjects without MDD. By combining these datasets, they tested whether SNT modulates either functional connectivity, directed signaling, or both in rs-fMRI (beyond placebo effects in sham SNT).

In an analysis of both functional connectivity and directed signaling patterns in rs-fMRI before and after SNT treatment, Mitra and colleagues found that SNT specifically modulates the directional flow of rs-fMRI activity between the dorsal

ACC and the salience network, which has previously been implicated in emotional processing (Hamilton et al. 2013, 2016; Seeley 2019). Specifically, they found that in the normally functioning salience network, activity tends to travel from the anterior insula to the dorsal ACC. However, in patients with MDD, the direction of propagation was *reversed*, such that signals in spontaneous fMRI activity tended to travel from dorsal ACC into the rest of the salience network, including the anterior insula.

Interestingly, although this potential biomarker appeared in many of the patients in the sample, it was not apparent in *every* patient, suggesting that this putative mechanism is only responsible for one biotype of major depression. Additionally, there was no statistically detectable shift in functional connectivity in a whole-brain screen, suggesting that the magnitude of functional connectivity shifts, if present, were not large enough to be detected in this sample. Finally, in addition to finding that SNT treatment modulates the propagation of rs-fMRI activity between dorsal ACC and the rest of the salience network, they also found that the magnitude of the temporal effect corresponds with treatment efficacy. Thus, SNT appears to correct a baseline aberrancy in directed rs-fMRI signals and that baseline rs-fMRI temporal structure predicts both individual depression severity and treatment response.

These findings suggest that there may exist an MDD biotype in which the pathophysiology is marked by aberrantly early signaling by the dorsal ACC within the salience network. However, several questions remain. First, given work highlighting successful treatment of MDD through alternative brain regions (Mayberg et al. 2005; Kirkby et al. 2018; Scangos et al. 2021b), further studies must determine whether the dorsal ACC-salience mechanism found here is a unique biotype of MDD, or a part of an overlapping set of neurophysiologic causes of MDD (Drysdale et al. 2017). Second, this hypothesis regarding the mechanistic function of SNT derives from a novel treatment paradigm involving a five-day course of spaced intermittent theta-burst stimulation. Future work is required to determine the role of administration frequency and pulse parameters in how rTMS modulates the temporal structure of rs-fMRI. Third, the patient cohort in this study consisted of individuals with treatment-resistant depression; thus, it remains unclear so far whether these results generalize to the wider population of non-treatment-resistant depression patients. Indeed, given the severe symptoms of the psychiatric inpatients included in the above work, it is possible that these findings may be more representative of severe, treatment-resistant MDD than a ‘typical’ case of MDD.

Finally, the importance of the temporal relationship between dorsal ACC and anterior insula in these results raises the intriguing possibility that modulation of the anterior insula may be an alternative means for normalizing the temporal flow of activity in the salience network. Recent work suggests it is possible to non-invasively modulate insular activity (Dinur-Klein et al. 2014; Zangen et al. 2021) (but see also: (Spagnolo et al. 2019)). However, future investigations are required to determine whether modulation of the anterior insula can be utilized for antidepressant effect. Other non-DLPFC regions, however, have indeed been employed as targets for TMS in depression. Let us now turn to these targets, to examine whether they have distinctive therapeutic mechanisms, or whether they instead share a final common mechanism with DLPFC-TMS in treating depression.

3.2 Non-DLFPC TMS Targets in Depression: Common or Distinctive Mechanisms?

While the DLPFC is by far the most commonly employed therapeutic target for TMS in depression, other targets have also been studied in clinical populations. The dorsomedial prefrontal cortex (DMPFC), for example, has shown effectiveness against depression in open-label case series (Downar et al. 2014; Bakker et al. 2015) and superiority to sham stimulation in some (Kreuzer et al. 2015) but not all (Dunlop et al. 2020) randomized trials. Recently, a randomized trial in 169 patients with major depression established non-inferiority for DMPFC versus DLPFC stimulation (Zangen et al. 2023), leading to formal FDA approval of DMPFC-TMS in this indication. Notably, baseline EEG and clinical-symptom profiles differed between the DLPFC- and DMPFC-responsive patients in this study, suggesting that these two interventions might act via slightly different neural mechanisms in treating depression.

In support of this possibility, as noted earlier, Drysdale and colleagues identified 4 distinct 'bio-types' of depression patients, each with distinctive symptom profiles as well as distinctive patterns of whole-brain resting-state functional connectivity (Drysdale et al. 2017). Two of these biotypes showed high rates of response to DMPFC-TMS (60–80%), while two showed poor response rates (20–30%) (Fig. 3a). Notably, one of the poor-responding biotypes was characterized by a prominent nexus of aberrant functional connectivity in the right lateral orbitofrontal cortex (RLOFC), raising the question of whether this structure might itself present a useful target for at least some patients with MDD (Fig. 3b).

The RLOFC has drawn interest in recent years as a key region in the brain's 'non-reward' pathway – a counterpart to the better-known medial orbitofrontal 'reward pathway', except focused on negative rather than positive incentives and reinforcers (Rolls and Deco 2016). Rolls' 'non-reward attractor theory of depression' posits that the RLOFC and its associated cortico-striatal-thalamic loop through the ventral striatum can enter a recurrent, self-reinforcing pattern of activity (Rolls 2016). In this attractor state, perseverative rumination and fixation on negative themes would ensue, while lateral inhibition of the reward pathway would abolish hedonic functioning and/or pursuit of positive incentives (Rolls 2019).

Studies of depressed patients prior to DBS implantation, using intracortical recordings and stimulation, have suggested that LOFC regions do indeed play a causal role in supporting rumination on negative themes (Rao et al. 2018), and closed-loop stimulation of ventral striatal pathways feeding LOFC appears to rapidly abolish incipient negative thoughts and relieve depression symptoms (Scangos et al. 2021a). These observations raise the question of whether non-invasive stimulation of

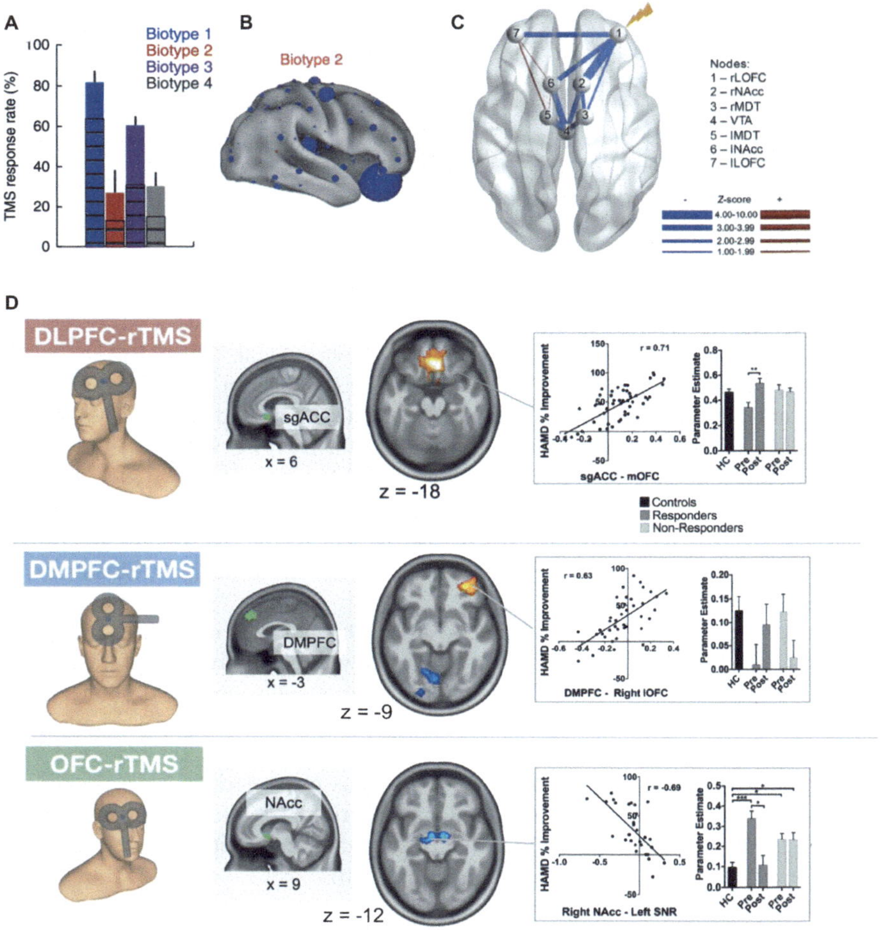

Fig. 3 Distinctive mechanisms of action for DLPFC, DMPFC, and RLOFC-TMS in depression. (**a**) Among four 'biotypes' of MDD patients identified by Drysdale et al. (2017) from whole-brain functional connectivity on resting-state fMRI, types 1 and 3 showed strong responses to DMPFC-TMS while types 2 and 4 showed poor response. (**b**) Biotype 2, a poor responder type to DMPFC-TMS, showed a prominent and distinctive nexus of abnormal connectivity in the RLOFC (large blue sphere), suggesting that this target might be more effective in some DMPFC-TMS non-responders. (**c**) Fettes et al. (2017a, b) showed a distinctive mechanism of action at work in a patient who achieved remission on RLOFC-TMS after failing to respond to courses of DLPFC- and DMPFC-TMS. This mechanism did not involve classic salience-network nodes but instead involved reductions in connectivity in a cortico-striatal-thalamic loop circuit that also incorporated the substantia-nigra, the RLOFC target of stimulation, and its left-sided counterpart. (**d**) Fettes, Mazza et al. (2024) examined the correlates of response to DLFPC-TMS ($N = 50$), DMPFC-TMS ($N = 40$), and RLOFC-TMS ($N = 30$) in depression and found that they were each quite different from the others: while DLPFC-TMS altered connectivity from the sgACC to more anterior medial prefrontal regions, RLOFC-TMS reduced connectivity between the nucleus accumbens, thalamus, and substantia nigra, and DMPFC-TMS reduced connectivity from DMPFC to DLPFC while increasing connectivity from DMPFC to RLOFC. No final common correlate of response was identified across the three TMS targets

the RLOFC could similarly relieve depression symptoms (Fettes et al. 2017a), particularly in cases where DLPFC- and/or DMPFC-TMS has failed.

An increasing number of studies have begun to investigate the effectiveness of RLOFC-TMS in depression, along with some hints at potential mechanisms. Nauczyciel et al. (2014) studied 1 Hz RLOFC-TMS using a double-cone coil in OCD patients, showing that this intervention reduced metabolic activity in the RLOFC target on FDG-PET (Nauczyciel et al. 2014). Feffer et al. then reported that, among 42 patients with depression who had already failed to improve on DMPFC-TMS, a 20–30 session course of 1 Hz RLOFC-TMS achieved response in 35% and remission in 24% of individuals (Feffer et al. 2018). Subsequently, in another sample of 41 patients who had failed to respond to DLPFC-TMS, Prentice et al. reported a similar 34% response and 27% remission rate following a mean 24 sessions of 1 Hz RLOFC-TMS (Prentice et al. 2023). In another independent report, Tadayonnejad et al. (2023) reported that, among 33 patients unresponsive to DLPFC-TMS, an additional 15 sessions of add-on 1 Hz RLOFC-TMS achieved response in 24% of patients (Tadayonnejad et al. 2023). Notably, these patients were characterized by higher scores on the Ruminative Responses Scale (RRS), and these scores improved significantly following the addition of RLOFC-TMS to the treatment regimen. Finally, a case report by Fettes et al. used resting-state functional MRI to characterize the neural correlates of remission on RLOFC-TMS in a patient with previous failure to respond to courses of DLPFC-TMS and then DMFPC-TMS (Fettes et al. 2017b). Remission was associated with reductions in connectivity between the RLOFC and its associated cortico-striatal-thalamic loop partners, including nucleus accumbens, mediodorsal thalamus, and ventral tegmental area (Fig. 3c).

Given that individual patients with depression appear to be more responsive to DLPFC, DMPFC, and/or RLOFC stimulation, the question therefore arises as to whether there is a distinctive neural mechanism of action that is unique to each of these targets, or whether instead there is a final common pathway that is required for antidepressant effect, regardless of target – for example, a reduction in sgACC-default mode network connectivity, or an increase in salience-network connectivity. To date, no formal randomized trial has directly addressed this question, although several are underway. However, preliminary evidence presented here suggests that there may indeed be distinctive mechanisms of action for each target in depression, evident in terms of unique functional connectivity and symptom profiles for responders versus nonresponders, and in terms of the types of changes in functional connectivity that are associated with clinical improvement.

Fettes, Mazza et al. have examined pre- and post-treatment resting-state fMRI data in 120 patients who received a therapeutic course of TMS targeting either DLPFC ($N = 50$), DMPFC ($N = 40$), or OFC ($N = 30$). In this sample, sgACC-DLPFC connectivity was not predictive of treatment outcome for any of the three groups. However, sgACC-to-target-site connectivity was predictive of treatment outcome for DMPFC-TMS (in agreement with the previous work of (Salomons et al. 2014)) but not for DLPFC- or RLOFC-TMS. Instead, both the neural predictors and correlates of clinical improvement appeared to be quite distinctive for each target (Fig. 3d). For DLPFC-TMS, patients with lower baseline sgACC-VMPFC connectivity showed greater response, and similarly, clinical improvement was correlated to increases in sgACC-VMPFC connectivity. For DMPFC-TMS, patients with low baseline connectivity between DMPFC and RLOFC showed greater response, and similarly, clinical improvement was correlated with increases in DMPFC-RLOFC connectivity. For RLOFC-TMS, clinical improvement was correlated to decreases in connectivity between the ventral striatum and the thalamus and substantia nigra (in agreement with the previous report of (Fettes et al. 2017b)).

While further work in larger samples is certainly required, the literature to date does appear to suggest that there may be more than one pathway to remission for a patient with depression who undergoes TMS. DLPFC-, DMPFC-, and RLOFC-TMS all appear to be effective for certain subpopulations of depression patients, and it may be possible in time to identify these patients prior to treatment using a combination of clinical features and functional MRI or other markers of neural activity. Furthermore, the changes in whole-brain neural activity that correlate to clinical improvement appear to be rather distinct for each type of intervention. Thus, the heterogeneity of depression at both the clinical and the neural level may need to be addressed by a heterogeneous suite of TMS interventions, capable of normalizing aberrant activity in different pathways within the functioning brain as a whole.

4 Synaptic and Neuroplastic Mechanisms of TMS in Depression

So far we have focused on identifying particular brain regions and large-scale networks that are effective targets for TMS in depression. Locating and targeting these circuits is essential to achieving therapeutic effects against depression specifically: if we reposition the coil over auditory cortex, for example, the therapeutic effects of TMS now apply not to depression but to tinnitus; reposition again over supplementary motor cortex, and the therapeutic effects now apply instead to Parkinson's disease, and so forth. Thus, the first half of the story of the antidepressant mechanism of TMS plays out in the spatial domain, at the level of large-scale networks, via mechanisms of inter-regional connectivity.

Yet there is an equally important second half to the story, playing out at the microscopic level of synapses and receptors, via the mechanisms of neuroplasticity. These mechanisms are required in order to explain why the antidepressant effects of TMS endure for weeks and months beyond the acute course of treatment. It is also equally important to understand why the antidepressant effects of TMS do *not* ensue within minutes of treatment, as they do for certain rapid-acting interventions such as intravenous ketamine (Zarate et al. 2006), intravenous scopolamine (Furey and Drevets 2006), and the closed-loop DBS case presented at the beginning of this chapter (where, notably, therapeutic effects ensue within seconds, not minutes) (Scangos et al. 2021a).

It is thus a key observation that TMS has a very different profile of 'pharmacodynamics' than DBS, ketamine, or scopolamine – almost the reciprocal, in fact. TMS sessions typically have *no* immediate marked effect on mood, and instead exert a gradually cumulative effect over many sessions of treatment. Indeed, with highly accelerated regimens in particular, the maximal therapeutic effect is sometimes not achieved until several days or weeks after treatment is completed (Cole et al. 2022a). Understanding *why* TMS has this cumulative, delayed, durable response trajectory is essential to a full account of the antidepressant effects of TMS. Let us therefore now turn to examine how TMS engages the mechanisms of neuroplasticity at the synaptic level.

4.1 TMS Effects on Plasticity in Local Microcircuits

TMS is widely assumed to act by facilitating synaptic plasticity in microcircuits within the prefrontal cortical target site. As reviewed above, TMS uses potent, focal, temporally patterned magnetic field pulses to generate electric currents non-invasively in spatially restricted volumes of brain tissue (Hallett 2007). Early models of TMS mechanisms focused on local effects on the dorsolateral prefrontal cortical target site, which may be hypoactive in some patients with depression, based on PET findings and case studies of emergent depression after a prefrontal stroke (George et al. 1995; Hallett 2007; Pascual-Leone et al. 1996b; Bench et al. 1992; Morris et al. 1996).

iTBS and other high-frequency protocols are assumed to rescue this dysfunction by potentiating neurons and forming new synapses in the PFC target site, based largely on studies in primary motor cortex. For example, in one such study (Pascual-Leone et al. 1994), rTMS targeting the primary motor cortex induced progressively larger motor evoked potentials (MEP) in adjacent hand muscles, consistent with an accumulating spread of excitation with repeated stimulation, as well as a period of increased excitability as indexed by increased MEP amplitude that lasted for 3–4 min after a single stimulation session. In contrast, continuous TBS (cTBS) and low-frequency protocols appear to have the opposite effect, reducing cortical excitability for at least 15 min after a single session in one study (Chen et al. 1997). These effects are believed to be driven at the cellular level by LTP- and LTD-like processes in cortical neurons, as they are blocked by co-treatment with the NMDA receptor antagonist, memantine (Huang et al. 2007). Conversely, a recent clinical trial showed that co-treating with iTBS and D-cycloserine – an NMDA receptor partial agonist that may facilitate synaptic plasticity – enhances antidepressant outcomes in patients (Cole et al. 2022b).

Other studies indicate that rTMS may also induce synapse formation (Hallett 2007; Thomson et al. 2020), a potential common mechanism shared by multiple antidepressants (Duman and Aghajanian 2012). Ketamine, psilocybin, and other antidepressants drive synaptogenesis in the prefrontal cortex (Li et al. 2010; Autry et al. 2011; Shao et al. 2021; Kwan et al. 2022), and synapse formation is required for the long-term maintenance of antidepressant-like behavioral effects in mice (Moda-Sava et al. 2019). Whether rTMS has similar effects on synapse formation is unclear at this time, but indirect evidence suggests that it might. For example, in both rodent and human neuronal culture systems, high-frequency rTMS activates the BDNF-Trk and EGR1 signaling pathways – which are linked respectively with synapse formation and LTP – in a dose-dependent manner, consistent with potential effects on synaptogenesis.

4.2 TMS Effects on Plasticity in Distributed Networks

Growing evidence indicates that TMS acts not only via facilitating plasticity in local microcircuits but also through indirect effects on downstream networks. Pioneering early work showed that antidepressant responses to rTMS may relate to individual differences in fMRI measures of functional connectivity between the DLPFC target site and the subgenual anterior cingulate (Fox et al. 2012; Drysdale et al. 2017; Siddiqi et al. 2021b; Cash et al. 2019; Elbau et al. 2023), a target for therapeutic deep brain stimulation in depression (Mayberg et al. 2005). Subsequently, multiple large-scale studies have highlighted an association between treatment outcomes and functional connectivity with a network of brain regions including the anterior cingulate cortex, anterior insula cortex, and orbitofrontal cortex, among other areas (Fox et al. 2012; Drysdale et al. 2017; Siddiqi et al. 2021b; Cash et al. 2019; Elbau et al. 2023; Dunlop et al. 2024; Liston et al. 2014; Chen et al. 2013). Furthermore, lesion network mapping analyses have revealed how a lesion's connectivity with a network of depression-related brain regions is more tightly associated with depression risk than the location of the lesion in isolation, again consistent with the hypothesis that TMS acts in part through downstream effects on distributed networks (Siddiqi et al. 2021a, 2022; Taylor et al. 2023).

More recently, interleaved TMS/fMRI and TMS/EEG studies have begun to identify networks of brain regions that may be activated indirectly by TMS pulses delivered to a restricted area of cortex (Webler et al. 2022; George et al. 2023; Bestmann et al. 2003; Bohning et al. 1999; Noda et al. 2018; Daskalakis et al. 2008). For example, concurrent TMS/fMRI studies have shown how stimulating the left dorsolateral prefrontal cortex induces neuronal activity responses not only in the local target site but also in a network of structures including the anterior insula, anterior cingulate, somatosensory, and auditory cortex (Nahas et al. 2001; Tik et al. 2023b). Whether and how these effects on distributed networks contribute to the therapeutic effects of rTMS is a critical outstanding question.

4.3 Modeling TMS in Rodents

To date, most studies of TMS-induced plasticity have relied either on measures of plasticity in motor cortex in humans or on effects in neuronal culture systems. As reviewed above, these studies have revealed a host of important insights into the neurobiological mechanisms underlying TMS effects in clinical settings, but these approaches also have some limitations. Studies of plasticity in motor cortex may not generalize to other brain regions that are more common therapeutic targets – a problem exacerbated in neuronal culture systems, which are not intended to model the complex connectivity patterns observed in the living brain. Furthermore, most studies in humans are by necessity non-invasive and inherently low-throughput.

Preclinical rodent models could provide a potentially transformative bridge spanning multiple levels of analysis. They have the potential to enable direct, high-throughput assessments of synaptic plasticity in local microcircuits and distributed brain networks. They could also facilitate a relatively rapid, systematic optimization of an extremely large parameter space that must be navigated in therapeutic TMS protocols, involving multi-dimensional variables such as targeting, pulse dose, train/inter-train duty cycle, duration, and network engagement, among others. Highlighting the potential of this approach, preclinical studies using rodent-compatible TMS coils have shown that low-intensity rTMS upregulates BDNF expression in the mouse brain, and that iTBS but not cTBS has rapid effects on myelination in the mouse cortex in vivo (Cullen et al. 2019; Tang et al. 2016; Makowiecki et al. 2014). In another study, iTBS or cTBS was delivered non-focally (covering the whole brain) in a rat using a 70-mm coil, and the authors found that both protocols had wide-spread effects on c-Fos expression – a marker of neuronal activity – that diminished with repeated stimulation sessions (Volz et al. 2013). New, mouse-compatible TMS coils capable of delivering focal, high-potency stimulation sufficient to drive neuronal spiking in a circumscribed area of cortex are on the horizon. Likewise, a powerful and rapidly developing suite of circuit neuroscience tools – including two-photon calcium imaging, high-density silicon probes, optogenetics, tissue clearing, and rapid whole-brain imaging of neuronal activity markers – have been optimized for use in rodents and have the potential to facilitate TMS modeling and reveal new mechanistic insights in future studies.

5 Neuropharmacological Mechanisms of TMS in Depression

Although TMS is a non-pharmacological treatment for depression, the durability of its therapeutic effects nonetheless depends on mechanisms of neuroplasticity that involve specific neurotransmitter and receptor systems. These receptor systems are amenable to pharmacological manipulation, potentially with beneficial effects in boosting or accelerating the therapeutic response. Let us therefore briefly review what is known about the neuropharmacological mechanisms of TMS, and how they offer opportunities for pharmacological 'augmentation' of the antidepressant effects of TMS effects.

5.1 Dopaminergic Mechanisms

As early as 2001, Strafella and colleagues used PET imaging to investigate the effects of TMS on dopaminergic neurotransmission (Strafella et al. 2001). Applying a series of 10 Hz trains of TMS to the left DLPFC, and using 11C-raclopride to assess D2 receptor occupancy, they observed a very focal reduction in binding potential (indicative of dopamine release) localized specifically to the head of the left caudate nucleus, at a location shown elsewhere to participate in the cortico-striatal-thalamic loop circuit serving the left DLPFC (Dowdle et al. 2018). Interestingly, no such effect was apparent in the corresponding right caudate structure, or in other striatal structures such as the putamen or nucleus accumbens.

In a follow-up study from the same group, Cho et al. used PET to study the effects of 10 Hz TMS targeting the DMPFC rather than the DLPFC, using a slightly different tracer with D2 and D3 affinity, [11C] + PHNO. In this case, a behavioral readout was available in the form of performance on a delay discounting test. This time, focal reductions in binding potential for bilateral DMPFC stimulation occurred not in the head of the left caudate but bilaterally in the dorsal putamen and globus pallidus (Cho et al. 2015). Interestingly, the relationship between the change in binding potential and the change in the behavioral readout (change in delay discounting coefficient) was U-shaped rather than monotonic – an observation whose significance we shall revisit in the next section.

The question of which dopamine receptor subtypes are most relevant to TMS-induced neuroplasticity was addressed in an interesting 2015 study by Enomoto and colleagues. This study used quadripulse stimulation (QPS), a burst pattern of stimulation in which short intervals between the pulses of a burst (e.g., 5 ms) have a consistent facilitatory effect while long intervals (e.g., 50 ms) have a consistent excitatory effect. Using motor evoked potentials (MEP) as the readout, the authors compared the effects of facilitatory QPS5 versus inhibitory QPS50 in healthy volunteers who were pretreated with either placebo, or pramipexole, or L-DOPA. While L-DOPA is metabolized to dopamine to provide agonism at all receptor subtypes D1-5, pramipexole is primarily an agonist at the D2/3/4 receptor subfamily, with little activity at the D1/5 subfamily. Pramipexole, as with placebo, showed minimal modulatory effects upon the MEP inhibition of QPS50 or the MEP facilitation of QPS5. However, L-DOPA potently enhanced both the facilitation of QPS5 and the inhibition of QPS50, suggesting a general effect in potentiating TMS-induced neuroplasticity, regardless of direction (Enomoto et al. 2015).

These preclinical findings imply that agonism or antagonism at the D2 receptor subfamily alone should have little effect on the therapeutic outcomes of TMS. However, agonism at the D1 (potentially in combination with the D2) subfamily should enhance the clinical effects of TMS – an effect that might be achieved more practically with psychostimulants than with L-DOPA.

In fact, this is precisely what has been observed in clinical samples. Schulze et al. found that, in a sample of 105 patients undergoing DMPFC-TMS for depression, treatment outcomes were no worse in patients whose regimens included antipsychotics; in fact, those on higher dose-equivalents trended toward having slightly better outcomes (Schulze et al. 2017). Conversely, Hunter et al. (2019), reviewing a sample of 181 patients undergoing DLPFC-TMS for depression, found that patients on concomitant psychostimulant medications showed close to double the rates of response and remission (Hunter et al. 2019). A subsequent replication review in 90 patients, by the same group, found once again that patients on psychostimulants had better outcomes, with the notable finding of a negative dose-response relationship for patients on lisdexamfetamine, reminiscent of the U-shaped behavioral effects of dopamine release in the preclinical study of Cho et al. discussed earlier.

Deploying psychostimulants in clinical populations undergoing TMS may therefore prove to be a simple and effective means of enhancing the therapeutic effect. However, even more commonplace agents might also be worth considering. Frick and colleagues, examining outcomes in patients undergoing DMPFC-TMS for depression, found that higher doses of habitual caffeine use potentiated the antidepressant effects of active TMS, but not sham TMS (Frick et al. 2021).

In summary, preclinical evidence suggests that therapeutic-like TMS protocols cause focal dopamine release in the striatal projection zones of the cortical target, and that both the neuroplastic effects of single-session TMS and the therapeutic effects of a full course of stimulation may be enhanced via dopaminergic manipulations, possibly via D1- rather than D2-dependent mechanisms. Future randomized controlled trials may clarify the therapeutic value of caffeine, psychostimulants, and L-DOPA, particularly in accelerated TMS regimens where only a few days of augmentation may be needed.

5.2 Glutamatergic Mechanisms

Both LTP and LTD are thought to be mediated in major part by glutamatergic mechanisms involving the NMDA receptor at the post-synaptic membrane. For this reason, it is of interest to determine whether glutamatergic mechanisms also play a role in the therapeutic effects of TMS. Preclinically, as noted earlier, the neuroplastic effects of commonly used theta-burst TMS protocols are blocked by the NMDA antagonist memantine (Huang et al. 2007), indicating a key role for this receptor system. In clinical populations, the effects of potentiating NMDA-receptor-dependent mechanisms of plasticity have been investigated via d-cycloserine, a well-tolerated agent with a benign and long-established safety profile. D-cycloserine acts as a partial antagonist at the NMDA receptor in high doses (~1,000 mg), and as a partial agonist at low doses (~100 mg). Cole et al. (2022a, b) recently published encouraging results from a randomized clinical trial in 50 patients undergoing once-daily DLPFC-TMS for depression, and taking either placebo or 100 mg of d-cycloserine prior to each of the first 10 sessions of treatment. Response and remission rates more than doubled in the d-cycloserine group, to an impressive 73.9% response and 39.1% remission despite the relatively short 20-session course of treatment (Cole et al. 2022b).

Of note, the augmenting effects of d-cycloserine may also prove to generalize well to other targets and indications: the same research group has recently found more than a two-fold enhancement of improvement in Y-BOCS scores for patients with obsessive-compulsive disorder undergoing DMPFC-TMS (Wrightson et al. 2023). Potentiation of glutamatergic mechanisms of LTP and LTD induction is therefore both a promising strategy for improving clinical outcomes of TMS treatment, and a window on the neuropharmacology of TMS itself.

6 Unresolved Questions on the Mechanisms of TMS in Depression

6.1 Where Are All the TMS 'Anti-responders'?

We opened this chapter with an illustrative case of a patient in whom brain stimulation produced an immediate and strongly negative effect on her mood state. Happily, such outcomes appear to be extremely rare for therapeutic TMS in depression. Studies of the trajectories of outcome for TMS reveal groups of rapid responders, medium responders, slow responders, and nonresponders (Fig. 4a), but importantly, no group of 'rapid anti-responders' and indeed no group of 'anti-responders' at all (Kaster et al. 2019; Chen et al. 2024). Likewise, studies of the distribution of TMS outcomes in large naturalistic samples reveal a bimodal distribution of responders and non-responders, but no similarly-large group of anti-responders (Fig. 4b) (Fitzgerald et al. 2016). Indeed, the percentage of patients who show clinically significant deterioration on TMS appears surprisingly low, perhaps on the order of ~5–10%, and the magnitude of the deterioration in these cases appears relatively small (Fitzgerald et al. 2016).

But *why* should this be? Why, after nearly 3 decades of clinical experience with TMS, have we yet to encounter a large percentage of 'anti-responders', whose symptoms get progressively worse as we apply more sessions of treatment? So far, nothing in either the larger-scale and the smaller-scale mechanisms we have reviewed above should forbid the presence of a large group of anti-responders. For example, casual network mapping identifies regions of the brain that are positively rather than negatively linked to depression severity (Siddiqi et al. 2021a) (Fig. 2c). When using standard scalp heuristics such as the 5 cm rule or the BeamF3 heuristic, a substantial proportion of patients end up receiving an entire course of TMS delivered not to the desired network but rather to its anti-network (Cash et al. 2021). So why do such patients merely show non-response, rather than a dramatic 'anti-response' as in the opening case of this chapter?

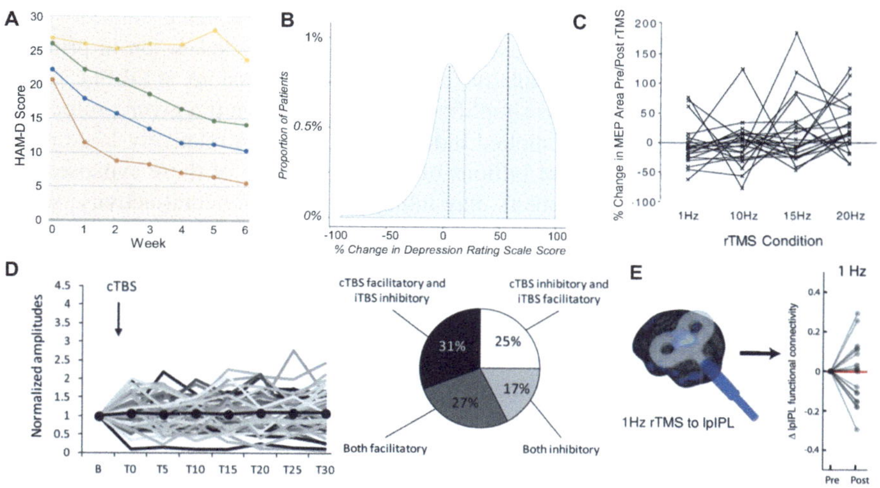

Fig. 4 The relative lack of 'anti-responders' to a full course of clinical TMS stands in stark contrast to the wide range of effects seen for single sessions of TMS in individuals. (**a**) Trajectories of response to a 30-session course of TMS in the THREE-D study (Kaster et al. 2019) reveal fast, medium, slow, and non-responders but no group of 'anti-responders'. (**b**) The bimodal population distribution of clinical response to TMS (Fitzgerald et al. 2016) includes a responder and a non-responder mode, but no third 'anti-responder' mode. (**d**) Maeda et al. (2000) offered an early demonstration of the marked variability in the effect of low-frequency and high-frequency TMS across individuals, with some showing strong facilitation and others showing strong inhibition, in contrast to the classical description of low-frequency protocols as inhibitory and high-frequency protocols as excitatory. (**d**) Hamada et al. (2013) likewise found tremendous inter-individual variability in the effects of nominally 'inhibitory' cTBS and nominally 'excitatory' iTBS; indeed, only 25% of individuals showed this classically described profile, and 31% of individuals actually showed the opposite profile. (**e**) Looking beyond motor cortex, Eldaief et al. (2011) similarly showed that nominally 'inhibitory' 1 Hz stimulation in fact elicited a wide variety of effects across individuals, ranging from strong increases to strong decreases, in functional connectivity between the target inferior partial lobule (IPL) and medial prefrontal cortex adjoining the sgACC

6.2 *Why Are TMS Clinical Outcomes Not as Heterogeneous as TMS Single-Session Effects?*

The same question may be asked regarding the many different patterns of stimulation that have been used therapeutically: high-frequency, low-frequency, intermittent theta-burst, continuous theta-burst, and so on. Few clinical trials to date have made any attempt to personalize the frequency or pattern of stimulation. So why do we not inadvertently create a large cohort of 'anti-responders', rather than merely non-responders, through inadvertently delivering the wrong pattern or protocol?

Classically, TMS protocols are described in blanket terms as being 'inhibitory' or 'excitatory' (usually based on the short-term effects of a single session in primary motor cortex, rather than the cumulative effects of 30–50 sessions at DLPFC). Yet there are two difficulties with this simplified model. First, even if a given protocol is indeed 'excitatory', it is being applied indiscriminately via a relatively large electromagnetic field to a mixture of billions of synapses. Some of these synapses are playing excitatory roles while others play inhibitory roles on neural activity; some participate in the 'direct' corticostriatal pathway and others in the opposed 'indirect' pathway. It is not immediately clear why applying large, high-frequency magnetic pulses to this heterogeneous mixture of effectors should exert a beneficial effect, or indeed any net effect at all.

Secondly, on closer study, most protocols seem to have tremendous inter-individual heterogeneity in their neuroplastic effects. As early as 2000, Maeda et al. reported both 1 Hz and 10 Hz stimulation to yield a fairly even mix of inhibition in some individuals and facilitation in others (Fig. 4c) (Maeda et al. 2000). Similarly, in 2013, Hamada et al. found that only 25% of individuals showed the classical pattern of excitation of iTBS and inhibition for cTBS; 31% showed exactly the reverse pattern, and for the remainder, both protocols were inhibitory or both protocols were excitatory (Fig. 4d) (Hamada et al. 2013). On the whole-brain level, using functional MRI, Eldaief et al. (2011) found that the effects of lateral parietal 1 Hz TMS on parietal-subgenual ACC functional connectivity were not unidirectional but instead yielded a range of effects from strong increases in some individuals to strong decreases in others (Fig. 4e) (Eldaief et al. 2011).

If the neuroplastic effects of our standard TMS protocols are so heterogeneous across individuals (facilitatory in some, excitatory in others, in roughly even proportion), then once again, why do we not see roughly similar numbers of 'anti-responders' to TMS, alongside our treatment responders? Our neuroanatomical, network-level models do not intrinsically forbid this; indeed, they propose both target networks and anti-networks. Likewise, currently available findings on TMS-induced plasticity also do not intrinsically forbid this; indeed, they identify the roughly symmetrical number of individuals who show facilitation versus inhibition on 10 Hz, 1 Hz, iTBS, and cTBS. Some yet-undescribed mechanism appears to be protecting our TMS patients from anti-response, even when we practitioners inadvertently stimulate the anti-network of the intended network, and even when we neglect to determine whether our intended 'excitatory' protocol is actually excitatory in the particular individual being treated.

6.3 *What Are the Implications of Meta-Plasticity for Therapeutic TMS Protocols?*

The safety mechanism shielding patients from 'anti-response' could potentially involve the synaptic mechanisms of *meta-plasticity* (Hulme et al. 2014). This term was first proposed in 1996 (Abraham and Bear 1996), based on observations in animal models in which the effects of a given protocol of stimulation (e.g., whether 10 Hz stimulation elicited LTP or LTD) were not fixed but rather depended upon the prior exposure of the synapse to LTP or LTP induction. The dynamic range of a synapse is not unlimited – indeed, each individual synapse appears to have built-in mechanisms to prevent self-reinforcing cycles of 'runaway facilitation' or 'runaway inhibition' from driving the overall network into non-functionality. Instead, synapses appear to be best positioned to adapt to changing network needs if they can maintain themselves relatively near the *middle* of their dynamic range, rather than at the high end or the low end.

Meta-plasticity acts as a stabilizing force in this regard. For synapses that have already been exposed to previous cycles of LTP induction, further cycles of LTP become harder to induce, and indeed, in such synapses, pulse patterns that would normally produce LTP can instead produce LTD. Conversely, for synapses that have already been driven close to the bottom of their dynamic range, further induction of LTD becomes difficult, and patterns that would normally produce LTD instead produce LTP (Hulme et al. 2014).

Mathematically, this phenomenon was modeled in the 1980s, in the Bienenstock, Cooper, and Munro (BCM) model of synaptic plasticity (Bienenstock et al. 1982). In the BCM model, for a neuron in the middle of its dynamic range, low-frequency stimulation would induce LTD and high-frequency stimulation would induce LTP. However, if a neuron is exposed to faciliatory stimulation, moving up toward the maximum of its dynamic range, then the threshold between LTD and LTP induction shifts upward as well, so that more frequencies now produce net inhibition. Conversely, if a neuron is exposed to inhibitory stimulation, moving down toward the minimum of its dynamic range, then the threshold for LTD shifts downward as well, so that more frequencies now produce net facilitation (Fig. 5a).

Intriguingly, the predictions of this model have been borne out quite neatly, in TMS studies in human primary motor cortex. In 2010, Hamada and colleagues used QPS (see Sect. 5.2) to map out the threshold between facilitation and inhibition, using changes in motor-evoked potential as the readout. In the absence of prior TMS stimulation, high-frequency bursts (intervals of 1.5–10 ms) produced MEP facilitation, while low-frequency bursts (intervals of 30–1,250 ms) produced MEP inhibition. However, if the motor cortex had previously been primed with high-frequency QPS, the threshold shifted so that almost all frequencies were now either neutral or inhibitory. If instead the motor cortex had previously been primed with low-frequency QPS, then conversely, the threshold shifted so that all frequencies were now either neutral or excitatory (Hamada and Ugawa 2010). The MEP amplitude plots for these findings bear a striking resemblance to the BCM curves originally proposed in 1982 (Fig. 5b).

How might we expect the mechanisms of meta-plasticity to filter the effects of nominally 'excitatory' high-frequency or 'inhibitory' low-frequency TMS protocols, when applied to a large population of neurons in the target volume of brain tissue? Among the billions of synapses stimulated by each TMS pulse, each one will be at a different preexisting position within its dynamic range: some near the minimum,

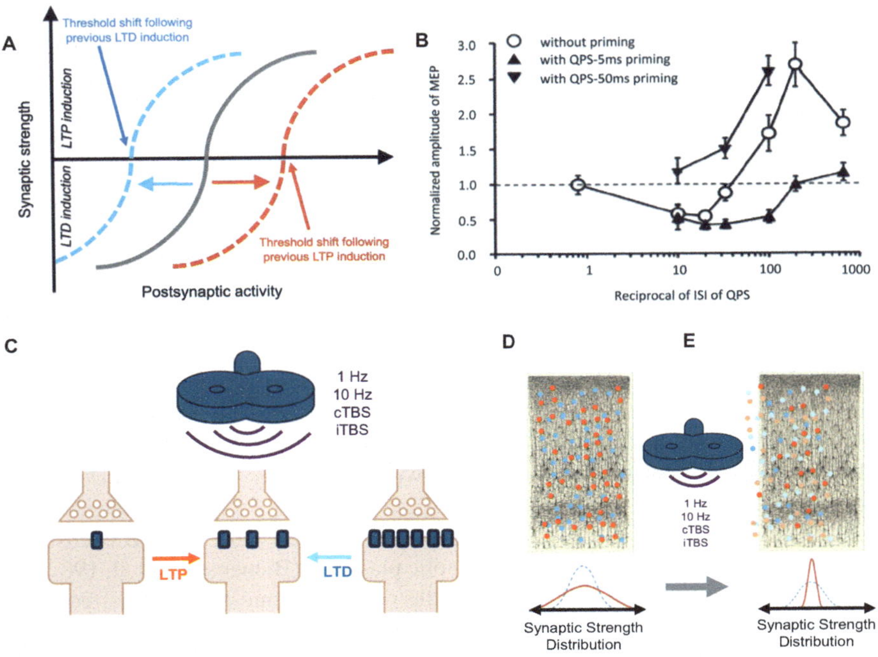

Fig. 5 Meta-plasticity, and its implications for population-level effects of TMS protocols on synaptic strength distributions in target brain regions. (**a**) In the Bienenstock-Cooper-Munro model, a threshold between LTD versus LTP induction exists for each synapse. Synapses that have previously undergone LTD induction have a compensatory reduced threshold for LTP induction, while synapses that have previously undergone LTP induction have an increased threshold for further LTP induction, thus providing a homeostatic mechanism to avoid 'runaway plasticity'. (**b**) TMS studies in motor cortex using quadripulse stimulation (Hamada and Ugawa 2010) provide experimental demonstration of this mechanism of 'meta-plasticity': if the motor cortex has previously undergone excitatory priming with high-frequency QPS5, most protocols now exert an inhibitory effect; if, instead, inhibitory priming with lower-frequency QPS50 is provided, most protocols now exert an excitatory effect. (**c**) When TMS is applied to a volume of tissue, regardless of protocol, the neuroplastic effects at each individual synapse will depend on its prior state: those already near the minimum of their dynamic range (bottom left) will tend to undergo LTP, while those already near the maximum of their dynamic range (bottom right) will tend to undergo LTD. (**d**) At the population level, synapses that have previously been driven by re-entrant activity toward the minimum or the maximum of their dynamic range will lie in the tails of the local distribution. Pathological attractor states, such as those postulated in depression, may therefore be associated with a platykurtic distribution of synaptic strengths in the affected neural pathways. (**e**) When these pathways are targeted with TMS (regardless of protocol), successive sessions will drive synapses near the maximum or minimum of their dynamic range toward the middle, thus exerting a pro-kurtotic effect on the population distribution as a whole. With fewer synapses stuck near the extremes of their dynamic range, re-entrant circuit activity becomes harder to sustain, thus allowing the system to escape from the pathological attractor states associated with depression

primed for facilitation; others already near the maximum, primed for inhibition. Thus, considering the population of synapses as a whole, the overall effect of any given TMS protocol might be best characterized not as 'excitatory' or 'inhibitory', but rather as 'moderating'. Synapses near the maximum of their dynamic range will progressively move closer to the middle, as will synapses near the minimum of their dynamic range (Fig. 5c, d). The exact final 'settling point' of each synapse may be less relevant than the simple result that successive sessions of TMS should just generally reduce the proportion of synapses that are stuck near the top or the bottom of their dynamic range, and increase the proportion of synapses that are nearer to the middle of their dynamic range.

6.4 Synaptic Pro-Kurtosis (SPrK) as a Therapeutic Mechanism for TMS

In statistical terms, the effect of most TMS protocols, over many sessions, would be not to shift the *mean* strength of the targeted synapse population toward excitation or inhibition, but rather to increase the *kurtosis* of the overall synapse population. In other words, synapses near the tails (i.e., those that have previously been driven near to the maximum or minimum of their dynamic range) are returned closer to the middle of their dynamic range, where they may contribute more effectively to the functioning of the overall network. For convenience, we will call this effect 'synaptic pro-kurtosis' (SPrK), as the effect of a given TMS protocol on a large population of synapses would be best understood as shifting the tails toward the mean, rather than simply as blanket 'excitation' or 'inhibition' (Fig. 5d).

Note that SPrK is intrinsically a *unidirectional* statistical phenomenon: For any synapses out in the tails of the distribution, stuck near the edge of their dynamic range, either LTP or LTD is strongly favored over its opposite, regardless of what protocol is delivered. These synapses will gradually regress toward the mean with each TMS session delivered. However, once a given synapse reaches some equilibrium between LTD and LTP, additional sessions will no longer exert much net effect. At this point, maximum therapeutic benefit will have been reached. On clinical scales, the trajectories of improvement toward this equilibrium should appear fairly similar *regardless of protocol*, since they reflect a population-level migration rather than the effect at a single synapse. Indeed, this is precisely what has been observed in direct comparisons of iTBS, 10 Hz, conventional once-daily, and accelerated regimens (Kaster et al. 2019, 2020).

If the population-level effect of TMS on synaptic strength is primarily one of pro-kurtosis rather than excitation or inhibition, this may account for why courses of TMS courses exert a therapeutic effect that is mostly unidirectional (i.e., either beneficial or neutral for most individuals), without any evident large cohort of anti-responders. Indeed, one prediction of the SPrK model is that the trajectories of TMS response should not be linear. Instead, the rate of improvement should be rapid at first, when more synapses are near the extremes of their dynamic range, but should then 'decay' exponentially as fewer and fewer synapses remain out in the tails of the overall distribution. Indeed, an exponential decay function has recently been shown to be a better fit than a linear function to the trajectories of clinical improvement for TMS (Berlow et al. 2023).

By the same token, the SPrK model may also help to explain why clinical practitioners do not routinely see patients continue to progress linearly beyond euthymia into hypomania and mania as additional sessions are given. Instead, patients appear to reach a plateau of improvement, and extending the course when this plateau has been reached appears to simply yield diminishing returns rather than progressively worsening adverse effects (Chen et al. 2024; McDonald et al. 2011). This fortunate yet overlooked feature of TMS dosing is precisely what one would expect from a treatment whose mechanism involves progressive pro-kurtosis, rather than progressive excitation or inhibition.

6.5 Implications of the SPrK Model for Interpreting the Neuropathology of Psychiatric Disorders

Note that the SPrK model also makes some testable predictions about the state of the population of synapses affected by major depression, prior to any treatment. Specifically, if one were to examine synapses in the affected circuits (e.g., those in RLOFC cortico-striatal-thalamic loop circuits (Fettes et al. 2017a) sustaining the 'non-reward attractor state' proposed by Rolls (2016)), there should be relatively more synapses stuck at either the high end or the low end of their dynamic range. This would be the end result of re-entrant, self-reinforcing activity in certain circuits, driving synapses progressively away from the mean until they run up against the biological safeguards that prevent full-blown runaway facilitation or inhibition.

OCD, PTSD, and other TMS-treatable conditions would presumably also involve a similar pathological process at the synaptic-population level. Aberrant, re-entrant neural activity in a given circuit would progressively drive its synapses to their extremes, locking the circuit's activity in a self-reinforcing loop until some external perturbation allows an escape from this attractor state. The escape would be rapid but relatively brief in the case of rapid-acting pharmacological manipulations of neurotransmission, such as opioids, ketamine, or scopolamine. However, in the case of TMS, SPrK effects would accumulate over many sessions, and the extreme synaptic weights supporting the attractor state would return to normality, eventually removing the synaptic substrate of the attractor state itself. Therapeutic effects would therefore ensue cumulatively, and potentially after the end of the treatment course; however, they would also persist for some time even in the absence of continuing treatment.

Further work will be required both preclinically and in humans to address the unresolved questions raised in this section. Preclinical work in animal models will also certainly be needed to test the hypothesis that the therapeutic effects of TMS involve pro-kurtotic effects at the level of the entire synaptic population under stimulation, rather than simply blanket excitation or inhibition. However, understanding the effects of TMS on entire populations of synapses in vivo, rather than relying on simple models of excitation or inhibition, will be essential to a future unified theory that would explain why TMS is effective across so many nominally distinct clinical disorders, ranging from depression to OCD to tinnitus to chronic pain. A decade from now, perhaps these disorders will all prove to share a common element of a pathological attractor state, supported by 'almost-runaway plasticity' in the affected circuits. Perhaps TMS will prove to act against each of these disorders in a common way, by flattening the extreme synapses that support these attractor states back to the mean: less of a scalpel, and more of an iron.

7 Transdiagnostic and Evolutionary Perspectives on the Mechanisms of TMS in Depression

We have at this point reviewed a set of neuroplastic mechanisms by which TMS may achieve durable changes in the activity and synaptic connectivity of a target brain region and its associated whole-brain network. Prior to this, we have reviewed how this process, when applied to a specific set of brain regions, may allow TMS to be used as a potent and rapid treatment for major depression specifically, as opposed to other known indications such as tinnitus, movement disorders, chronic pain, and so forth.

Yet the account so far leaves unanswered an important larger question: what is it about these particular brain regions that make them useful targets in depression? More concretely: why should a patient with terminal cancer and existential despair (Downar et al. 2024), or a patient with longstanding major depression under the burden of pervasive psychosocial stressors, somehow be able to return to euthymia and regular daily functioning, *even without any change in the underlying external stressors*, by the mere dint of undergoing a few sessions of TMS?

Our account of the therapeutic mechanism of TMS would be incomplete (at least in the eyes of either patients or practitioners) without at least some consideration of why durable changes in the activity of the DLPFC, sgACC, etc. somehow translate into the actual patient's restored capacity to reclaim their life: to emerge from bed, leave the home, engage with other people, fulfill roles, pursue occupations, imagine meaningful futures, and work to achieve them.

Bridging this gap requires an ethological, evolutionary perspective on the link between depression and the brain. In this perspective, from first principles, the brain supports survival by constantly tuning behavior to the needs of the organism and the constraints of the environment. Although the set of all possible behaviors may seem intractably large, evolutionary processes seem to have found a convenient way to dimension-reduce this complexity down to a relatively small number of behavioral 'modes', which, deployed skillfully, enable any organism to survive and flourish. These modes are usually taught as a series of 'F's: Foraging and Feeding are useful modes of behavior in the absence of a threat, while threats to survival are best managed with a universal Swiss Army knife of threat-defense modes, including active modes such as Fleeing or Fighting, and passive modes such as Freezing or Folding (comprising depression, defeat, and sickness modes).

Although every species has its own unique behavior set, these underlying 'modes' of behavior have changed very little in the half-billion years since the emergence of bilateral organisms with a brain at the forward end, because each of these modes solves a specific problem in survival faced by every species. Likewise, although every species has its own unique neuroanatomy, some regions (in the brainstem, for example) are highly conserved throughout the vertebrate world, and it is in one of these regions that neurons appear to be laid out in a 'mode-o-topic' fashion that is convenient to our discussion.

The periaqueductal gray (PAG) of human beings is structured in a manner remarkably similar to that of other mammals, and even that of non-mammalian vertebrates. Such conservation is understandable, considering that the enormous complexity of human behavior still dimension-reduces down fairly well to the time-tested fundamental 'modes' of foraging, feeding, fighting, fleeing, freezing, folding, and so on.

Anatomically, the PAG is a tubular gray matter structure in the midbrain, whose neurons are organized into columns, arranged in a clockwise fashion around the cerebral aqueduct (Benarroch 2012) (Fig. 6). Each column appears to coordinate autonomic and behavioral aspects of a particular behavioral mode, via descending projections to the brainstem as well as ascending projections to the amygdala, hypothalamus, and beyond (Depaulis and Bandler 2012; Carrive 1993). For example, neurons in the ventrolateral PAG (vlPAG) column support passive threat defense mechanisms associated with defeat or sickness behavior, such as immobility, quiescence, bradycardia, and hypotension (Carrive et al. 1997; Lovick 1992). In contrast, neurons in the lateral and dorsolateral (l/dlPAG) columns promote active defense mechanisms including escape behaviors and active confrontational defense behaviors, alongside tachycardia and increased respiration rates (Zhang et al. 1990; Morgan et al. 1998).

In addition, a relatively newly identified column in the rostral aspect of the lateral PAG (rlPAG) contains neurons that are active during foraging and hunting behavior (Comoli et al. 2003; Mota-Ortiz et al. 2009; Yu et al. 2021) (Fig. 6a). The behaviors promoted by these columns stand in opposition to those of active or passive threat defense, and activity in the rlPAG appears to be mutually inhibitory to activity in activity in the other aforementioned columns of the vlPAG, both in animal models (Comoli et al. 2003; Silva and McNaughton 2019), and in high-field resting-state neuroimaging studies of the PAG in humans (Fischbach et al. 2024).

The inputs to the rlPAG and vlPAG columns (Yin et al. 2020) are of particular interest for our purposes. In rodents, the rlPAG has prominent telencephalic inputs from the anterior cingulate and insular cortex, as well as the centromedial amygdala (Mota-Ortiz et al. 2009). Social defeat appears to suppress the information flow

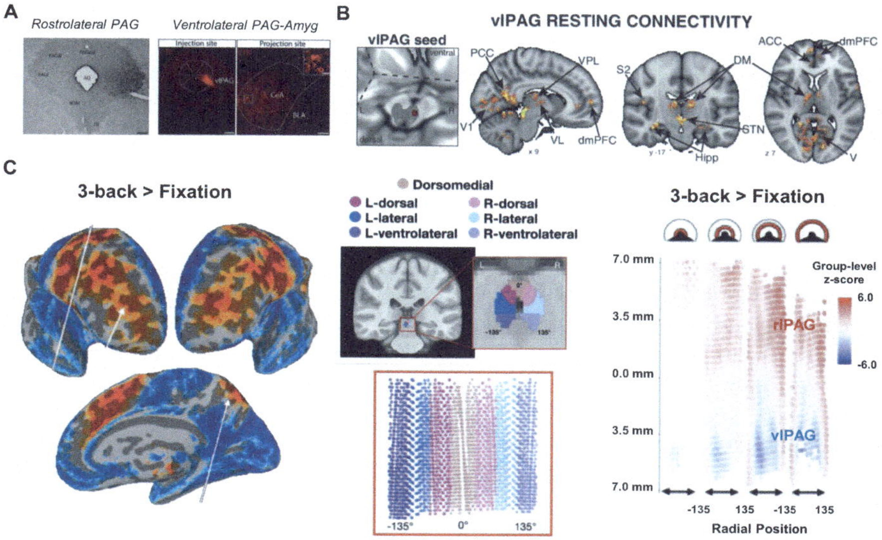

Fig. 6 Reciprocal roles of the rostrolateral and ventrolateral periaqueductal gray in foraging/hunting versus passive threat defense. (**a**) The rostrolateral PAG, rlPAG, (left) contains neurons that are active specifically during foraging and hunting behavior in rodents (adapted from Mota-Ortiz et al. 2009). The ventrolateral periaqueductal gray, vlPAG, (right), taking inputs from the centromedian amygdala, contains neurons active during passive threat-defense behaviors in response to pain, social defeat, and other threats for which active evasion is ineffective (adapted from Yin et al. 2020). (**b**) In humans, the vlPAG is co-active with the default-mode network during the resting state on functional MRI (Faull and Pattison 2017). (**c**) During the performance of goal-oriented behavior requiring cognitive control (e.g., the N-back task), activity is apparent in salience-network regions such as the DLPFC and posterior lateral parietal cortex, as well as the rlPAG in this high-field (7T) study of Fischbach et al. (2024); note also the reciprocal deactivation of the vlPAG and default-mode regions during goal-oriented task performance

between the medial prefrontal cortex and the dmPAG (Franklin et al. 2017). In humans, recent work using 7T functional MRI has revealed resting-state co-activation of the vlPAG with default-mode regions including posterior cingulate cortex, medial prefrontal cortex, and dorsomedial thalamus (Faull and Pattinson 2017) (Fig. 6b). The rlPAG, by contrast, is activated on-task during cognitive control with regions corresponding closely to the salience network: the left and right DLPFC, dorsal anterior cingulate and SMA, anterior insula, and inferior parietal lobule, while deactivations are seen in the vlPAG (Kragel et al. 2019) (Fig. 6c).

It will be interesting in future work to assess whether the whole-brain network/anti-network causally associated with depression in lesion and stimulation studies in Sect. 2.2 (Fig. 2c) corresponds to the set of areas that are coactive with the rlPAG and the vlPAG, respectively. If so, this would help to provide an ethological grounding to explain why stimulation or lesioning of these areas might drive a given patient into, or out of, the mode of passive threat defense at the expense of the daily activities that are analogous to 'foraging and hunting' behaviors in modern human societies. In essence, a course of TMS targeting DLPFC would strengthen the tendency of the central nervous system to maintain itself in the foraging/hunting attractor state, and reduce its tendency to become mired in the attractor state of passive threat defense. In this ethological context, depression would be best understood as a 'mode disorder' rather than a 'mood disorder': a nervous system that is spending too much time locked in the mode of passive threat defense, and not enough time locked in the mode of active foraging and hunting.

From an ethological perspective, our goal in treating depression is to enable the patient to spend less time stuck in the passive threat defense mode of 'fold', and more time in either active threat defense modes (so that stressors may be resolved) or better yet, in the 'foraging modes' of behavior that allow a person to find resources, allies, mates, and opportunities, and build a life worth living. Social workers and community activists, and others may pursue this goal by seeking change in the external environment. Medical specialists equipped with TMS coils (or other biological interventions) may instead pursue this goal by trying to tip the balance of activity of the nervous system itself.

The strategy, in overview, is to suppress the activity of brain circuits that coordinate the passive threat-defense modes of 'fold' and 'freeze', and instead promote the activity of competing circuits that coordinate the active resource-pursuing modes of 'foraging', perhaps together with nuanced deployment of active threat-defense modes such as 'fighting' and 'fleeing'. Because of the unmatched complexity of the set of specific behaviors that humans can deploy in service of these modes, every major level of the nervous system is involved. However, the specific viable targets at each level should be identifiable from their direct or indirect connectivity to particular columns of the PAG.

In particular, brain regions providing input to the rostrolateral PAG (which have been delineated in recent 7T imaging work, and which appear to correspond to the salience network (Kragel et al. 2019)), should serve as useful targets for excitatory stimulation in major depression, via TMS or other modalities. In parallel, brain regions providing input to the ventrolateral PAG (which are predicted to correspond to a non-reward network including the RLOFC and sgACC) may also serve as useful targets for inhibitory stimulation in major depression, via TMS, ECT, DBS, or other modalities. Indeed, since these two behavioral modes are opposed, we may expect inhibitory connections between their neural substrates at every level of the nervous system. In the PAG, high-field imaging would reveal anticorrelation of activity between the opposed rlPAG and vlPAG columns (Fischbach et al. 2024). In the cortex, a rlPAG-covarying network and a vlPAG-covarying anti-network of regions would show opposed resting-state activity (cf. Sect. 2.2, Fig. 2c), and by tipping the balance between these opposed networks, we would expect lesions, DBS, or TMS to generally promote the foraging mode or suppress the passive-defense folding mode.

In summary, the antidepressant effect of TMS might therefore be best understood as merely one aspect of a more general shift of the activity of the entire central nervous system: weakening the attractor states associated with passive-defense modes, and potentiating the attractor states associated with the foraging/hunting mode that is a necessary ingredient for survival in any normally functioning organism.

8 Summary and Conclusions

Let us conclude this review with one final illustrative case study, which we will assign to a date 5 years in the future. A 29-year-old woman presents at a TMS clinic with a 3-year history of increasingly severe depression, unresponsive to previous trials of the SSRI sertraline and the SNRI duloxetine, and now disabling her from work as a grocery store cashier over the last 9 months. We note a precipitating factor of escalating workplace harassment by her manager beginning 3 years ago, alongside the end of a relationship with her common-law partner 2 years ago, precipitating financial insecurity which led to loss of housing 1 year ago. Now living with her parents, she took medical leave from work 9 months ago and has rarely left the house since this time. We note also a family history of unipolar major depression in her elder brother, mother, and grandmother.

From an ethological standpoint, we may interpret her onset of illness as the entry into a passive mode of threat defense, triggered appropriately by aggression from a socially dominant conspecific, against whom active modes of threat defense have been ineffective in removing the threat. Neurobiologically, we would understand this passive threat-defense mode as sustained by the emergence of an attractor state of self-reinforcing activity in a whole-brain network of regions whose common element is structural and/or functional connectivity to the vlPAG. This attractor state, via mutually inhibitory pathways at multiple levels from brainstem to diencephalon to telencephalon, acts to suppress the reciprocal attractor state of self-reinforcing activity in the rlPAG and associated regions supporting the foraging/hunting mode, including salience-network regions such as the dorsal anterior cingulate, anterior insula, DLPFC, and DMPFC. Given the family history of depression in this individual polygenic factors may have structured the PAG columns and their afferents so as to generally favor the passive threat-defense mode attractor state over the foraging-mode attractor state, even at baseline.

Pharmacological efforts to tip the balance back in favor of the rlPAG and its associated regions appear to have been ineffective in this case. However, the TMS provider offers an alternative: having determined TMS to be safe and indicated for this patient, the recommendation is to undergo a course of 30–50 sessions of treatment, with the coil placed over the DLPFC, so as to engage the network of regions linked to the rlPAG and its associated set of foraging/hunting functions. A PET scan, if performed, might have shown these regions to be generally less active than usual in this depressed patient (cf. Fig. 1a). However, the TMS provider instead offers the option of resting-state functional MRI to fine-tune the exact DLPFC target for maximal engagement of the rlPAG-associated network and maximal inhibition of the vlPAG-associated network (which would include the sgACC and default-mode-network regions).

Tipping the central nervous system from the vlPAG-centered attractor state to the rlPAG-centered attractor state does not ensue within minutes or hours, as it might with other interventions such as ketamine or scopolamine. Instead, with each pulse, billions of synapses are stimulated with every pulse, including those of inhibitory and excitatory neurons, and the net effect on the circuit is neither grossly excitatory nor grossly inhibitory. Instead, the net effect is best understood as a *pro-kurtotic* effect at the population level: synapses near the maximum of their dynamic range undergo LTD induction, while those near the minimum of their dynamic range undergo LTP, and both types are driven closer to the middle of their range. Over many sessions, as the synapses at these extrema are driven gradually toward the middle, they no longer form a substrate to sustain the self-reinforcing, pathological, vlPAG-associated attractor state associated with depression. Once this attractor state collapses, it ceases to inhibit the opposed rlPAG-associated attractor state associated with foraging and hunting behaviors, as well as other prosocial behaviors associated with the dmPAG.

With a successful course of TMS, the balance of behavioral modes will thus be tipped away from passive threat defense, and back toward prosocial and foraging modes of behavior. Clinically, successful TMS treatment will yield improvements on standardized symptom scales, ideally to below an arbitrary threshold for 'remission'. More broadly, however, remission will be apparent on the symptom domains tracked on functional scales (e.g., SF-36) and disability scales (e.g., SDS, WHODAS 2.0): self-care, engagement with friends, pursuit of relationships, participation in work, education, or home life, and re-engagement with a larger community of fellow human beings: all among the human analogs of the core behavioral modes to be pursued in the absence of survival threats.

How long will her remission last without re-treatment? The answer depends partly on neuroplasticity, and partly on broader social and environmental factors. Although TMS treatment may have effaced the vlPAG-centric attractor states associated with depression, the synapses that once sustained them are capable of returning once again from the middle back toward the extrema of their dynamic range. If they do so, the depressed attractor state will eventually re-emerge, and exert an escalating inhibitory effect upon the competing rlPAG-associated modes of foraging and prosocial behavior. Eventually, a tipping point may be reached once again, at which point the patient will relapse into depression.

External factors play heavily into this process, even in rodent models. Repeated exposure to social defeat, inescapable threat, or other similar factors (Franklin et al. 2017) will drive the system back to the vlPAG-associated attractor state of passive defense – this is simply the organism's only remaining available response to threats that are (i) persistent, (ii) must be faced without allies, and (iii) cannot be ignored, outrun, or outfought.

Among humans, the story is no different. If our patient successfully completes a course of TMS treatment, only to return to an environment of economic insecurity, social defeat, threats to survival, and no obvious allies or means of escape to be found, then relapse is the most likely, if not inevitable, outcome. In this respect, the words used by nineteenth century medical pioneer, Sir William Osler, to describe tuberculosis might apply equally well today to depression: "a social disease with a medical aspect". On the one hand, we are fortunate to find ourselves in the midst of an age of rapid progress for neuromodulation treatments in depression, with major advances occurring almost yearly. On the other hand, for individual patients, successful treatment may offer only a transient escape from the neural consequences of a precarious existence – a problem afflicting more and more of the population at large in recent years. Progress against the stubbornly high prevalence of depression will require steady progress on both these fronts, in the years to come.

References

Abraham WC, Bear MF (1996) Metaplasticity: the plasticity of synaptic plasticity. Trends Neurosci 19:126–130

Argyelan M et al (2023) Electroconvulsive therapy-induced volumetric brain changes converge on a common causal circuit in depression. Mol Psychiatry. https://doi.org/10.1038/s41380-023-02318-2

Autry AE et al (2011) NMDA receptor blockade at rest triggers rapid behavioural antidepressant responses. Nature 475:91–95

Avissar M et al (2017) Functional connectivity of the left DLPFC to striatum predicts treatment response of depression to TMS. Brain Stimul 10:919–925

Baeken C, Duprat R, Wu G-R, De Raedt R, van Heeringen K (2017) Subgenual anterior cingulate-medial orbitofrontal functional connectivity in medication-resistant major depression: a neurobiological marker for accelerated intermittent theta burst stimulation treatment? Biol Psychiatry Cogn Neurosci Neuroimaging 2:556–565

Bakker N et al (2015) rTMS of the dorsomedial prefrontal cortex for major depression: safety, tolerability, effectiveness, and outcome predictors for 10 Hz versus intermittent theta-burst stimulation. Brain Stimul 8:208–215

Baxter LR Jr et al (1989) Reduction of prefrontal cortex glucose metabolism common to three types of depression. Arch Gen Psychiatry 46:243–250

Belluck P (2021) A 'Pacemaker for the Brain': no treatment helped her depression – until this. The New York Times

Benarroch EE (2012) Periaqueductal gray: an interface for behavioral control. Neurology 78:210–217

Bench CJ et al (1992) The anatomy of melancholia--focal abnormalities of cerebral blood flow in major depression. Psychol Med 22:607–615

Bergmann TO, Hartwigsen G (2021) Inferring causality from noninvasive brain stimulation in cognitive neuroscience. J Cogn Neurosci 33:195–225

Berlow YA et al (2023) Modeling the antidepressant treatment response to transcranial magnetic stimulation using an exponential decay function. Sci Rep 13:7138

Bestmann S, Baudewig J, Siebner HR, Rothwell JC, Frahm J (2003) Subthreshold high-frequency TMS of human primary motor cortex modulates interconnected frontal motor areas as detected by interleaved fMRI-TMS. Neuroimage 20:1685–1696

Bienenstock EL, Cooper LN, Munro PW (1982) Theory for the development of neuron selectivity: orientation specificity and binocular interaction in visual cortex. J Neurosci 2:32–48

Blumberger DM et al (2018) Effectiveness of theta burst versus high-frequency repetitive transcranial magnetic stimulation in patients with depression (THREE-D): a randomised non-inferiority trial. Lancet 391:1683–1692

Blumberger DM et al (2021) A randomized sham controlled comparison of once vs twice-daily intermittent theta burst stimulation in depression: a Canadian rTMS treatment and biomarker network in depression (CARTBIND) study. Brain Stimul 14:1447–1455

Blumberger DM et al (2022) Effectiveness of standard sequential bilateral repetitive transcranial magnetic stimulation vs bilateral theta burst stimulation in older adults with depression: the FOUR-D randomized noninferiority clinical trial. JAMA Psychiatry 79:1065–1073

Bohning DE et al (1999) A combined TMS/fMRI study of intensity-dependent TMS over motor cortex. Biol Psychiatry 45:385–394

Bolt T et al (2022) A parsimonious description of global functional brain organization in three spatiotemporal patterns. Nat Neurosci 25:1093–1103

Carrive P (1993) The periaqueductal gray and defensive behavior: functional representation and neuronal organization. Behav Brain Res 58:27–47

Carrive P, Leung P, Harris J, Paxinos G (1997) Conditioned fear to context is associated with increased Fos expression in the caudal ventrolateral region of the midbrain periaqueductal gray. Neuroscience 78:165–177

Cash RFH et al (2019) Subgenual functional connectivity predicts antidepressant treatment response to transcranial magnetic stimulation: independent validation and evaluation of personalization. Biol Psychiatry 86:e5–e7

Cash RFH et al (2021) Using brain imaging to improve spatial targeting of transcranial magnetic stimulation for depression. Biol Psychiatry 90:689–700

Chen R et al (1997) Depression of motor cortex excitability by low-frequency transcranial magnetic stimulation. Neurology 48:1398–1403

Chen AC et al (2013) Causal interactions between fronto-parietal central executive and default-mode networks in humans. Proc Natl Acad Sci U S A 110:19944–19949

Chen X et al (2024) Depressive symptom trajectories with prolonged rTMS treatment. Brain Stimul. https://doi.org/10.1016/j.brs.2024.04.010

Cho SS et al (2015) Investing in the future: stimulation of the medial prefrontal cortex reduces discounting of delayed rewards. Neuropsychopharmacology 40:546–553

Cole EJ et al (2020) Stanford accelerated intelligent neuromodulation therapy for treatment-resistant depression. Am J Psychiatry 177:716–726

Cole EJ et al (2022a) Stanford neuromodulation therapy (SNT): a double-blind randomized controlled trial. Am J Psychiatry 179:132–141

Cole J et al (2022b) Efficacy of adjunctive D-cycloserine to intermittent theta-burst stimulation for major depressive disorder: a randomized clinical trial. JAMA Psychiatry 79:1153–1161

Comoli E, Ribeiro-Barbosa ER, Canteras NS (2003) Predatory hunting and exposure to a live predator induce opposite patterns of Fos immunoreactivity in the PAG. Behav Brain Res 138: 17–28

Croarkin PE et al (2024) Accelerated 1 Hz dorsomedial prefrontal transcranial magnetic stimulation for generalized anxiety disorder in adolescents and young adults: a case series. Brain Stimul 17: 269–271

Cullen CL et al (2019) Low-intensity transcranial magnetic stimulation promotes the survival and maturation of newborn oligodendrocytes in the adult mouse brain. Glia 67:1462–1477

Daskalakis ZJ et al (2008) Long-interval cortical inhibition from the dorsolateral prefrontal cortex: a TMS-EEG study. Neuropsychopharmacology 33:2860–2869

Depaulis A, Bandler R (2012) The midbrain periaqueductal gray matter: functional, anatomical, and neurochemical organization. Springer Science & Business Media

Devinsky O, Morrell MJ, Vogt BA (1995) Contributions of anterior cingulate cortex to behaviour. Brain 118(Pt 1):279–306

Dinga R et al (2019) Evaluating the evidence for biotypes of depression: methodological replication and extension of. Neuroimage Clin 22:101796

Dinur-Klein L et al (2014) Smoking cessation induced by deep repetitive transcranial magnetic stimulation of the prefrontal and insular cortices: a prospective, randomized controlled trial. Biol Psychiatry 76:742–749

Dong K et al (2022) Comparative efficacy of transcranial magnetic stimulation on different targets in Parkinson's disease: a Bayesian network meta-analysis. Front Aging Neurosci 14:1073310
Dowdle LT, Brown TR, George MS, Hanlon CA (2018) Single pulse TMS to the DLPFC, compared to a matched sham control, induces a direct, causal increase in caudate, cingulate, and thalamic BOLD signal. Brain Stimul 11:789–796
Downar J et al (2014) Anhedonia and reward-circuit connectivity distinguish nonresponders from responders to dorsomedial prefrontal repetitive transcranial magnetic stimulation in major depression. Biol Psychiatry 76:176–185
Downar J et al (2024) Accelerated transcranial magnetic stimulation for psychological distress in advanced cancer: a phase 2a feasibility and preliminary efficacy clinical trial. Palliat Med 38: 485–491
Drevets WC (2000) Neuroimaging studies of mood disorders. Biol Psychiatry 48:813–829
Drevets WC et al (1997) Subgenual prefrontal cortex abnormalities in mood disorders. Nature 386: 824–827
Drysdale AT et al (2017) Resting-state connectivity biomarkers define neurophysiological subtypes of depression. Nat Med 23:28–38
Duman RS, Aghajanian GK (2012) Synaptic dysfunction in depression: potential therapeutic targets. Science 338:68–72
Dunlop K et al (2020) Dorsomedial prefrontal cortex repetitive transcranial magnetic stimulation for treatment-refractory major depressive disorder: a three-arm, blinded, randomized controlled trial. Brain Stimul 13:337–340
Dunlop K et al (2024) Dimensional and categorical solutions to parsing depression heterogeneity in a large single-site sample. Biol Psychiatry. https://doi.org/10.1016/j.biopsych.2024.01.012
Elbau IG et al (2023) Functional connectivity mapping for rTMS target selection in depression. Am J Psychiatry 180:230–240
Eldaief MC, Halko MA, Buckner RL, Pascual-Leone A (2011) Transcranial magnetic stimulation modulates the brain's intrinsic activity in a frequency-dependent manner. Proc Natl Acad Sci U S A 108:21229–21234
Enomoto H et al (2015) Effects of L-Dopa and pramipexole on plasticity induced by QPS in human motor cortex. J Neural Transm 122:1253–1261
F3 Measurement System. https://clinicalresearcher.org/F3/calculate.php.
Faull OK, Pattinson KT (2017) The cortical connectivity of the periaqueductal gray and the conditioned response to the threat of breathlessness. eLife 6:e21749. https://doi.org/10.7554/eLife.21749
Feffer K et al (2018) 1Hz rTMS of the right orbitofrontal cortex for major depression: safety, tolerability and clinical outcomes. Eur Neuropsychopharmacol 28:109–117
Feinsod M, Kreinin B, Chistyakov A, Klein E (1998) Preliminary evidence for a beneficial effect of low-frequency, repetitive transcranial magnetic stimulation in patients with major depression and schizophrenia. Depress Anxiety 7:65–68
Fettes P, Mazza F et al (2024) Predictors and correlates of outcome for dorsolateral, dorsomedial, and orbitofrontal rTMS in major depression. Res Sq.. https://doi.org/10.21203/rs.3.rs-4484095/v1
Fettes P, Schulze L, Downar J (2017a) Cortico-striatal-thalamic loop circuits of the orbitofrontal cortex: promising therapeutic targets in psychiatric illness. Front Syst Neurosci 11:25
Fettes P, Peters S, Giacobbe P, Blumberger DM, Downar J (2017b) Neural correlates of successful orbitofrontal 1 Hz rTMS following unsuccessful dorsolateral and dorsomedial prefrontal rTMS in major depression: a case report. Brain Stimul 10:165–167
Fischbach AK et al (2024) 7-Tesla evidence for columnar and rostral-caudal organization of the human periaqueductal gray response in the absence of threat: a working memory study. J Neurosci. https://doi.org/10.1523/JNEUROSCI.1757-23.2024
Fitzgerald PB et al (2009) A randomized trial of rTMS targeted with MRI based neuro-navigation in treatment-resistant depression. Neuropsychopharmacology 34:1255–1262

Fitzgerald PB, Hoy KE, Anderson RJ, Daskalakis ZJ (2016) A study of the pattern of response to rTMS treatment in depression. Depress Anxiety 33:746–753
Folstein MF, Maiberger R, McHugh PR (1977) Mood disorder as a specific complication of stroke. J Neurol Neurosurg Psychiatry 40:1018–1020
Fox MD, Buckner RL, White MP, Greicius MD, Pascual-Leone A (2012) Efficacy of transcranial magnetic stimulation targets for depression is related to intrinsic functional connectivity with the subgenual cingulate. Biol Psychiatry 72:595–603
Fox MD et al (2014) Resting-state networks link invasive and noninvasive brain stimulation across diverse psychiatric and neurological diseases. Proc Natl Acad Sci U S A 111:E4367–E4375
Franklin TB et al (2017) Prefrontal cortical control of a brainstem social behavior circuit. Nat Neurosci 20:260–270
Frick A, Persson J, Bodén R (2021) Habitual caffeine consumption moderates the antidepressant effect of dorsomedial intermittent theta-burst transcranial magnetic stimulation. J Psychopharmacol 35:1536–1541
Furey ML, Drevets WC (2006) Antidepressant efficacy of the antimuscarinic drug scopolamine: a randomized, placebo-controlled clinical trial. Arch Gen Psychiatry 63:1121–1129
Ganos C et al (2022) A neural network for tics: insights from causal brain lesions and deep brain stimulation. Brain 145:4385–4397
George MS et al (1994) Regional brain activity when selecting a response despite interference: an H2 (15) O PET study of the stroop and an emotional stroop. Hum Brain Mapp 1:194–209
George MS et al (1995) Daily repetitive transcranial magnetic stimulation (rTMS) improves mood in depression. Neuroreport 6:1853–1856
George MS et al (1996) Changes in mood and hormone levels after rapid-rate transcranial magnetic stimulation (rTMS) of the prefrontal cortex. J Neuropsychiatry Clin Neurosci 8:172–180
George MS et al (1997) Mood improvement following daily left prefrontal repetitive transcranial magnetic stimulation in patients with depression: a placebo-controlled crossover trial. Am J Psychiatry 154:1752–1756
George MS et al (2010) Daily left prefrontal transcranial magnetic stimulation therapy for major depressive disorder: a sham-controlled randomized trial. Arch Gen Psychiatry 67:507–516
George MS et al (2023) EEG synchronized left prefrontal transcranial magnetic stimulation (TMS) for treatment resistant depression is feasible and produces an entrainment dependent clinical response: a randomized controlled double blind clinical trial. Brain Stimul 16:1753–1763
Hallett M (2007) Transcranial magnetic stimulation: a primer. Neuron 55:187–199
Hamada M, Ugawa Y (2010) Quadripulse stimulation – a new patterned rTMS. Restor Neurol Neurosci 28:419–424
Hamada M, Murase N, Hasan A, Balaratnam M, Rothwell JC (2013) The role of interneuron networks in driving human motor cortical plasticity. Cereb Cortex 23:1593–1605
Hamilton JP, Chen MC, Gotlib IH (2013) Neural systems approaches to understanding major depressive disorder: an intrinsic functional organization perspective. Neurobiol Dis 52:4–11
Hamilton JP et al (2016) Effects of salience-network-node neurofeedback training on affective biases in major depressive disorder. Psychiatry Res Neuroimaging 249:91–96
Herbsman T et al (2009) More lateral and anterior prefrontal coil location is associated with better repetitive transcranial magnetic stimulation antidepressant response. Biol Psychiatry 66:509–515
Holtzheimer PE et al (2017) Subcallosal cingulate deep brain stimulation for treatment-resistant depression: a multisite, randomised, sham-controlled trial. Lancet Psychiatry 4:839–849
Huang Y-Z, Chen R-S, Rothwell JC, Wen H-Y (2007) The after-effect of human theta burst stimulation is NMDA receptor dependent. Clin Neurophysiol 118:1028–1032
Hulme SR, Jones OD, Raymond CR, Sah P, Abraham WC (2014) Mechanisms of heterosynaptic metaplasticity. Philos Trans R Soc Lond B Biol Sci 369:20130148
Hunter AM et al (2019) Concomitant medication use and clinical outcome of repetitive transcranial magnetic stimulation (rTMS) treatment of major depressive disorder. Brain Behav 9:e01275

Isserles M et al (2021) Deep transcranial magnetic stimulation combined with brief exposure for posttraumatic stress disorder: a prospective multisite randomized trial. Biol Psychiatry 90:721–728
Jing Y et al (2020) Pregenual or subgenual anterior cingulate cortex as potential effective region for brain stimulation of depression. Brain Behav 10:e01591
Joutsa J et al (2018) Identifying therapeutic targets from spontaneous beneficial brain lesions. Ann Neurol 84:153–157
Joutsa J et al (2022) Brain lesions disrupting addiction map to a common human brain circuit. Nat Med 28:1249–1255
Kaster TS et al (2019) Trajectories of response to dorsolateral prefrontal rTMS in major depression: a THREE-D study. Am J Psychiatry 176:367–375
Kaster TS et al (2020) Depressive symptom trajectories associated with standard and accelerated rTMS. Brain Stimul 13:850–857
Kirkby LA et al (2018) An amygdala-hippocampus subnetwork that encodes variation in human mood. Cell 175:1688–1700.e14
Kito S, Fujita K, Koga Y (2008) Regional cerebral blood flow changes after low-frequency transcranial magnetic stimulation of the right dorsolateral prefrontal cortex in treatment-resistant depression. Neuropsychobiology 58:29–36
Klein E et al (1999) Therapeutic efficacy of right prefrontal slow repetitive transcranial magnetic stimulation in major depression: a double-blind controlled study. Arch Gen Psychiatry 56:315–320
Kragel PA et al (2019) Functional involvement of human periaqueductal gray and other midbrain nuclei in cognitive control. J Neurosci 39:6180–6189
Kreuzer PM et al (2015) The ACDC pilot trial: targeting the anterior cingulate by double cone coil rTMS for the treatment of depression. Brain Stimul 8:240–246
Kryatova MS, Seiner SJ, Brown JC, Siddiqi SH (2024) Older age associated with better antidepressant response to H1-coil transcranial magnetic stimulation in female patients. J Affect Disord 351:66–73
Kwan AC, Olson DE, Preller KH, Roth BL (2022) The neural basis of psychedelic action. Nat Neurosci 25:1407–1419
Li N et al (2010) mTOR-dependent synapse formation underlies the rapid antidepressant effects of NMDA antagonists. Science 329:959–964
Li Y et al (2023) A vertigo network derived from human brain lesions and brain stimulation. Brain Commun 5:fcad071
Liang Z et al (2020) Repetitive transcranial magnetic stimulation on chronic tinnitus: a systematic review and meta-analysis. BMC Psychiatry 20:547
Liston C et al (2014) Default mode network mechanisms of transcranial magnetic stimulation in depression. Biol Psychiatry 76:517–526
Lovick TA (1992) Inhibitory modulation of the cardiovascular defence response by the ventrolateral periaqueductal grey matter in rats. Exp Brain Res 89:133–139
Lozano AM et al (2008) Subcallosal cingulate gyrus deep brain stimulation for treatment-resistant depression. Biol Psychiatry 64:461–467
Maeda F, Keenan JP, Tormos JM, Topka H, Pascual-Leone A (2000) Interindividual variability of the modulatory effects of repetitive transcranial magnetic stimulation on cortical excitability. Exp Brain Res 133:425–430
Makowiecki K, Harvey AR, Sherrard RM, Rodger J (2014) Low-intensity repetitive transcranial magnetic stimulation improves abnormal visual cortical circuit topography and upregulates BDNF in mice. J Neurosci 34:10780–10792
Mayberg HS (1997) Limbic-cortical dysregulation: a proposed model of depression. J Neuropsychiatry Clin Neurosci 9:471–481
Mayberg HS et al (2002) The functional neuroanatomy of the placebo effect. Am J Psychiatry 159: 728–737

Mayberg HS et al (2005) Deep brain stimulation for treatment-resistant depression. Neuron 45:651–660

McDonald WM et al (2011) Improving the antidepressant efficacy of transcranial magnetic stimulation: maximizing the number of stimulations and treatment location in treatment-resistant depression. Depress Anxiety 28:973–980

Mir-Moghtadaei A et al (2015) Concordance between BeamF3 and MRI-neuronavigated target sites for repetitive transcranial magnetic stimulation of the left dorsolateral prefrontal cortex. Brain Stimul 8:965–973

Mitra A, Raichle ME (2016) How networks communicate: propagation patterns in spontaneous brain activity. Philos Trans R Soc Lond B Biol Sci 371

Mitra A, Snyder AZ, Hacker CD, Raichle ME (2014) Lag structure in resting-state fMRI. J Neurophysiol 111:2374–2391

Mitra A, Snyder AZ, Blazey T, Raichle ME (2015a) Lag threads organize the brain's intrinsic activity. Proc Natl Acad Sci U S A 112:E2235–E2244

Mitra A, Snyder AZ, Tagliazucchi E, Laufs H, Raichle ME (2015b) Propagated infra-slow intrinsic brain activity reorganizes across wake and slow wave sleep. eLife 4

Mitra A et al (2016) Human cortical-hippocampal dialogue in wake and slow-wave sleep. Proc Natl Acad Sci U S A 113:E6868–E6876

Mitra A et al (2018) Spontaneous infra-slow brain activity has unique spatiotemporal dynamics and laminar structure. Neuron 98:297–305.e6

Mitra A, Raichle ME, Geoly AD, Kratter IH, Williams NR (2023) Targeted neurostimulation reverses a spatiotemporal biomarker of treatment-resistant depression. Proc Natl Acad Sci U S A 120:e2218958120

Moda-Sava RN et al (2019) Sustained rescue of prefrontal circuit dysfunction by antidepressant-induced spine formation. Science 364

Morgan MM, Whitney PK, Gold MS (1998) Immobility and flight associated with antinociception produced by activation of the ventral and lateral/dorsal regions of the rat periaqueductal gray. Brain Res 804:159–166

Morris PL, Robinson RG, Raphael B, Hopwood MJ (1996) Lesion location and poststroke depression. J Neuropsychiatry Clin Neurosci 8:399–403

Mota-Ortiz SR, Sukikara MH, Felicio LF, Canteras NS (2009) Afferent connections to the rostrolateral part of the periaqueductal gray: a critical region influencing the motivation drive to hunt and forage. Neural Plast 2009:612698

Mylius V et al (2013) Definition of DLPFC and M1 according to anatomical landmarks for navigated brain stimulation: inter-rater reliability, accuracy, and influence of gender and age. Neuroimage 78:224–232

Nahas Z et al (2001) Unilateral left prefrontal transcranial magnetic stimulation (TMS) produces intensity-dependent bilateral effects as measured by interleaved BOLD fMRI. Biol Psychiatry 50:712–720

Nauczyciel C et al (2014) Repetitive transcranial magnetic stimulation over the orbitofrontal cortex for obsessive-compulsive disorder: a double-blind, crossover study. Transl Psychiatry 4:e436

Noda Y et al (2018) Impaired neuroplasticity in the prefrontal cortex in depression indexed through paired associative stimulation. Depress Anxiety 35:448–456

Notzon S et al (2015) Psychophysiological effects of an iTBS modulated virtual reality challenge including participants with spider phobia. Biol Psychol 112:66–76

Oathes DJ et al (2021) Combining transcranial magnetic stimulation with functional magnetic resonance imaging for probing and modulating neural circuits relevant to affective disorders. Wiley Interdiscip Rev Cogn Sci 12:e1553

Padmanabhan JL et al (2019) A human depression circuit derived from focal brain lesions. Biol Psychiatry 86:749–758

Pascual-Leone A, Valls-Solé J, Wassermann EM, Hallett M (1994) Responses to rapid-rate transcranial magnetic stimulation of the human motor cortex. Brain 117(Pt 4):847–858

Pascual-Leone A, Catalá MD, Pascual-Leone Pascual A (1996a) Lateralized effect of rapid-rate transcranial magnetic stimulation of the prefrontal cortex on mood. Neurology 46:499–502
Pascual-Leone A, Rubio B, Pallardó F, Catalá MD (1996b) Rapid-rate transcranial magnetic stimulation of left dorsolateral prefrontal cortex in drug-resistant depression. Lancet 348:233–237
Prentice A et al (2023) 1Hz right orbitofrontal TMS benefits depressed patients unresponsive to dorsolateral prefrontal cortex TMS. Brain Stimul 16:1572–1575
Rao VR et al (2018) Direct electrical stimulation of lateral orbitofrontal cortex acutely improves mood in individuals with symptoms of depression. Curr Biol 28:3893–3902.e4
Raut RV, Mitra A, Snyder AZ, Raichle ME (2019) On time delay estimation and sampling error in resting-state fMRI. Neuroimage 194:211–227
Raut RV et al (2020) Organization of propagated intrinsic brain activity in individual humans. Cereb Cortex 30:1716–1734
Reich MM et al (2022) A brain network for deep brain stimulation induced cognitive decline in Parkinson's disease. Brain 145:1410–1421
Ressler KJ, Mayberg HS (2007) Targeting abnormal neural circuits in mood and anxiety disorders: from the laboratory to the clinic. Nat Neurosci 10:1116–1124
Rolls ET (2016) A non-reward attractor theory of depression. Neurosci Biobehav Rev 68:47–58
Rolls ET (2019) The orbitofrontal cortex and emotion in health and disease, including depression. Neuropsychologia 128:14–43
Rolls ET, Deco G (2016) Non-reward neural mechanisms in the orbitofrontal cortex. Cortex 83:27–38
Sackeim HA et al (2020) Clinical outcomes in a large registry of patients with major depressive disorder treated with transcranial magnetic stimulation. J Affect Disord 277:65–74
Salomons TV et al (2014) Resting-state cortico-thalamic-striatal connectivity predicts response to dorsomedial prefrontal rTMS in major depressive disorder. Neuropsychopharmacology 39:488–498
Scangos KW et al (2021a) Closed-loop neuromodulation in an individual with treatment-resistant depression. Nat Med 27:1696–1700
Scangos KW, Makhoul GS, Sugrue LP, Chang EF, Krystal AD (2021b) State-dependent responses to intracranial brain stimulation in a patient with depression. Nat Med 27:229–231
Schaper FLWVJ et al (2023) Mapping lesion-related epilepsy to a human brain network. JAMA Neurol 80:891–902
Schulze L et al (2017) Effect of antipsychotic pharmacotherapy on clinical outcomes of intermittent theta-burst stimulation for refractory depression. J Psychopharmacol 31:312–319
Schutter DJLG, Laman DM, van Honk J, Vergouwen AC, Koerselman GF (2009) Partial clinical response to 2 weeks of 2 Hz repetitive transcranial magnetic stimulation to the right parietal cortex in depression. Int J Neuropsychopharmacol 12:643–650
Seeley WW (2019) The salience network: a neural system for perceiving and responding to homeostatic demands. J Neurosci 39:9878–9882
Seminowicz DA et al (2004) Limbic-frontal circuitry in major depression: a path modeling metanalysis. Neuroimage 22:409–418
Shao L-X et al (2021) Psilocybin induces rapid and persistent growth of dendritic spines in frontal cortex in vivo. Neuron 109:2535–2544.e4
Siddiqi SH et al (2020) Distinct symptom-specific treatment targets for circuit-based neuromodulation. Am J Psychiatry 177:435–446
Siddiqi SH et al (2021a) Brain stimulation and brain lesions converge on common causal circuits in neuropsychiatric disease. Nat Hum Behav 5:1707–1716
Siddiqi SH, Weigand A, Pascual-Leone A, Fox MD (2021b) Identification of personalized transcranial magnetic stimulation targets based on subgenual cingulate connectivity: an independent replication. Biol Psychiatry 90:e55–e56
Siddiqi SH, Kording KP, Parvizi J, Fox MD (2022) Causal mapping of human brain function. Nat Rev Neurosci 23:361–375
Siddiqi SH et al (2023) Lesion network localization of depression in multiple sclerosis. Nat Ment Health 1:36–44

Siddiqi S et al (2024) Causal network localization of brain stimulation targets for trait anxiety. Res Sq. https://doi.org/10.21203/rs.3.rs-4221074/v1
Silva C, McNaughton N (2019) Are periaqueductal gray and dorsal raphe the foundation of appetitive and aversive control? A comprehensive review. Prog Neurobiol 177:33–72
Spagnolo PA et al (2019) Lack of target engagement following low-frequency deep transcranial magnetic stimulation of the anterior insula. Neuromodulation 22:877–883
Stefurak T et al (2003) Deep brain stimulation for Parkinson's disease dissociates mood and motor circuits: a functional MRI case study. Mov Disord 18:1508–1516
Strafella AP, Paus T, Barrett J, Dagher A (2001) Repetitive transcranial magnetic stimulation of the human prefrontal cortex induces dopamine release in the caudate nucleus. J Neurosci 21:RC157
Sylvester CM et al (2012) Functional network dysfunction in anxiety and anxiety disorders. Trends Neurosci 35:527–535
Tadayonnejad R et al (2023) Right lateral orbitofrontal cortex inhibitory transcranial magnetic stimulation for treatment of refractory mood and depression. Brain Stimul 16:1374–1376
Tang AD et al (2016) Construction and evaluation of rodent-specific rTMS coils. Front Neural Circuits 10:47
Taylor JJ et al (2023) A transdiagnostic network for psychiatric illness derived from atrophy and lesions. Nat Hum Behav 7:420–429
Teferi M et al (2023) Continuous theta-burst stimulation to the right dorsolateral prefrontal cortex may increase potentiated startle in healthy individuals. Biol Psychiatry Glob Open Sci 3:470–479
Thomson AC et al (2020) Transcranial magnetic stimulation-induced plasticity mechanisms: TMS-related gene expression and morphology changes in a human neuron-like cell model. Front Mol Neurosci 13:528396
Tik M et al (2017) Towards understanding rTMS mechanism of action: stimulation of the DLPFC causes network-specific increase in functional connectivity. Neuroimage 162:289–296
Tik M et al (2023a) Concurrent TMS/fMRI reveals individual DLPFC dose-response pattern. Neuroimage 282:120394
Tik M et al (2023b) Acute TMS/fMRI response explains offline TMS network effects – an interleaved TMS-fMRI study. Neuroimage 267:119833
Trapp NT, Pace BD, Neisewander B, Ten Eyck P, Boes AD (2023) A randomized trial comparing beam F3 and 5.5 cm targeting in rTMS treatment of depression demonstrates similar effectiveness. Brain Stimul 16:1392–1400
Volz LJ, Benali A, Mix A, Neubacher U, Funke K (2013) Dose-dependence of changes in cortical protein expression induced with repeated transcranial magnetic theta-burst stimulation in the rat. Brain Stimul 6:598–606
Webler RD et al (2022) DLPFC stimulation alters working memory related activations and performance: an interleaved TMS-fMRI study. Brain Stimul 15:823–832
Weigand A et al (2018) Prospective validation that subgenual connectivity predicts antidepressant efficacy of transcranial magnetic stimulation sites. Biol Psychiatry 84:28–37
Williams NR et al (2018) High-dose spaced theta-burst TMS as a rapid-acting antidepressant in highly refractory depression. Brain J Neurol 141:e18
Williamson J (2019) Establishing causal claims in medicine. Int Stud Philos Sci 32:33–61
Wrightson JG, Cole J, Sohn MN, McGirr A (2023) The effects of D-cycloserine on corticospinal excitability after repeated spaced intermittent theta-burst transcranial magnetic stimulation: a randomized controlled trial in healthy individuals. Neuropsychopharmacology 48:1217–1224
Wu JC et al (1992) Effect of sleep deprivation on brain metabolism of depressed patients. Am J Psychiatry 149:538–543
Wu X et al (2016) Dysfunction of the cingulo-opercular network in first-episode medication-naive patients with major depressive disorder. J Affect Disord 200:275–283
Yin W et al (2020) A central amygdala-ventrolateral periaqueductal gray matter pathway for pain in a mouse model of depression-like behavior. Anesthesiology 132:1175–1196

Yu H et al (2021) Periaqueductal gray neurons encode the sequential motor program in hunting behavior of mice. Nat Commun 12:6523
Zangen A et al (2021) Repetitive transcranial magnetic stimulation for smoking cessation: a pivotal multicenter double-blind randomized controlled trial. World Psychiatry 20:397–404
Zangen A et al (2023) Pursuing personalized medicine for depression by targeting the lateral or medial prefrontal cortex with Deep TMS. JCI Insight 8
Zarate CA Jr et al (2006) A randomized trial of an N-methyl-D-aspartate antagonist in treatment-resistant major depression. Arch Gen Psychiatry 63:856–864
Zhang SP, Bandler R, Carrive P (1990) Flight and immobility evoked by excitatory amino acid microinjection within distinct parts of the subtentorial midbrain periaqueductal gray of the cat. Brain Res 520:73–82
Zhou M et al (2017) Intrinsic cerebral activity at resting state in adults with major depressive disorder: a meta-analysis. Prog Neuropsychopharmacol Biol Psychiatry 75:157–164

Mechanism of Action of ECT in Depression

Jean-Baptiste Belge, Philip van Eijndhoven, and Peter C. R. Mulders

Contents

Abstract Electroconvulsive therapy (ECT) remains the most potent antidepressant treatment available for patients with major depressive disorder (MDD). ECT is highly effective, achieving a response rate of 70–80% and a remission rate of 50–60% even in treatment-resistant patients. The underlying mechanisms of ECT are not fully understood, although several hypotheses have been proposed, including the monoamine hypothesis, anticonvulsive hypothesis, neuroplastic effects, and immunomodulatory properties. In this paper, we provide an overview of magnetic resonance imaging evidence that addresses the neuroplastic changes that occur after ECT at the human systems level and elaborate further on ECTs potent immunomodulatory properties. Despite a growing body of evidence that suggests ECT may normalize many of the structural and functional changes in the brain associated with severe depression, there is a lack of convergence between neurobiological changes and the robust clinical effects observed in depression. This may be due to sample

J.-B. Belge (✉)
Department of Psychiatry, Collaborative Antwerp Psychiatric Research Institute (CAPRI), Faculty of Medicine and Health Sciences, University of Antwerp, Wilrijk, Belgium

Department of Psychiatry, Radboud University Medical Centre, Nijmegen, The Netherlands
e-mail: jeanbaptiste.belge@uantwerpen.be; jean-baptiste.belge@radboudumc.nl

P. van Eijndhoven and P. C. R. Mulders
Department of Psychiatry, Radboud University Medical Centre, Nijmegen, The Netherlands

Donders Institute for Brain, Cognition and Behavior, Centre for Medical Neuroscience, Nijmegen, The Netherlands

Curr Topics Behav Neurosci (2024) 66: 279–296
https://doi.org/10.1007/7854_2023_450
Published Online: 15 November 2023

sizes used in ECT studies being generally small and differences in data processing and analysis pipelines. Collaborations that acquire large datasets, such as the GEMRIC consortium, can help translate ECT's clinical efficacy into a better understanding of its mechanisms of action.

1 Introduction

Electroconvulsive therapy (ECT), which has been in use since the 1940s, remains the strongest antidepressant treatment available today (UK ECT Review Group 2003). The technique involves the application of an electrical current through the brain, usually to the temporal region, with the aim of inducing a generalized epileptic seizure. A typical course of ECT involves several weeks of treatment, with two to three sessions per week. It is effective for various psychiatric disorders, including bipolar disorder, schizophrenia, Parkinson's disease, and catatonia, but is most often used to treat major depressive disorder (MDD) that has not responded to other treatments (Kho et al. 2003; Kellner et al. 2020).

As a matter of fact, while both psychotherapy and medication can be effective in treating depression, it often takes a relatively long time before the effects become apparent. Unfortunately, about 30% of patients do not experience improvement even after multiple treatment attempts. ECT on the contrary achieves a remarkable 70–80% response rate and a 50–60% remission rate, even in treatment-resistant patients (Hussain et al. 2004; Kho et al. 2003). Comparative studies have shown that ECT has higher response rates and better outcomes compared to traditional antidepressant drugs (Kellner et al. 2020). Additionally, ECT is one of the quickest-acting treatments for affective disorders, sometimes achieving antidepressant effects within days to weeks instead of the 4–6 weeks for expected results from pharmacological treatments for depression (Kho et al. 2003). Due to its greater efficacy and rapid clinical response, it should be emphasized that even though ECT is often recommended when other treatment protocols have had little or no effect, ECT is not a last resort treatment and should be considered as a first-line treatment whenever a rapid clinical response is necessary. Examples include patients who are severely agitated, psychotic, actively suicidal, or whose health is seriously compromised due to severe depression (Kellner et al. 2020).

Despite its effectiveness, the underlying (neuro)biological mechanisms of ECT remain unclear. Throughout its clinical use, hypotheses of ECTs mechanism of action have been proposed that are important to briefly discuss. Firstly the monoamine hypothesis states that, similar to psychopharmacology, ECT works through increases in monoamine systems such as serotonin and dopamine. This hypothesis is based on both our understanding of the depressed brain and the finding that ECT affects monoamine systems and receptor availability (Merkl et al. 2009). Secondly, the anticonvulsive hypothesis, which states that ECT, by eliciting seizures, drives the brain to become more resistant to convulsions through increasing inhibitory processes, as evidenced by increases in the inhibitory neurotransmitter gamma-amino butyric acid (GABA, Sanacora et al. 2003) and the clinical observation that patients develop an increased seizure threshold during the course of treatment (Duthie et al. 2015). Thirdly, one of the main mechanisms of action behind ECTs powerful therapeutic effects appears to be the neuroplastic effects of the treatment (Madsen et al. 2000) which could explain the widespread changes in both the structure and functioning of the brain (Bolwig 2014). Finally, a new wave of ECT research indicates that ECT can induce changes in both central and peripheral inflammatory pathways which is particularly interesting with the inflammatory hypothesis of depression in mind (Yrondi et al. 2018, 2019).

These hypotheses are not mutually exclusive and, ultimately, it is unlikely that any one hypothesis fully explains the broad effects ECT has, but instead each of these (and likely others) may play an important role and interact to create the strong antidepressant response we see in patients. For the purpose of this chapter, we will provide an overview of magnetic resonance imaging evidence that addresses the neuroplastic changes that occur after ECT at the human systems level and elaborate further on ECTs potent immunomodulatory properties.

2 Volumetric Brain Changes

Neuroimaging studies into psychiatric disorders have had an early emphasis on structural and volumetric changes. A variety of methods aimed at determining brain structure have become available over the years, with the most popular derived measures being cortical thickness, cortical volume, or regional volumes, for example calculated using voxel-based morphometry (VBM). While these metrics all aim to extract a different measure to relate to treatment, the core premise is that structure relates (strongly) to function, and consequently that changes in brain structure are indicative of changes in brain function. Important to consider is that in neuroimaging studies in general, but structural neuroimaging studies in particular, the available results from any given analysis are determined by the method employed, and more precisely the areas under investigating. Broad explorative studies might be under-powered to detect small but relevant changes, while focused region-of-interest approaches might uncover significant changes but cannot determine the specificity of the regional effects. Especially as ECT research studies have, until recently, been predominantly performed in smaller samples, such choices have shaped our current understanding of what happens in the brain during and after treatment.

Depression has been linked to significant reductions in gray matter volume, with the most prominent reductions occurring in the medial prefrontal cortex, specifically the anterior cingulate cortex, as well as the lateral prefrontal cortex, encompassing the dorsolateral and orbitofrontal regions, and the medial temporal lobe, which includes the hippocampus and amygdala (Koolschijn et al. 2009; Bora et al. 2012).

Regarding the impact of electroconvulsive therapy (ECT) on the brain, available evidence suggests an increase in the volume of the medial temporal lobe, which is consistent with the changes observed in depression. The degree of volumetric change varies across different studies, with some studies reporting changes that encompass the entire medial temporal lobe, including adjacent structures. However, the increase in volume is most consistently reported for both the hippocampus and, to a lesser extent, the amygdala, insula, and temporal poles.

The combined findings of neuroimaging studies in patients receiving electroconvulsive therapy (ECT) for depression have led to a current consensus that ECT leads to a significant volumetric increase of approximately 5% in both the left and right hippocampus compared to baseline (Dukart et al. 2014; Nuninga et al. 2020; Nordanskog et al. 2014; Oltedal et al. 2018; Tendolkar et al. 2013). These volumetric changes may potentially reflect an increase in angiogenesis, neurogenesis, and/or gliagenesis. It is noteworthy that reduced hippocampal volume is a hallmark feature of depression, and this decrease is more pronounced in patients with a longer disease duration, multiple depressive episodes, or poor treatment response.

The clinical implications of hippocampal enlargement induced by ECT remain unclear, as it appears to have an ambiguous association with the clinical effects of ECT (Oltedal et al. 2018). This suggests that the hippocampal volume increase may be an epiphenomenon of ECT, rather than being responsible for the therapeutic outcome of the intervention itself (Oltedal et al. 2018). However, some studies have found an association between hippocampal volume increase and therapeutic efficacy (Dukart et al. 2014; Nuninga et al. 2020; Ota et al. 2015), with Nuninga et al. (2020) observing a significant and specific association between an increase in the dentate gyrus, the primary site of hippocampal neurogenesis, and an improvement in depressive symptoms. Nevertheless, the follow-up period was insufficient to determine whether these effects were long-lasting or only temporary (Nuninga et al. 2020). Most studies reporting an increase in medial temporal lobe volume did not find any association with clinical outcome, while there was in general a relation between the number of ECT sessions and the increase in volume (Oltedal et al. 2018).

Moreover, it is essential to note that time courses differ, with the antidepressant effects of ECT mostly lasting beyond the volumetric increase, which generally returns to baseline within several weeks to months (19, 20). This may emphasize the unrelatedness of the biological (hippocampal volume increase) and clinical effects of ECT (Oltedal et al. 2018). In fact, repeated findings of large-scale increases in the volume of the medial temporal lobe may be biased by studies that have focused exclusively on those regions based on previous work.

Importantly, a recent study by the Global ECT-MRI Research Collaboration (GEMRIC) identified a multivariate discriminative pattern of structural change relevant to clinical response to ECT, which did not include any of mediotemporal lobe regions (Mulders et al. 2020).

Based on the premise that the hippocampus is a critical cognitive center in the temporal lobe, Van Oostrom et al. (2018) found that the transient increase in hippocampal volume during and after electroconvulsive therapy (ECT) is linked to the transient cognitive side effects of the treatment. This finding was recently validated and replicated by Argyelan et al. (2021), who observed that an enlargement of the hippocampus was associated with poorer cognitive outcomes across two distinct cohorts with varying diagnoses, electrode placements, and numbers of ECT sessions. The notion that cognitive deficits may arise not only from hippocampal atrophy, but also from sudden neurogenesis, has been extensively documented in the literature. The authors propose that a rapid proliferation of new neurons and synapses with diverse electrophysiological properties is likely to influence the functionality of the entire hippocampal circuitry, resulting in a negative impact on cognitive functioning. Over time, the maturation of the new cells, combined with the synaptic pruning of aberrant connections, could ultimately lead to proper integration with the existing circuits, thereby explaining the transient nature of cognitive impairment following ECT (Van Oostrom et al. 2018).

Studies have shown that the brain changes induced by electroconvulsive therapy (ECT) are not limited to the medial temporal lobe (MTL), as other brain regions involved in depression have also demonstrated structural changes post-ECT (Mulders et al. 2020; Ousdal et al. 2022; Wade et al. 2016). These findings suggest that ECT engages brain plasticity mechanisms beyond local neurogenesis. Notably, ECT has been found to increase the volume of several subregions of the anterior cingulate cortex (ACC), while depression is generally related to a volumetric decrease in this region (Drevets et al. 2008). Although the changes in the ACC are less pronounced, they are more strongly associated with clinical status, both in the depressed state and following ECT treatment. Among the subregions of the ACC, the subgenual area has been established as an important target in depression, as demonstrated by successful deep-brain stimulation and repetitive transcranial magnetic stimulation studies (Mayberg et al. 2005; Baeken et al. 2014). Multiple studies have reported an increase in subgenual ACC volume following ECT, with this result being the most consistently reported finding after changes in the MTL. Furthermore, a parcellation-based study revealed an increase in ACC thickness that correlated with clinical improvement (Pirnia et al. 2016).

Finally, the striatum consists of several regions that govern, among others, emotion and reward processes. The nucleus accumbens, caudate nucleus, and the putamen all have been reported as being affected by ECT and, as with other regions, reports mostly indicate an increase in volume. All of these areas have been implicated in the pathophysiology of depression and may relate to changes in reward sensitivity or anhedonia (Koolschijn et al. 2009) and are also particularly affected in treatment-resistant depression. While the clinical relevance of these neuroplastic changes following ECT is yet unclear, a study by Bouckaert et al. (2016a, b) observed that after ECT a volumetric increase in the bilateral caudate nucleus correlated significantly with an improvement of psychomotor retardation (Bouckaert et al. 2016a, b; Herrero et al. 2002). In the same line, Belge and colleagues (2019) found that the volume increase of the nucleus accumbens region correlated significantly with an improvement on psychomotor retardation subscale (Belge et al. 2020b). Finally, intriguingly, Wade et al. demonstrated that baseline volume and shape metrics of the basal ganglia predicted overall response to ECT with up to 89% accuracy (Wade et al. 2016), thus suggesting that these structures, first and foremost the subcortical components of the motor system, might be crucial in the pathophysiology of depression and in the antidepressant effect of ECT.

Overall, it seems unlikely that clinical effects of ECT can be directly tied to volumetric increases in a single brain region. Large studies show that volumetric changes occur in almost all brain regions after ECT and that this effect accumulates as the number of ECT treatment received increases. The patterns within these changes do relate to clinical outcome, but whether that is due to specific changes within those that respond, or that these patterns are present in those more likely to respond, remains undecided. Due to the changes occurring across the brain, collaboration is required to validate clinical association in large enough samples to overcome the heterogeneity in treatment effects and clinical depression.

3 Structural Connectivity Changes

The structural connectivity between different brain regions is underpinned by the architectural and microstructural properties of the white matter (WM) (Jbabdi et al. 2015). Diffusion tensor imaging (DTI) measures the restriction of freely moving water molecules in brain tissue and uses these restrictions to infer WM microstructural properties (van Velzen et al. 2020). If modeled with an ellipsoid shape, represented by a tensor, metrics like fractional anisotropy (FA), mean (MD), radial (RD), and axial (AD) diffusivity can be derived (Le Bihan 2003; Jbabdi et al. 2015). Of these metrics, FA is the most commonly used. With higher FA values thought to represent directionally constrained diffusion and greater uniformity and compactness of the WM fiber bundles due to more intact myelin, FA is often used as an index of WM integrity (van Velzen et al. 2020; Lyden et al. 2014; Jbabdi et al. 2015).

The use of DTI has consistently revealed alterations in white matter microstructure among patients with depression (van Velzen et al. 2020). These investigations have predominantly reported reductions in FA in frontal, limbic, and striatal connections, specifically in the anterior limb of the internal capsule, inferior longitudinal fasciculus, posterior thalamic radiation, superior longitudinal fasciculus, and fronto-occipital fasciculus (van Velzen et al. 2020). More significant reductions in FA have been found to be associated with inadequate treatment response (de Diego-Adeliño et al. 2014). Recently, the Neuroimaging Genetics through Meta-Analysis (ENIGMA) consortium's major depressive disorder working group analyzed 20 international cohorts and confirmed these alterations, revealing subtle but widespread lower FA in patients compared to controls, with the largest differences observed in the corpus callosum and corona radiata, suggesting structural dysconnectivity in patients with depression (van Velzen et al. 2020).

Interestingly, electroconvulsive therapy (ECT) appears to influence white matter (WM) microstructural properties and architecture (Bai et al. 2019a, b; Gryglewski et al. 2020; Lyden et al. 2014; Nickl-Jockschat et al. 2016; Nobuhara et al. 2004; Repple et al. 2020; Zeng et al. 2015). It has been proposed that ECT's potent neuroplastic properties may specifically affect WM (Lyden et al. 2014; Zeng et al.

2015). While sparsely investigated, ECT has been shown to induce changes in various diffusion metrics. For example, in a pilot study, Nobuhara et al. (2004) reported an increase in FA after ECT in frontal white matter. In a subsequent study, Lyden et al. (2014) observed an increase in FA in the anterior cingulum, forceps minor, and left superior longitudinal fasciculus, which was positively associated with treatment response. However, Nickl-Jockschat et al. (2016) did not detect any white matter alterations associated with ECT. Finally, in a more recent study, Gryglewski et al. (2020) observed an increase in axial diffusivity (AD), an indirect measure of axonal integrity, in the posterior limb of the internal capsule of the right hemisphere, but changes of FA were not observed. The inconsistency of these results could be due to the limited power and methodological differences between these studies, which makes it difficult to highlight consistent patterns of change in diffusion metrics following ECT.

Importantly, a multi-center harmonization approach revealed an increase in FA in the splenium following ECT (Belge et al. 2022). The splenium is the most posterior part of the corpus callosum, composed of reciprocal fibers from the temporal association and parietal association, connecting regions of the parietal and occipital lobes through the forceps major (Wakana et al. 2004; Nolte 2009). Reductions in FA are consistently reported in the corpus callosum of patients with depression (van Velzen et al. 2020; Chen et al. 2016; Liao et al. 2013; Murphy and Frodl 2011) and a decrease of myelin in the axons of the splenium has been demonstrated in a post-mortem sample of patients with depression (Williams et al. 2019). The splenium is known to synchronize interhemispheric oscillatory activity through the balanced excitation or suppression of the contralateral homotopic and heterotopic cortical areas (Knyazeva 2013). This is important as depression is related to interhemispheric functional synchronization deficits (Grimm et al. 2008; Guo et al. 2013; Xu et al. 2013). Interestingly, ECT has been shown to selectively modulate this interhemispheric functional synchronization (Wei et al. 2014). These results suggest that ECT restores interhemispheric connectivity through its neuroplastic properties on splenial white matter and its efferent fiber tracts. Finally, an increase in white matter consistency of the superior longitudinal fasciculus has been repeatedly demonstrated following ECT (Lyden et al. 2014; Belge et al. 2022). Interestingly, the superior longitudinal fasciculus participates in the modulation of the dorsolateral prefrontal cortex (DLPFC). WM dysconnectivity of this tract appears to drive an attenuated top-down cognitive control from the DLPFC of limbic hyperactivity, instigating ruminative or repetitive negative thinking, a core behavior of the depressive syndrome (Zuo et al. 2012; Pisner et al. 2019). It is thus tempting to propose that the microstructural changes of the superior longitudinal fasciculus by ECT could enhance top-down cognitive control over emotional states to regulate mood, constituting in essence a core mechanism behind the therapeutic effects of ECT. This corroborates moreover functional imaging findings demonstrating an enhanced connectivity of the DLPFC following ECT (Belge et al. 2021).

4 Functional Connectivity Changes

Functional connectivity refers to a popular neuroimaging analysis that, instead of investigation increased brain activity in response to a stimulus, instead looks at co-occurring fluctuations in brain activity during a task or during rest (i.e., resting-state functional connectivity). Such studies show how brain regions are functionally connected, and how they are organized within large-scale functional brain networks. These networks are spread throughout the brain, with the main networks of interest in the context of psychiatry being the "default mode network" (DMN), which is most active during rest, the "cognitive executive network" (CEN), which is most active during cognitively demanding tasks, and the "salience" or "affective network" (SN), which is mostly implicated during emotionally salient stimuli. The studies exploring these metrics, while fewer in number, indicate that ECT appears to restore aberrant functional connectivity present at baseline in treatment-resistant depression.

In addition to being a region-of-interest when looking at volumetric changes, an important region with reported changes in functional connectivity is the ACC. The ACC is an important monitoring hub, contributing to attributing valence to stimuli as well as cognitive, social, and arousal processes (Holroyd and Umemoto 2016). In major depressive disorder most functional studies find evidence for hyperactive "emotional" limbic areas such as the amygdala and ACC and hypoactive "controlling" lateral areas such as the lateral prefrontal cortex (Mayberg 1997; Northoff et al. 2011). Because of the consistent hyperactivity within the ACC, deep-brain stimulation for treatment-resistant depression also targets this area (Mayberg et al. 2005).

Resting-state connectivity studies also support changes primarily within and between these core regions, with connectivity being increased between anterior parts of the default mode network (medial prefrontal cortex) and parts of the ACC and the salience network (including the amygdala), while functional connectivity between cognitive executive regions is decreased (Mulders et al. 2015). Following ECT, functional connectivity changes from the ACC to the temporal pole have also been reported, although whether ECT increases or decreases the strength of this functional connection is undecided (Liu et al. 2015). Functional connectivity from the subgenual ACC to the temporal pole and parahippocampal cortex is increased in one study (Liu et al. 2015), while connectivity from the subcallosal cingulate cortex to the same areas is decreased in another study (Argyelan et al. 2016). A study using the amygdala as a seed-region reports a decrease in connectivity to the subgenual ACC (Cano et al. 2016). These results could indicate locally specific changes in ACC-connectivity, but also highlight the significant impact small variations in seed-region can have on retrieved connectivity patterns. Of note, a study reporting high subcallosal cingulate cortex activity at baseline (compared to healthy controls) also found that this increased activity predicted treatment efficacy (Argyelan et al. 2016).

Furthermore, studies have shown that electroconvulsive therapy (ECT) can restore connectivity between prefrontal areas and the dorsal nexus, a node of the anterior default mode network (DMN) that exhibits increased depression-associated connectivity with other networks implicated in depression (Sheline et al. 2010).

Moreover, Bai et al. (2019a, b) have found that ECT-induced enhanced connectivity between the dorsal medial prefrontal cortex (dmPFC) and the precuneus (PCu) cortex was positively correlated with clinical improvement.

Similarly, studies have revealed that ECT can normalize network connectivity between the posterior and the anterior default mode network, as well as between the posterior default mode network and the left central executive network (Leaver et al. 2016; Perrin et al. 2012). Wei et al. (2018) have also observed a post-ECT increase in the functional connectivity strength between the left angular gyrus, a functional node of the anterior DMN, and the whole brain. Belge et al. (2021) have replicated and extended this finding by demonstrating a significant post-ECT increase in connectivity between the left CEN and the left angular gyrus, a central region in depression that processes information by connecting distinct, functionally specialized regions related to emotion regulation (Achard et al. 2006). This supports a recent whole-brain functional connectome study reporting a globally decreased connectivity of the left angular gyrus in patients with depression (Lai et al. 2017).

The angular gyrus is also a functional node of the posterior DMN, and decreased connectivity between the CEN and the posterior DMN has specifically been reported in depressed patients (Mulders et al. 2015). The normalization of this connectivity following ECT is consistent with findings reported by Abbott et al. (2013). This reduced CEN-posterior-DMN connectivity may explain the difficulty that patients with depression experience in switching between a "default-state," in which the DMN is dominant and attention is focused on internal mental events, and an "executive state," in which the CEN is dominant and attention is directed toward external stimuli (Mulders et al. 2015). This deficiency in flexibility results in excessive ruminations and other self-referential activities. Therefore, by increasing the connectivity between the CEN and the posterior DMN, ECT could improve the ability to switch between these two states.

Patients with depression consistently show reduced connectivity between the dorsolateral prefrontal cortex and other regions of the cognitive control network (CEN), as reported in various studies (Alexopoulos et al. 2012; Liston et al. 2014; Ye et al. 2012). This suggests that the dorsolateral prefrontal cortex may have weakened top-down control over subcortical limbic hyperactivity, based on the hypothesis proposed by Mayberg (1997) and Northoff et al. (2011). According to Beck's cognitive model of depression, increased bottom-up reactivity in subcortical limbic regions, such as the amygdala, coupled with decreased top-down control from the dorsolateral prefrontal cortex, generates and maintains depressive symptoms (Disner et al. 2011). Studies have shown that ECT increases intra-network connectivity within the CEN and also enhances connectivity between the left CEN and the left-middle frontal gyrus, the anatomical location of the dorsolateral prefrontal cortex (Wang et al. 2018; Belge et al. 2021). Additionally, a recent systematic review has identified the modulation of the superior and middle frontal gyri and the subgenual anterior cingulate cortex (ACC) as key mechanisms of action for ECT, highlighting the importance of the CEN and default mode network, respectively (Porta-Casteràs et al. 2021).

In summary, changes in functional connectivity induced by ECT have primarily been reported in regions implicated in depression, i.e. the default mode, cognitive executive and salience networks. Limited evidence suggests that ECT restores network organization in depressed subjects.

5 Immune Effects

A significant body of research has established that at least a subset of individuals afflicted with depression exhibit signs of an inflammatory response (Dantzer et al. 2008; Miller et al. 2009; Raison et al. 2006). Notably, analyses of inflammatory markers in patients with depression as compared to healthy controls have shown elevated levels of various pro-inflammatory cytokines, including IL-6, IL1-β, and TNF-α, in both peripheral blood and cerebrospinal fluid (Dantzer et al. 2008; Miller et al. 2009; Raison et al. 2006). In animal models, sickness behavior induced by experimental infection manifests as a cluster of behavioral symptoms that are frequently observed in depression (Dantzer et al. 2008). It is posited that these symptoms are triggered by the production of peripheral pro-inflammatory cytokines that subsequently interact with specific brain regions through intricate immune-to-brain pathways (Dantzer et al. 2008). These mechanisms also appear to be relevant in humans (Dantzer et al. 2008). Elevated production of pro-inflammatory cytokines can inhibit neurotrophic support and neurogenesis in brain regions implicated in depression by activating microglial cells, leading to the release of a barrage of oxidative and neurotoxic mediators (Dantzer et al. 2008; Miller et al. 2009; Raison et al. 2006).

When examining the immunological aspects of electroconvulsive therapy (ECT), it is crucial to differentiate its acute and long-term effects. A solitary ECT session can cause a sudden surge in the expression of pro-inflammatory cytokines, such as TNF-α, IL-1β, and IL-6, but this effect is transient and lasts for only 15–30 min. Conversely, multi-session ECT may eventually lead to the normalization of peripheral blood cytokine concentrations. Studies have demonstrated that TNF-α decreases after a full course of ECT, while IL-6 reduction was observed by Freire et al. (2017), Järventausta et al. (2017), and Belge et al. (2019). Rotter et al. (2013) also observed a decrease in IL-5 levels. A recent meta-analysis by Gay et al. (2021) confirmed these separate findings by demonstrating that ECT can induce an initial acute increase in IL-6 levels and a potential reduction in TNF-α levels. While ECT appears to exert peripheral immunomodulatory effects, the underlying mechanisms behind these properties and their precise relationship to inflammatory changes at the brain level require further systematic investigation.

Bioque et al. (2019) revealed in a recent study that a single ECT session causes a decrease in the nuclear transcription factor κB (NFκB) pathway, a central regulatory circuit of the immune system that is upregulated in depressed patients and directly controls the expression of various inflammatory cytokines. It is tempting to suggest that the modulation of this important transcription factor is a common mechanism underlying ECT-induced peripheral immune changes.

Furthermore, ECT induces immune changes at the central level as well. In an experimental autoimmune encephalomyelitis mice model of neuroinflammation, electroconvulsive seizures suppressed microglial neurotoxicity by reducing the expression of various inflammatory markers and oxidative stress (Goldfarb et al. 2020). Whether ECT also impacts neuroinflammation in humans is yet to be investigated, including approaches such as positron emission tomography to visualize microglial activation.

Given the cytokine model of depression, it is reasonable to hypothesize that the observed ECT-induced immune changes may be significant for the clinical effects of ECT. Although the relationship between changes in peripheral inflammatory markers and clinical symptom domains after ECT has not been extensively investigated, some evidence indicates that changes in peripheral inflammatory cytokines are associated with significant changes in clinical symptom domain following ECT. Specifically, a decrease in peripheral IL-6 levels appears to be linked to an improvement in psychomotor retardation after ECT. Järventausta et al. (2017) and Belge et al. (2020a, b) reported a trend toward a reduction in peripheral IL-6 in ECT responders. Moreover, Kranaster et al. (2018) demonstrated that IL-6 was reduced after ECT in remitters but not in non-remitters when studying immune markers in cerebrospinal fluid. Although Kruse et al. (2018) found that higher levels of IL-6 before ECT predicted a more pronounced decrease in HDRS-17 scores, other studies did not replicate these findings.

6 Conclusion

In general, the reason for the strong efficacy of ECT is still not completely understood. It is clear that ECT induces a wealth of neurobiological changes at different system levels. A growing body of evidence indicates that ECT may especially normalize many of the structural and functional changes in the brain associated with severe depression and the treatment appears also to have potent immunomodulatory properties which could be relevant with the immune hypothesis of depression in mind. Yet, it remains unknown how these changes at different levels relate to each other and which are most essential for symptom improvement. There are various potential reasons for the lack of convergence between neurobiological changes and the robust clinical effects observed in depression. Firstly, the sample sizes used in ECT studies are generally small and, while they may be sufficient to report longitudinal changes in a paired fashion, even with high response rates ranging from 60% to 70%, these studies may be underpowered to reliably identify associations between treatment and neurobiology. Another explanation for this lack of convergence is that attempting to link clinical improvement to longitudinal changes in individual structures or connections may be an overly simplistic approach, as depression is associated with widespread alterations in both structure and function across the brain and beyond. Moreover, existing studies of ECT's neurobiological underpinnings are characterized by clinical and demographic heterogeneity and by differences in data processing and analysis pipelines. To address this, collaborations that acquire large datasets, such as the GEMRIC consortium, can help translate neuroimaging findings into clinical settings. Additionally, using consistent methodologies can improve the reproducibility of results. Advances in multimodal imaging, which combine different types of imaging data, and multivariate techniques such as machine learning, may offer further insights into the therapeutic mechanisms of ECT. Finally, insight into the complex neurobiology of ECT has until today been obtained by scrutinizing all system levels sequentially to then integrate interactions between or among specific variables at each level. Ideally, future research should be aimed at integrating findings from each level into one supra-model. This comprehensive model would consider as many variables as possible and look for common patterns through methods such as modality reduction (Groves et al. 2011).

References

Abbott CC, Lemke NT, Gopal S et al (2013) Electroconvulsive therapy response in major depressive disorder: a pilot functional network connectivity resting state FMRI investigation. Front Psych 4:10. Published 2013 Mar 1. https://doi.org/10.3389/fpsyt.2013.00010

Achard S, Salvador R, Whitcher B, Suckling J, Bullmore E (2006) A resilient, low-frequency, small-world human brain functional network with highly connected association cortical hubs. J Neurosci 26(1):63–72. https://doi.org/10.1523/JNEUROSCI.3874-05.2006

Alexopoulos GS, Hoptman MJ, Kanellopoulos D, Murphy CF, Lim KO, Gunning FM (2012) Functional connectivity in the cognitive control network and the default mode network in late-life depression. J Affect Disord 139(1):56–65. https://doi.org/10.1016/j.jad.2011.12.002

Argyelan M, Lencz T, Kaliora S, Sarpal DK, Weissman N, Kingsley PB et al (2016) Subgenual cingulate cortical activity predicts the efficacy of electroconvulsive therapy. Transl Psychiatry 6(4):e789

Argyelan M, Lencz T, Kang S, Ali S, Masi PJ, Moyett E, Joanlanne A, Watson P, Sanghani S, Petrides G, Malhotra AK (2021) ECT-induced cognitive side effects are associated with hippocampal enlargement. Transl Psychiatry 11(1):516. https://doi.org/10.1038/s41398-021-01641-y

Baeken C, Marinazzo D, Wu G-R, Van Schuerbeek P, De Mey J, Marchetti I et al (2014) Accelerated HF-rTMS in treatment-resistant unipolar depression: insights from subgenual anterior cingulate functional connectivity. World J Biol Psychiatry 15:286

Bai T, Wei Q, Xie W et al (2019a) Hippocampal-subregion functional alterations associated with antidepressant effects and cognitive impairments of electroconvulsive therapy. Psychol Med 49(8):1357–1364

Bai T, Wei Q, Zu M et al (2019b) Functional plasticity of the dorsomedial prefrontal cortex in depression reorganized by electroconvulsive therapy: validation in two independent samples. Hum Brain Mapp 40(2):465–473. https://doi.org/10.1002/hbm.24387

Belge JB, van Diermen L, Sabbe B, Parizel P, Morrens M, Coppens V, Constant E, de Timary P, Sienaert P, Schrijvers D, van Eijndhoven P (2020a) Inflammation, hippocampal volume, and therapeutic outcome following electroconvulsive therapy in depressive patients: a pilot study. Neuropsychobiology 79(3):222–232. https://doi.org/10.1159/000506133. Epub 2020 Feb 28

Belge JB, Van Diermen L, Schrijvers D et al (2020b) The basal ganglia: a central hub for the psychomotor effects of electroconvulsive therapy. J Affect Disord 265:239–246. https://doi.org/10.1016/j.jad.2020.01.033

Belge JB, Mulders PCR, Oort JV, Diermen LV, Poljac E, Sabbe B, de Timary P, Constant E, Sienaert P, Schrijvers D, van Eijndhoven P (2021) Movement, mood and cognition: preliminary insights into the therapeutic effects of electroconvulsive therapy for depression through a resting-state connectivity analysis. J Affect Disord 290:117–127. https://doi.org/10.1016/j.jad.2021.04.069. Epub 2021 May 3

Belge JB, Mulders PCR, Van Diermen L, Schrijvers D, Sabbe B, Sienaert P, Oudega ML, Tendolkar I, Dols A, van Eijndhoven P (2022) White matter changes following electroconvulsive therapy for depression: a multicenter ComBat harmonization approach. Transl Psychiatry 12(1):517. https://doi.org/10.1038/s41398-022-02284-3. PMCID: PMC9758171

Bioque M, Mac-Dowell KS, Meseguer A et al (2019) Effects of electroconvulsive therapy in the systemic inflammatory balance of patients with severe mental disorder. Psychiatry Clin Neurosci 73(10):628–635. https://doi.org/10.1111/pcn.12906

Bolwig TG (2014) Neuroimaging and electroconvulsive therapy: a review. J ECT 30(2):138–142. Available from: http://www.ncbi.nlm.nih.gov/pubmed/24800687

Bora E, Fornito A, Pantelis C, Yücel M (2012) Gray matter abnormalities in major depressive disorder: a meta-analysis of voxel based morphometry studies. J Affect Disord 138(1–2):9–18. https://doi.org/10.1016/j.jad.2011.03.049

Bouckaert F, De Winter FL, Emsell L et al (2016a) Grey matter volume increase following electroconvulsive therapy in patients with late life depression: a longitudinal MRI study. J Psychiatry Neurosci 41(2):105–114. https://doi.org/10.1503/jpn.140322

Bouckaert F, Dols A, Emsell L, De Winter F-L, Vansteelandt K, Claes L et al (2016b) Relationship between hippocampal volume, serum BDNF and depression severity following electroconvulsive therapy in late-life depression. Neuropsychopharmacol Off Publ Am Coll Neuropsychopharmacol:2741–2748

Cano M, Cardoner N, Urretavizcaya M, Martínez-Zalacaín I, Goldberg X, Via E et al (2016) Modulation of limbic and prefrontal connectivity by electroconvulsive therapy in treatment-resistant depression: a preliminary study. Brain Stimul 9(1):65–71. https://doi.org/10.1016/j.brs.2015.08.016

Chen G, Hu X, Li L et al (2016) Disorganization of white matter architecture in major depressive disorder: a meta-analysis of diffusion tensor imaging with tract-based spatial statistics. Sci Rep 6:21825

Dantzer R, O'Connor JC, Freund GG, Johnson RW, Kelley KW (2008) From inflammation to sickness and depression: when the immune system subjugates the brain. Nat Rev Neurosci 9(1):46–56. https://doi.org/10.1038/nrn2297. PMCID: PMC2919277

de Diego-Adeliño J, Pires P, Gómez-Ansón B et al (2014) Microstructural white-matter abnormalities associated with treatment resistance, severity and duration of illness in major depression. Psychol Med 44(6):1171–1182

Disner SG, Beevers CG, Haigh EA, Beck AT (2011) Neural mechanisms of the cognitive model of depression. Nat Rev Neurosci 12(8):467–477. Published 2011 Jul 6. https://doi.org/10.1038/nrn3027

Drevets W, Savitz J, Trimble M (2008) The subgenual anterior cingulate cortex in mood disorders. CNS Spectr 13(8):663–681. Available from: http://www.ncbi.nlm.nih.gov/pmc/articles/pmc2729429/

Dukart J, Regen F, Kherif F et al (2014) Electroconvulsive therapy-induced brain plasticity determines therapeutic outcome in mood disorders. Proc Natl Acad Sci U S A 111(3):1156–1161. https://doi.org/10.1073/pnas.1321399111

Duthie AC, Perrin JS, Bennett DM, Currie J, Reid IC (2015) Anticonvulsant mechanisms of electroconvulsive therapy and relation to therapeutic efficacy. J ECT 31(3):173–178

Freire TFV, Rocha NSD, Fleck MPA (2017) The association of electroconvulsive therapy to pharmacological treatment and its influence on cytokines. J Psychiatr Res 92:205–211. https://doi.org/10.1016/j.jpsychires.2017.05.004
Gay F, Romeo B, Martelli C, Benyamina A, Hamdani N (2021) Cytokines changes associated with electroconvulsive therapy in patients with treatment-resistant depression: a meta-analysis. Psychiatry
Goldfarb S, Fainstein N, Ben-Hur T (2020) Electroconvulsive stimulation attenuates chronic neuroinflammation. JCI Insight 5(17):e137028. Published 2020 Sep 3. https://doi.org/10.1172/jci.insight.137028
Grimm S, Beck J, Schuepbach D et al (2008) Imbalance between left and right dorsolateral prefrontal cortex in major depression is linked to negative emotional judgment: an fMRI study in severe major depressive disorder. Biol Psychiatry 63(4):369–376
Groves AR, Beckmann CF, Smith SM, Woolrich MW (2011) Linked independent component analysis for multimodal data fusion. Neuroimage 54(3):2198–2217. https://doi.org/10.1016/j.neuroimage.2010.09.073
Gryglewski G, Seiger R, Baldinger-Melich P et al (2020) Changes in white matter microstructure after electroconvulsive therapy for treatment-resistant depression. Int J Neuropsychopharmacol 23(1):20–25
Guo W, Liu F, Xue Z et al (2013) Decreased interhemispheric coordination in treatment-resistant depression: a resting-state fMRI study. PloS One 8(8):e71368
Herrero MT, Barcia C, Navarro JM (2002) Functional anatomy of thalamus and basal ganglia. Childs Nerv Syst 18(8):386–404. https://doi.org/10.1007/s00381-002-0604-1
Holroyd CB, Umemoto A (2016) The research domain criteria framework: the case for anterior cingulate cortex. Neurosci Biobehav Rev 71:418–443. https://doi.org/10.1016/j.neubiorev.2016.09.021
Husain SS, Kevan IM, Linnell R, Scott AIF (2004) Electroconvulsive therapy in depressive illness that has not responded to drug treatment. J Affect Disord 83(2–3):121–126. Available from: http://www.ncbi.nlm.nih.gov/pubmed/15555704
Jbabdi S, Sotiropoulos SN, Haber SN, Van Essen DC, Behrens TE (2015) Measuring macroscopic brain connections in vivo. Nat Neurosci 18:1546–1555
Järventausta K, Sorri A, Kampman O et al (2017) Changes in interleukin-6 levels during electroconvulsive therapy may reflect the therapeutic response in major depression. Acta Psychiatr Scand 135(1):87–92. https://doi.org/10.1111/acps.12665
Kellner CH, Obbels J, Sienaert P (2020) When to consider electroconvulsive therapy (ECT). Acta Psychiatr Scand 141(4):304–315. https://doi.org/10.1111/acps.13134
Kho KH, van Vreeswijk MF, Simpson S, Zwinderman AH (2003) A meta-analysis of electroconvulsive therapy efficacy in depression. J ECT 19(3):139–147. https://doi.org/10.1097/00124509-200309000-00005
Knyazeva MG (2013) Splenium of corpus callosum: patterns of interhemispheric interaction in children and adults. Neural Plast 2013:639430
Koolschijn PCMP, Van Haren NEM, Lensvelt-Mulders GJLM, Hulshoff Pol HE, Kahn RS (2009) Brain volume abnormalities in major depressive disorder: a meta-analysis of magnetic resonance imaging studies. Hum Brain Mapp 30(11):3719–3735
Kranaster L, Hoyer C, Aksay SS et al (2018) Antidepressant efficacy of electroconvulsive therapy is associated with a reduction of the innate cellular immune activity in the cerebrospinal fluid in patients with depression. World J Biol Psychiatry 19(5):379–389. https://doi.org/10.1080/15622975.2017.1355473
Kruse JL, Congdon E, Olmstead R et al (2018) Inflammation and improvement of depression following electroconvulsive therapy in treatment-resistant depression. J Clin Psychiatry 79(2):17m11597. https://doi.org/10.4088/JCP.17m11597
Lai CH, Wu YT, Hou YM (2017) Functional network-based statistics in depression: theory of mind subnetwork and importance of parietal region. J Affect Disord 217:132–137. https://doi.org/10.1016/j.jad.2017.03.073

Leaver AM, Espinoza R, Pirnia T, Joshi SH, Woods RP, Narr KL (2016) Modulation of intrinsic brain activity by electroconvulsive therapy in major depression. Biol Psychiatry Cogn Neurosci Neuroimaging 1(1):77–86. https://doi.org/10.1016/j.bpsc.2015.09.001
Le Bihan D (2003) Looking into the functional architecture of the brain with diffusion MRI. Nat Rev Neurosci 4(6):469–480
Liao Y, Huang X, Wu Q et al (2013) Is depression a disconnection syndrome? Meta-analysis of diffusion tensor imaging studies in patients with MDD. J Psychiatry Neurosci 38(1):49–56
Liston C, Chen AC, Zebley BD et al (2014) Default mode network mechanisms of transcranial magnetic stimulation in depression. Biol Psychiatry 76(7):517–526. https://doi.org/10.1016/j.biopsych.2014.01.023
Liu Y, Du L, Li Y, Liu H, Zhao W, Liu D et al (2015) Antidepressant effects of electroconvulsive therapy correlate with subgenual anterior cingulate activity and connectivity in depression. Medicine 94(45):e2033. Available from: http://journals.lww.com/md-journal/Fulltext/2015/11110/Antidepressant_Effects_of_Electroconvulsive.65.aspx
Lyden H, Espinoza RT, Pirnia T et al (2014) Electroconvulsive therapy mediates neuroplasticity of white matter microstructure in major depression. Transl Psychiatry 4(4):e380
Madsen TM, Treschow A, Bengzon J, Bolwig TG, Lindvall O, Tingström A (2000) Increased neurogenesis in a model of electroconvulsive therapy. Biol Psychiatry 47(12):1043–1049. Available from: http://www.ncbi.nlm.nih.gov/pubmed/10862803
Mayberg HS (1997) Limbic-cortical dysregulation: a proposed model of depression. J Neuropsychiatry Clin Neurosci 9(3):471–481. Available from: http://psychiatryonline.org/doi/abs/10.1176/jnp.9.3.471
Mayberg HS, Lozano AM, Voon V, HE MN, Seminowicz D, Hamani C et al (2005) Deep brain stimulation for treatment-resistant depression. Neuron 45(5):651–660. Available from: http://www.ncbi.nlm.nih.gov/pubmed/15748841
Merkl A, Heuser I, Bajbouj M (2009) Antidepressant electroconvulsive therapy: mechanism of action, recent advances and limitations. Exp Neurol 219(1):20–26. Available from: http://www.ncbi.nlm.nih.gov/pubmed/19426729
Miller AH, Maletic V, Raison CL (2009) Inflammation and its discontents: the role of cytokines in the pathophysiology of major depression. Biol Psychiatry 65(9):732–741. https://doi.org/10.1016/j.biopsych.2008.11.029. Epub 2009 Jan 15. PMCID: PMC2680424
Mulders PC, van Eijndhoven PF, Schene AH, Beckmann CF, Tendolkar I (2015) Resting-state functional connectivity in major depressive disorder: a review. Neurosci Biobehav Rev 56:330–344. Available from: http://www.sciencedirect.com/science/article/pii/S0149763415001992
Mulders PCR, Llera A, Beckmann CF et al (2020) Structural changes induced by electroconvulsive therapy are associated with clinical outcome. Brain Stimul 13(3):696–704. https://doi.org/10.1016/j.brs.2020.02.020
Murphy ML, Frodl T (2011) Meta-analysis of diffusion tensor imaging studies shows altered fractional anisotropy occurring in distinct brain areas in association with depression. Biol Mood Anxiety Disord 1(1):3
Nickl-Jockschat T, Palomero Gallagher N, Kumar V et al (2016) Are morphological changes necessary to mediate the therapeutic effects of electroconvulsive therapy? Eur Arch Psychiatry Clin Neurosci 266(3):261–267
Nobuhara K, Okugawa G, Minami T et al (2004) Effects of electroconvulsive therapy on frontal white matter in late-life depression: a diffusion tensor imaging study. Neuropsychobiology 50(1):48–53
Nolte J (2009) The human brain: an introduction to its functional anatomy, 6th edn. Mosby/Elsevier, Philadelphia
Nordanskog P, Larsson MR, Larsson EM, Johanson A (2014) Hippocampal volume in relation to clinical and cognitive outcome after electroconvulsive therapy in depression. Acta Psychiatr Scand 129(4):303–311
Northoff G, Wiebking C, Feinberg T, Panksepp J (2011) The "resting-state hypothesis" of major depressive disorder-a translational subcortical-cortical framework for a system disorder.

Neurosci Biobehav Rev 35(9):1929–1945. Available from: http://www.ncbi.nlm.nih.gov/pubmed/21192971
Nuninga JO, Mandl RCW, Boks MP et al (2020) Volume increase in the dentate gyrus after electroconvulsive therapy in depressed patients as measured with 7T. Mol Psychiatry 25(7): 1559–1568. https://doi.org/10.1038/s41380-019-0392-6
Oltedal L, Narr KL, Abbott C et al (2018) Volume of the human hippocampus and clinical response following electroconvulsive therapy. Biol Psychiatry 84(8):574–581. https://doi.org/10.1016/j.biopsych.2018.05.017
Ota M, Noda T, Sato N et al (2015) Effect of electroconvulsive therapy on gray matter volume in major depressive disorder. J Affect Disord 186:186–191. https://doi.org/10.1016/j.jad.2015.06.051
Ousdal OT, Brancati GE, Kessler U et al (2022) The neurobiological effects of electroconvulsive therapy studied through magnetic resonance: what have we learned, and where do we go? Biol Psychiatry 91(6):540–549. https://doi.org/10.1016/j.biopsych.2021.05.023
Perrin JS, Merz S, Bennett DM et al (2012) Electroconvulsive therapy reduces frontal cortical connectivity in severe depressive disorder. Proc Natl Acad Sci U S A 109(14):5464–5468. https://doi.org/10.1073/pnas.1117206109
Pirnia T, Joshi SH, Leaver AM, Vasavada M, Njau S, Woods RP et al (2016) Electroconvulsive therapy and structural neuroplasticity in neocortical, limbic and paralimbic cortex. Transl Psychiatry 6(6):e832
Pisner DA, Shumake J, Beevers CG, Schnyer DM (2019) The superior longitudinal fasciculus and its functional triple-network mechanisms in brooding. Neuroimage Clin 24:101935
Porta-Casteràs D, Cano M, Camprodon JA et al (2021) A multimetric systematic review of fMRI findings in patients with MDD receiving ECT. Prog Neuropsychopharmacol Biol Psychiatry 108:110178. https://doi.org/10.1016/j.pnpbp.2020.110178
Raison CL, Capuron L, Miller AH (2006) Cytokines sing the blues: inflammation and the pathogenesis of depression. Trends Immunol 27(1):24–31. https://doi.org/10.1016/j.it.2005.11.006. Epub 2005 Nov 28. PMCID: PMC3392963
Repple J, Meinert S, Bollettini I et al (2020) Influence of electroconvulsive therapy on white matter structure in a diffusion tensor imaging study. Psychol Med 50(5):849–856
Rotter A, Biermann T, Stark C et al (2013) Changes of cytokine profiles during electroconvulsive therapy in patients with major depression. J ECT 29(3):162–169. https://doi.org/10.1097/YCT.0b013e3182843942
Sanacora G, Mason GF, Rothman DL, Hyder F, Ciarcia JJ, Ostroff RB et al (2003) Increased cortical GABA concentrations in depressed patients receiving ECT. Am J Psychiatry 160(3): 577–579. Available from: http://journals.psychiatryonline.org/article.aspx?articleid=176102
Sheline YI, Price JL, Yan Z, Mintun MA (2010) Resting-state functional MRI in depression unmasks increased connectivity between networks via the dorsal nexus. Proc Natl Acad Sci U S A 107(24):11020–11025. https://doi.org/10.1073/pnas.1000446107
Tendolkar I, van Beek M, van Oostrom I, Mulder M, Janzing J, Voshaar RO et al (2013) Electroconvulsive therapy increases hippocampal and amygdala volume in therapy refractory depression: a longitudinal pilot study. Psychiatry Res 214(3):197–203
UK ECT Review Group (2003) Efficacy and safety of electroconvulsive therapy in depressive disorders: a systematic review and meta-analysis. Lancet 361(9360):799–808. Available from: http://www.ncbi.nlm.nih.gov/pubmed/12642045
Van Oostrom I, van Eijndhoven P, Butterbrod E et al (2018) Decreased cognitive functioning after electroconvulsive therapy is related to increased hippocampal volume: exploring the role of brain plasticity. J ECT 34(2):117–123. https://doi.org/10.1097/YCT.0000000000000483
van Velzen LS, Kelly S, Isaev D et al (2020) White matter disturbances in major depressive disorder: a coordinated analysis across 20 international cohorts in the ENIGMA MDD working group. Mol Psychiatry 25(7):1511–1525

Wade BS, Joshi SH, Njau S et al (2016) Effect of electroconvulsive therapy on striatal morphometry in major depressive disorder. Neuropsychopharmacology 41(10):2481–2491. https://doi.org/10.1038/npp.2016.48

Wakana S, Jiang H, Nagae-Poetscher LM, van Zijl PC, Mori S (2004) Fiber tract-based atlas of human white matter anatomy. Radiology 230(1):77–87

Wang J, Wei Q, Wang L et al (2018) Functional reorganization of intra- and internetwork connectivity in major depressive disorder after electroconvulsive therapy. Hum Brain Mapp 39(3):1403–1411. https://doi.org/10.1002/hbm.23928

Wei Q, Tian Y, Yu Y et al (2014) Modulation of interhemispheric functional coordination in electroconvulsive therapy for depression. Transl Psychiatry 4(9):e453

Wei Q, Bai T, Chen Y et al (2018) The changes of functional connectivity strength in electroconvulsive therapy for depression: a longitudinal study. Front Neurosci 12:661. Published 2018 Sep 25. https://doi.org/10.3389/fnins.2018.00661

Williams MR, Sharma P, Macdonald C, Pearce RKB, Hirsch SR, Maier M (2019) Axonal myelin decrease in the splenium in major depressive disorder. Eur Arch Psychiatry Clin Neurosci 269(4):387–395

Xu K, Jiang W, Ren L et al (2013) Impaired interhemispheric connectivity in medication-naive patients with major depressive disorder. J Psychiatry Neurosci 38(1):43–48

Ye T, Peng J, Nie B et al (2012) Altered functional connectivity of the dorsolateral prefrontal cortex in first-episode patients with major depressive disorder. Eur J Radiol 81(12):4035–4040. https://doi.org/10.1016/j.ejrad.2011.04.058

Yrondi A, Sporer M, Péran P, Schmitt L, Arbus C, Sauvaget A (2018) Electroconvulsive therapy, depression, the immune system and inflammation: a systematic review. Brain Stimul 11(1):29–51. https://doi.org/10.1016/j.brs.2017.10.013

Yrondi A, Péran P, Sauvaget A, Schmitt L, Arbus C (2019) Structural-functional brain changes in depressed patients during and after electroconvulsive therapy. Acta Neuropsychiatr 30(1):17–28. https://doi.org/10.1017/neu.2016.62

Zeng J, Luo Q, Du L et al (2015) Reorganization of anatomical connectome following electroconvulsive therapy in major depressive disorder. Neural Plast 2015:271674

Zuo N, Fang J, Lv X et al (2012) White matter abnormalities in major depression: a tract-based spatial statistics and rumination study. PloS One 7(5):e37561

Invasive Neuromodulation Techniques for Treatment-Resistant Depression

Peter Theiss, Francesco G. Pucci, and Konstantin V. Slavin

Contents

Abstract Surgically implanted neurostimulation devices for the treatment of depression have been studied for the last three decades. While the surgical risk associated with these treatment approaches clearly limits their use to the most severely impacted depressed patients, they offer a unique opportunity to better understand the impact of relatively localized alteration of neural activity in patient groups. As a result, these approaches provide a strict test of the role of individual neural structures or networks in mechanistic models of depression. In this chapter, we review the proposed mechanisms of action and evidence for clinical efficacy of vagal nerve stimulation, deep brain stimulation, and epidural cortical stimulation in patients with depression. The evidence for efficacy remains limited for all three modalities, but the long-term follow-up studies of treated patients have highlighted the importance of interactions between neural regions in determining therapeutic

P. Theiss and F. G. Pucci
Department of Neurosurgery, University of Illinois at Chicago, Chicago, IL, USA

K. V. Slavin (✉)
Department of Neurosurgery, University of Illinois at Chicago, Chicago, IL, USA

Neurology Section, Jesse Brown Veterans Administration Medical Center, Chicago, IL, USA
e-mail: kslavin@uic.edu

Curr Topics Behav Neurosci (2024) 66: 297–312
https://doi.org/10.1007/7854_2023_460
Published Online: 12 December 2023

response, and suggest that personalized approaches to stimulation are likely to be required.

Keywords Cortical stimulation · Deep brain stimulation · Refractory depression · Vagal nerve stimulation

1 Introduction

Treatment-resistant depression (TRD) comprises a significant subset of those affected by mood disorders worldwide, and represents a major challenge to patients, clinicians, and healthcare systems. The prevalence of depression in the general adult population in the United States is about 5%, of which 10–20% experience progressive symptoms despite psychotropic medications and psychotherapy (George et al. 2000). Once labeled treatment-resistant, many of these patients go on to receive electroconvulsive therapy or other procedural treatments, but their prospects for long-term symptom relief are generally poor. They remain on two or more psychotropic medications with significant adverse effects and interactions, often for life. Perhaps of all psychiatric disorders, therefore, treatment-resistant depression represents the most essential target for treatments that could yield population-level effects. In this chapter, we review the proposed mechanism of action and efficacy data for invasive neurosurgical neuromodulatory interventions for treatment-resistant depression.

2 Vagal Nerve Stimulation

With a deeper understanding of the neurobiological pathophysiology of depression have come deeper insights into potential targets for neuromodulation. Among the most rigorously investigated of these has been vagal nerve stimulation (VNS), which was first reported as a treatment for refractory epilepsy in 1990. VNS was made commercially available for this indication in Europe in 1994 and the United States in 1997, but more recent clinical and basic science data has supported its use in TRD as well. Here we discuss the putative mechanism by which VNS may influence the pathophysiology of depression, review the clinical evidence for its use in TRD, and comment on its current and future role in the treatment of mood disorders.

2.1 Mechanism of Action

The vagus nerve provides efferent parasympathetic innervation throughout the body, but also has a significant afferent component that converges on the nucleus of the solitary tract. In fact, sensory fibers in the nerve significantly outnumber motor fibers. The solitary nucleus, in turn, has wide projections to CNS regions associated with mood regulation, and the diversity of these connections may explain why the central effects of vagal stimulation may extend well beyond seizure suppression. Major projections from the solitary nucleus include ascending pathways to the forebrain via the parabrachial nucleus and the locus coeruleus (LC). LC is the primary locus of noradrenergic neurons in the brain, and animal studies have demonstrated increased cellular activity in the LC with VNS (Naritoku et al. 1995). This is consistent with current drug treatments for depression, selective serotonin and/or norepinephrine reuptake inhibitors (SSRI/SNRIs), which are aimed at increasing CNS levels of these neurotransmitters. Independent of this, animal studies have shown that VNS increased CSF levels of gamma-aminobutyric acid (GABA) and 5-hydroxyindoleacetic acid (5H1AA), a serotonin metabolite (Ben-Menachem et al. 1995; Carpenter et al. 2004; Hammond et al. 1992).

Beyond LC, the nucleus of the solitary tract projects to the basal forebrain and the limbic system, areas strongly implicated in the pathogenesis of mood disorders. These include the amygdala, hippocampus, cingulate gyrus, and hypothalamus. Early PET studies in patients implanted with VNS for seizure control showed decreased activity in the amygdala, which correlates with improvement in depression and anxiety and is also seen with effective SSRI treatment (Henry et al. 1998; Mayberg et al. 1997). Likewise, cerebral blood flow (CBF)-based functional imaging has revealed increased activity in the insular and inferior frontal cortices, and decreased activity in the hippocampi, amygdalae, and posterior cingulate cortices with vagal stimulation (Conway et al. 2018). In a 2005 PET study of patients who underwent VNS for TRD, CBF increases were identified in frontal and prefrontal regions and the anterior cingulate gyrus in all patients (Conway et al. 2018). These observations correlate well with the current understanding of the functional neuroanatomy of depression, which postulates decreased activity in the prefrontal cortex with complementary increases in the limbic system and amygdala.

2.2 *Clinical Experience*

Reports of the anti-epileptic effect on vagal nerve stimulation first arrived in 1990, but the first clinical trial-based evidence came in 1994 from the randomized controlled trial (RCT) of the NeuroCybernetic Prosthesis (NCP) system for adult patients with medication-resistant complex partial epilepsy (EO3), performed by the Vagal Nerve Stimulation Study Group (VNS Study Group 1995). Subjects in this trial reported mood improvements after VNS implantation, and this was confirmed with quantitative analysis of standard psychiatric rating scales and self-reported questionnaires. Specifically, 9 of 11 patients in this cohort demonstrated sustained mood improvements 6 months after implantation. This was compared to 2 of 11 patients demonstrating persistent seizure response, suggesting that the mood effects of VNS were independent of seizure control (Elger et al. 2000).

In 2000, Harden et al. also reported significant improvement in mood scores over time in 20 patients implanted with VNS for epilepsy. Their study did not find a significant difference in scores compared to a non-implanted control group, but their subjects had only mild depressive symptoms and the authors postulated that studies of patients with more severely impaired neuropsychiatric testing results may show a larger statistical yield (Harden et al. 2000).

These results, combined with the basic science data referenced above and the observation of mood improvements with anti-epileptic drugs such as carbamazepine, lamotrigine, and valproic acid, led to trial enrollments for VNS for TRD in the early 2000s. Rush et al. (2000) reported an open cohort of 30 patients with medication-resistant depression followed for 10 weeks after implantation (Rush et al. 2000). Using the 28-item Hamilton Depression Rating Scale ($HDRS_{28}$), the Clinical Global Impressions – Improvement Index (CGI), and the Montgomery-Åsberg Depression Rating Scale (MADRS), they reported mean improvements of 40%, 40%, and 50%, respectively. There were no significant predictors for lack of treatment response in this group, but patients who had previously failed to respond to electroconvulsive therapy (ECT) showed a trend toward lower VNS response, perhaps indicating more severe disease.

The encouraging results from this study led to the first randomized controlled trial for VNS in TRD, from Rush et al. in 2005. This study randomized 235 patients in 1:1 fashion to VNS implantation or a sham operation, with the primary outcome targeted as 50% improvement in $HDRS_{24}$ at 10 weeks. Interestingly, the final analysis did not reveal a significant difference between the VNS and sham groups, with only 15% of the subjects in the active group meeting the primary outcome compared to 10% in the control group. The authors proposed a number of explanations for this result, including the increased severity and prolonged disease course of their cohort (compared to the smaller studies and epilepsy cohorts referenced above) and the relatively short-term follow-up. The parameters of stimulation, namely current, frequency, and pulse width, were not observed to have any effect on mood symptoms in this study and were not considered a viable explanation. By comparison, however, a cohort of 60 patients with similar baseline characteristics, reported by Sackeim et al. in 2001 in a non-randomized efficacy study, found a 30% response rate at the same clinical endpoint after VNS implantation using the $HDRS_{28}$ (Sackeim et al. 2001). There, risk factors for poor treatment response were the extent of treatment resistance, manifested by the high number of failed antidepressant medications and, especially, previous ECT failure.

The results of the 2005 study by Rush et al. therefore, demonstrated a VNS response that was at least 50% worse than other non-randomized data in the literature performed on similar cohorts. Although the possibility of a placebo effect on the sham-operated patients may explain part of the statistical insignificance between the two groups, the 15% rate of treatment response in the treatment group remained unexplained. The same group therefore extended their observation of their VNS cohort, and in the same year published a comparison between their VNS plus treatment-as-usual (VNS-TAU) group with a similar treatment-as-usual (TAU) cohort (George et al. 2005). This study was not randomized (the control cohort was matched for disease severity), but added the power of a large sample size by including patients that had been originally placed in the sham VNS cohort but elected to cross over to active treatment after the completion of the study. Importantly, this study extended observations to 1 year post-treatment, comparing 205 VNS-TAU to 124 TAU subjects, and found a significant difference in response rate using the 30-item inventory of depressive symptomatology self-report ($IDS\text{-}SR_{30}$) and the $HDRS_{24}$ – 27% vs 13%, $p < 0.001$ (Ramsay et al. 1994). These results, although from non-randomized data, showed a benefit to VNS that appeared to grow over time and was closer to previously reported rates of mood improvement. To follow this up, the same group published a 5-year follow-up of 600 TRD patients, including their original cohort, demonstrating that 62% of VNS-TAU patients showed a measurable clinical response (including those with subsequent relapse), compared to 37% in the TAU group (Kumar et al. 2019).

Lastly, the safety profile of any surgical treatment, particularly for a psychiatric disease, should be reviewed. VNS has been extremely safe since its introduction, with adverse effects reported in the 1% range or less. While voice changes are relatively common, severe dysphagia or cardiac side effects are almost never reported. The most significant surgical risk is the risk of infection that comes with any implanted device, and surgical site infections have been reported at an acceptable rate in all published cohorts.

2.3 Conclusions and Future Directions

While the initial randomized trial for VNS for TRD yielded equivocal results, the follow-up study and concurrent and subsequent cohort studies provided consistent evidence that VNS improved mood symptoms over longer follow-up periods in at least a subset of TRD patients. Systematic reviews of the data mentioned above have likewise found a large variation in effect size between studies, which is likely due to a combination of differing degrees of disease severity between cohorts, placebo effect on sham controls, and variations in study design (Martin and Martín-Sánchez 2012). Recent research has focused on tailoring VNS therapy to maximize the response of those patients that may benefit from the intervention. In 2013, Aaronson et al. published a cohort of 331 patients randomized to low, medium, or high-amplitude stimulation and followed for a period of 50 weeks. All groups showed sustained response to treatment, with high-amplitude stimulation having the most significant effect. This cohort similarly lacked a negative control, but continues to suggest that VNS may have mood effects in TRD patients over longer follow-up times (Aaronson et al. 2013).

VNS was approved by the FDA for use in TRD in 2005, and is currently a viable option for patients who have exhausted all other clinical options. It has a strong scientific basis and a relatively large body of clinical data to support its use. Nevertheless, clinicians and patients considering VNS should be cautious that the treatment is not a panacea. The single sham-controlled, randomized trial showed no significant effect on the treatment group; and although there are many studies suggesting that VNS is effective over long-term follow-up, the effect is modest. Naturally, this is to be expected of any new treatment targeting TRD, and any effect on the natural history of the disease should be considered a clinical victory. Clinicians can expect a 50% improvement in disease severity scores in at least 15% of patients at 10 weeks post-treatment, and 22–37% of patients at 1 year. More recent follow-up on these cohorts at 5 years has indicated that treatment response may be as high as 60% (Kumar et al. 2019), suggesting that VNS does have a disease-modifying effect on TRD that takes place over years after implantation. Furthermore, VNS may yield both quality-of-life and cognitive benefits in TRD patients, manifested as improvements in both learning and memory tasks (Desbeaumes Jodoin et al. 2018). At the time of this publication, enrollment is ongoing for a follow-up multicenter RCT of VNS for TRD (the "RECOVER" study), which will randomize patients blindly to sham vs active stimulation and follow clinical outcomes for 12 months. The promise of higher enrollment and more extended clinical follow-up compared to earlier studies will warrant close attention to the results, which may add to our confidence in the longer-term mood effect of VNS implantation (clinicaltrials.gov).

More recently, VNS has been investigated as a treatment for anxiety disorders, given its demonstrated effect on the limbic circuits implicated in their pathogenesis. A non-randomized cohort of patients with obsessive-compulsive disorder, panic disorder, and post-traumatic stress disorder undergoing VNS showed some improvement in severity scores over time, but further studies will be needed (George et al. 2008).

3 Deep Brain Stimulation

The next frontier in the surgical treatment of depression will likely be deep brain stimulation (DBS), which has shown promising results in a subset of patients (Mohr et al. 2011). Subcortical stimulation and lesioning have been under earnest clinical investigation since the 1950s, primarily targeting movement disorders such as Parkinson's disease or essential tremor. Use of neurosurgical stereotactic frames and evolving imaging technology has enabled submillimetric targeting of deep structures, and these techniques have been extrapolated to the treatment of refractory psychiatric disorders and enabled the investigation of a variety of putative targets. In psychiatric disease, DBS was first examined in refractory obsessive-compulsive disorder (OCD) and Tourette syndrome and currently carries a Humanitarian Device Exemption (HDE) in the United States for use in OCD. In treatment-resistant depression, both gray and white matter regions have been investigated as potential targets for modulation of the limbic pathways associated with severe forms of the disease including the subcallosal cingulate gyrus, internal globus pallidus, ventral internal capsule, nucleus accumbens, thalamus, and habenula (Mohr et al. 2011).

3.1 Subcallosal Cingulate Gyrus (SCG)

Initial reports on DBS use in TRD primarily comprise vignettes of mood disorder improvement in patients implanted for movement disorder indications, which suggested that DBS could have lasting antidepressant effects outside the Parkinson's disease population. The first region specifically selected as a DBS target for TRD was the anterior portion of the cingulate gyrus ventral to the genu of the corpus callosum in the medial frontal lobe. Individuals carrying allelic variants of the serotonin transporter gene show decreased gray matter volume in both the subgenual cingulate region and the amygdala and have an associated increased propensity toward mood disorders (Pezavas et al. 2005). Initial PET studies of patients with refractory and recovering depression indicated that the SCG demonstrates increased FDG uptake in healthy volunteers experiencing sadness, whereas uptake is decreased or absent in symptomatic depressed patients. In patients undergoing successful SSRI or SNRI treatment, or successful electroconvulsive therapy or psychotherapy, hypermetabolism is often restored to this region (Mayberg 1997). Likewise, the subgenual prefrontal cortex demonstrates generalized volume loss in depressed patients based on both MRI studies and autopsy data (Dougherty et al. 2003). This fact has led some to question dogma that the SCG is hypofunctional in depressed patients and to propose, conversely, that given the associated volume loss in this cortex FDG uptake per tissue volume may increase in the setting of depression. Indeed, chronic hyperactivity of this region shown by blood flow studies has been associated with more treatment-resistant forms of depression (Mayberg 2009). The limits of current functional imaging techniques and variation in study results make it difficult to determine whether modulation or lesioning of this region is needed to disrupt the pathophysiology of TRD, but numerous studies have demonstrated that targeting this region produces sustained mood effects in a subset of patients. It is probably more correct to divide the anterior cingulate into dorsal, ventral, and rostral compartments, each of which projects to independent limbic areas and also contains internal connections; activity in the rostral component has been identified as an independent predictor of treatment response in TRD (Mayberg 1997). Importantly, however, this is a region of anatomic variability between patients and targeting of specific areas within the SCG challenges the accuracy of current DBS techniques.

While the SCG was established by multiple studies as a hub of pathophysiologic activity in TRD, precise lesioning of the cingulum and/or cingulate gyrus (cingulotomy) was proposed prior to the introduction of stimulation techniques. Cingulotomy produced a sustained clinical response in a subset of patients with a correlated modulation of FDG uptake in multiple limbic targets in the same patients (Dougherty et al. 2003). Specifically, more rostral lesions in the anterior cingulate were associated with more sustained clinical response after lesioning, consistent with imaging data and arguing against a placebo effect (Steele et al. 2008). Although randomized trials of clinical response to cingulotomy do not exist, case series matching imaging and clinic responses provided proof of concept that modulation of this region can interrupt limbic circuits in TRD. It is important to note that cingulotomy targets the white matter and therefore creates an axonal lesion, aimed at interrupting cortical circuits converging on the gray matter which seem to be reliably present in patients suffering from severe mood disorders. The SCG contains both afferent and efferent projections to nearly every limbic area associated with mood, memory, learning, and reward conditioning including the insula, orbitofrontal and prefrontal cortex, amygdala and hypothalamus, nucleus accumbens, and the brainstem (Mayberg 2009). Targeting the SCG therefore indirectly modulates activity in each of these regions.

Stimulation of the SCG was first reported in an unblinded study by Mayberg et al. in 2005 (Pezawas et al. 2005). This study reported the findings in six patients with severe depression refractory to at least four treatments who underwent implantation of bilateral stimulation leads to Brodmann Area 25 (SCG). These patients were followed clinically for 6 months and underwent stimulation at 130 Hz. At the end-of-study time point, 4 of the 6 patients showed a sustained clinic response, defined as a 50% decrease in HDRS-17 score from the pretreatment baseline, and three showed near complete resolution of their mood symptoms with the continuation of daily medication regimens. To address the possibility of a placebo affect the subject with the strongest clinic response was subsequently subjected to blind discontinuation of simulation; clinical effects of DBS persisted for 2 weeks but then subsided, with the return of depression symptoms. Blind resumption of stimulation resulted in rapid normalization of symptoms (Pezawas et al. 2005). Implantation was subsequently replicated in 14 additional patients with consistent results of 60% response rate at 6 months (Mayberg 2009). Concurrently, PET studies of responding subjects showed expected effects in both the SCG and connected limbic areas. In the SCG itself, hyperactivity was suppressed even below that of matched normal controls. In the long-term responders, CBF decreases were seen and sustained in the orbitofrontal cortex, hypothalamus, and insula among others. In the poorest responder of the initial 6 subjects, none of the CBF changes in associated limbic regions were observed (Mayberg et al. 2005). In the patients with the most sustained and dramatic responses to initial implantation, subjective reports of immediate resolution of dread, loss of hopelessness, increased motor speed and alertness were reported even while still in the operating room. Importantly, the safety profile of SCG DBS in these initial subjects was excellent and comparable to DBS for movement disorders, with no reported intracerebral hematomas or device infections.

The necessary next step was a placebo-controlled study of SCG stimulation. This was achieved in 2017, with the completion of a sham-controlled trial of 90 patients implanted with SCG DBS electrodes and randomized to 6 months of active ($n = 60$) or sham ($n = 30$) stimulation. Following the blinded period, subjects were followed for an additional 6 months of open-label stimulation (Holtzheimer et al. 2017). This multicenter study failed to find a significant clinical difference between active or sham stimulation at the 6-month time point, with only approximately 20% of subjects showing a sustained clinical response. This was a significantly lower response rate than expected based on the prior open-label studies, and the study authors attributed this to the longer pre-trial duration of depression in many patients enrolled in the study (12 years on average). In the subsequent open-label phase of this trial and on long-term follow-up at 2 years, nearly 50% of subjects showed antidepressant response and 25% demonstrated remission (Holtzheimer et al. 2017). Subjects from this and other trials have been followed for up to 8 years, and clinical response rates have been consistently reported around 50% (Crowell et al. 2019).

Subsequent retrospective studies of the SCG DBS responders have sought to further define the characteristics that predict a positive response and may guide patient selection in the future. Anatomic studies revealed no difference in placement of the active contacts within the SCG between responders and non-responders (Riva-Posse et al. 2014). While all responders showed PET changes in close proximity to the electrodes, changes in the frontal cortex and dorsal cingulate were equally reliable. Improving MR fiber tractography technology (activation volume tractography, AVT) allowed analysis of changes in white matter connections in those responding to SCG DBS. Three distinct fiber bundle pathways were identified as showing reliable modulation in DBS responders: the forceps minor in the genu of the corpus callosum, cingulum white matter bundles connecting the internal nuclei of the anterior cingulate gyri, and the medial uncinate fasciculus (Riva-Posse et al. 2014). Up to 2 years after DBS implantation, active contact adjustments resulting in blood flow alterations in all three of these white matter pathways were most associated with positive clinical response, and even with conversion of non-responders to responders. This supports the hypothesis that the white matter of the SCG is a hub of intersection among the white matter tracts that mediate TRD, and that precision modulation of this hub reliably alters the disease course. Non-responders may simply be those patients whose white matter tract anatomy differs broadly enough from normal, so that identically placed SCG leads do not adequately modulate this entire network.

This "connectomic" understanding of TRD pathophysiology calls for an individualized stimulation guided by tractography to increase the likelihood of a clinical response. This approach has been investigated in a prospective trial of SCG DBS, with patients undergoing pre-operative DTI MRI with precise target selection to guide lead placement to aim at the intersecting white matter tracts in this region (Riva-Posse et al. 2018). With precise bilateral targeting of the intersection of the forceps minor, uncinate fasciculus, cingulum fibers, and fronto-striatal fibers, nine of 11 patients demonstrated clinical response at 1 year and six demonstrated remission of their disease. These results provide an encouraging proof of concept and lend validity to our current understanding of the brain regions and tracts essential to the pathophysiology of TRD. Just as stimulation of the subthalamic nucleus disrupts the basal ganglia circuits that underlie movement disorders by altering an essential synaptic node, stimulation of the SCG alters the frontal white matter tracts that are aberrant in TRD. That STN DBS has been more clinically successful to date reflects differences in both the disease and the target. Movement disorder circuitry is well understood and localized to a relatively small number of gray matter regions, and the STN is a distinct gray matter target with little variation between patients and a specific microelectrode signature making it a familiar neurosurgical target. The SCG, on the other hand, is a white matter target with significant variability that lies at the intersection of a much larger interconnection of fibers that project to virtually the entire brain.

3.2 Other DBS Targets

Given the complexity of the circuitry of TRD and the moderate results in the randomized trial of SCG stimulation, it is reasonable to consider other targets. The ventral internal capsule (VC) has been an established target for ablative procedures (anterior capsulotomy) for refractory OCD since the 1960s. This target was gradually modified to the ventral capsule/ventral striatum (VC/VS), and clinical follow-up studies of patients undergoing stimulation of this region demonstrated that improvements in OCD symptoms were often accompanied by improvements in depressive symptoms, when present. Given that VC/VS stimulation was known to be safe and was already FDA-approved in the United States (for OCD, on HDE basis), extrapolation of this technique to open-label and later to randomized, sham-controlled trials for TRD was natural. In the first open-label trial of 15 patients, clinical response (measured by HDRS improvements, as in other DBS trials) was seen in 40% of patients at 6 months and 53% of patients at the longest follow-up, up to 4 years later (Malone et al. 2009). A subsequent randomized trial in 2014 enrolled 15 subjects in active and sham stimulation groups but was stopped after an early 16-week analysis demonstrated no significant difference and approximately a 20% response rate in each group (Dougherty et al. 2015). A subsequent Dutch study implanted 25 patients with VC leads who then underwent 1 year of open-label stimulation until a stable response state with tailored stimulation parameters was achieved. They then blinded the subjects to a period of sham vs active stimulation with a subsequent crossover phase. In the first phase, 10 of the 25 patients demonstrated initial clinical response; of those, there was a significant difference in active vs sham stimulation during the second phase (Bergfeld et al. 2016). In summary, VC/VS stimulation has demonstrated a moderate clinical response in open-label trials but has failed to show a significant benefit in its one true RCT, and so far lacks the supporting evidence from neuroimaging and connectome-based studies that is available for SCG DBS.

The nucleus accumbens (NA) is an essential component of the dopaminergic reward pathway and lies in anatomic proximity to both the ventral striatum and the SCG. Given the prevalence of anhedonia as a central feature of MDD, modulation of this region has also been proposed. An open-label analysis of 10 patients undergoing bilateral NA stimulation showed 50% response at 1 year from implantation and, importantly, decreased in FDG uptake in the SCG and orbitofrontal cortex in fMRI studies of responding subjects (Bewernick et al. 2010). The medial forebrain bundle connects the ventral tegmental area to the NA and has been investigated by direct stimulation in seven subjects in an open-label design. Interestingly, these subjects showed a high rate of rapid improvement in symptoms (within the first week), but by 3 months only 2 of 7 patients showed a sustained clinical response (Schlaepfer et al.

2013). Taken with the NA data, these results suggest that targeting of reward pathway-associated regions may temporarily correct anhedonia in some patients but may fail to produce lasting changes in the widespread depression circuitry that centers on the SCG. Other proposed DBS targets include the inferior thalamic peduncle, which has been investigated in a single patient, and the lateral habenula, which is overactive in severe depression but has only been investigated in humans in a single unpublished, terminated trial (Jiménez et al. 2005; Wang 2011).

3.3 Conclusions

In summary, as our understanding of the anatomic basis of the white matter circuitry improves and imaging and surgical techniques advance, DBS holds promise as a therapeutic option for patients with severe and resistant disease. Multiple independent studies, primarily targeting the SCG but also the VC/VS and NA, have found an approximately 50% response rate with a 20–25% long-term remission rate. Functional tractography has allowed further investigation of responding patients and implicated the white matter tracts converging on the SCG as essential modulation targets to achieve clinical response. Indeed, even stimulation of the nucleus accumbens requires functional changes in the closely associated SCG to be successful. It must be pointed out that open-label studies have reliably demonstrated a higher response rate than sham-controlled randomized trials, regardless of target, and furthermore that no randomized trial has met statistical success. For DBS to be widely adopted, a positive RCT will likely be required, and it appears that the complexity of the disease and individual variations in SCG anatomy and connections between patients make this difficult to achieve. The future of SCG stimulation will likely mirror advancements in neuroimaging that allow tailoring of lead placement with pre-operative tractography achieve the highest rate of clinic success. At the time of writing, 12 active clinical trials of DBS for TRD are recruiting subjects, including a further investigation of tractography-guided SCG stimulation and randomized trials of both medial forebrain bundle and VC/VS stimulation, among others. Whether DBS becomes a widely accepted treatment option for TRD may depend on the results of these studies (clinicaltrials.gov).

4 Epidural Cortical Stimulation

Epidural cortical stimulation (ECS) offers another neuromodulation option that theoretically carries lower risk of intracerebral hemorrhage than DBS. Epidural paddle-type leads can be tunneled through a cranial bur hole with a small scalp incision and connected to an implanted generator. ECS lacks the precise targeting available with the neurosurgical stereotactic techniques used in DBS, but implantation of a broader electrode array allows for multiple contacts to be stimulated to create a clinical effect. This technique is extrapolated from its established use as a treatment for chronic neuropathic pain including refractory trigeminal pain, post-stroke pain, or post-traumatic pain in patients with spinal cord injury or brachial plexus avulsion, although it remains an off-label treatment in the United States.

While the SCG and other deep targets are not directly accessible by epidural stimulation, the widespread cortical connections of the white matter pathways involved in depression raise the possibility of modulating the entire network by targeting the cortical endpoint. In the treatment of neuropathic pain, motor cortex stimulation has been shown to cause downstream functional changes in the lateral thalamus, periaqueductal gray, and the orbitofrontal cortex (Garcia-Larrea and Peyron 2007). The possible targets for ECS are far more limited than DBS and essentially comprise the supratentorial surface. To control neuropathic pain, ECS seems to employ the strong white matter connection between the motor cortex and the thalamus to indirectly modulate thalamic activity. As we have seen in our discussion of DBS targets, the white matter circuitry of depression is complex and can be modulated at multiple anatomic locations. The prefrontal cortex at the frontal pole has been selected for an open-label trial of ECS for TRD in the United States (Nahas et al. 2010). This region was selected for its extensive white matter connections to the default mode network and the possibility of downstream modulation of the orbitofrontal cortex, cingulate cortex (including the subgenual region), and precuneus (Raichle and Snyder 2007). Neuronavigation techniques were used to target lead placement over the frontal pole at Brodmann Areas 10 and 46 bilaterally in five patients with TRD, and subjects underwent intermittent stimulation with clinical follow-up for 7 months. Three of the five patients reached remission of their disease at this final time point, although one patient required lead explantation due to infection. These results are encouraging but have not been replicated or expanded to randomized or sham-controlled studies (Nahas et al. 2010).

5 Conclusions

Surgical treatments for TRD have now been present for over 30 years, and although extrapolated from techniques used to treat epilepsy, movement disorders, and chronic pain, they have been well-investigated in TRD with randomized, sham-controlled trials. These trials have failed to produce a strong positive result, and many psychiatrists remain reluctant to regularly refer treatment-refractory patients to neurosurgeons for this reason. But these results may reflect the complexity of the anatomical basis of the disease rather than the inefficacy of the treatments, as follow-up and individualized studies of DBS treatments targeting the white matter connections at the SCG have yielded significant results in most patients. Furthermore, post-implantation imaging studies of DBS patients have contributed significantly to our understanding of the neuroanatomy of depression. With improvement in neuroimaging techniques and advancement in implantable device technology, the patient selection and clinical results may improve in the near future. Many clinical trials of interventional treatments for TRD are currently underway.

References

Aaronson ST, Carpenter LL, Conway CR, Reimherr FW, Lisanby SH, Schwartz TL, Moreno FA, Dunner DL, Lesem MD, Thompson PM, Husain M, Vine CJ, Banov MD, Bernstein LP, Lehman RB, Brannon GE, Keepers GA, O'Reardon JP, Rudolph RL, Bunker M (2013) Vagus nerve stimulation therapy randomized to different amounts of electrical charge for treatment-resistant depression: acute and chronic effects. Brain Stimul 6(4):631–640

Ben-Menachem E, Hamberger A, Hedner T, Hammond EJ, Uthman BM, Slater J, Treig T, Stefan H, Ramsay RE, Wernicke JF et al (1995) Effects of vagus nerve stimulation on amino acids and other metabolites in the CSF of patients with partial seizures. Epilepsy Res 20(3): 221–227

Bergfeld IO, Mantione M, Hoogendoorn ML, Ruhé HG, Notten P, van Laarhoven J, Visser I, Figee M, de Kwaasteniet BP, Horst F, Schene AH, van den Munckhof P, Beute G, Schuurman R, Denys D (2016) Deep brain stimulation of the ventral anterior limb of the internal capsule for treatment-resistant depression: a randomized clinical trial. JAMA Psychiatry 73(5): 456–464

Bewernick BH, Hurlemann R, Matusch A, Kayser S, Grubert C, Hadrysiewicz B, Axmacher N, Lemke M, Cooper-Mahkorn D, Cohen MX, Brockmann H, Lenartz D, Sturm V, Schlaepfer TE (2010) Nucleus accumbens deep brain stimulation decreases ratings of depression and anxiety in treatment-resistant depression. Biol Psychiatry 67(2):110–116

Carpenter LL, Moreno FA, Kling MA, Anderson GM, Regenold WT, Labiner DM, Price LH (2004) Effect of vagus nerve stimulation on cerebrospinal fluid monoamine metabolites, norepinephrine, and gamma-aminobutyric acid concentrations in depressed patients. Biol Psychiatry 56(6):418–426

Conway CR, Kumar A, Xiong W, Bunker M, Aaronson ST, Rush AJ (2018) Chronic Vagus nerve stimulation significantly improves quality of life in treatment-resistant major depression. J Clin Psychiatry 79(5):18m12178

Crowell AL, Riva-Posse P, Holtzheimer PE, Garlow SJ, Kelley ME, Gross RE, Denison L, Quinn S, Mayberg HS (2019) Long-term outcomes of subcallosal cingulate deep brain stimulation for treatment-resistant depression. Am J Psychiatry 176(11):949–956

Desbeaumes Jodoin V, Richer F, Miron JP, Fournier-Gosselin MP, Lespérance P (2018) Long-term sustained cognitive benefits of Vagus nerve stimulation in refractory depression. J ECT 34(4): 283–290

Dougherty DD, Weiss AP, Cosgrove GR, Alpert NM, Cassem EH, Nierenberg AA, Price BH, Mayberg HS, Fischman AJ, Rauch SL (2003) Cerebral metabolic correlates as potential predictors of response to anterior cingulotomy for treatment of major depression. J Neurosurg 99(6):1010–1017

Dougherty DD, Rezai AR, Carpenter LL, Howland RH, Bhati MT, O'Reardon JP, Eskandar EN, Baltuch GH, Machado AD, Kondziolka D, Cusin C, Evans KC, Price LH, Jacobs K, Pandya M, Denko T, Tyrka AR, Brelje T, Deckersbach T, Kubu C, Malone DA Jr (2015) A randomized sham-controlled trial of deep brain stimulation of the ventral capsule/ventral striatum for chronic treatment-resistant depression. Biol Psychiatry 78(4):240–248

Elger G, Hoppe C, Falkai P, Rush AJ, Elger CE (2000) Vagus nerve stimulation is associated with mood improvements in epilepsy patients. Epilepsy Res 42(2–3):203–210

Garcia-Larrea L, Peyron R (2007) Motor cortex stimulation for neuropathic pain: from phenomenology to mechanisms. NeuroImage 37(Suppl 1):S71–S79

George MS, Sackeim HA, Rush AJ, Marangell LB, Nahas Z, Husain MM, Lisanby S, Burt T, Goldman J, Ballenger JC (2000) Vagus nerve stimulation: a new tool for brain research and therapy. Biol Psychiatry 47(4):287–295

George MS, Rush AJ, Marangell LB, Sackeim HA, Brannan SK, Davis SM, Howland R, Kling MA, Moreno F, Rittberg B, Dunner D, Schwartz T, Carpenter L, Burke M, Ninan P, Goodnick P (2005) A one-year comparison of vagus nerve stimulation with treatment as usual for treatment-resistant depression. Biol Psychiatry 58(5):364–373

George MS, Ward HE Jr, Ninan PT, Pollack M, Nahas Z, Anderson B, Kose S, Howland RH, Goodman WK, Ballenger JC (2008) A pilot study of vagus nerve stimulation (VNS) for treatment-resistant anxiety disorders. Brain Stimul 1(2):112–121

Hammond EJ, Uthman BM, Reid SA, Wilder BJ (1992) Electrophysiological studies of cervical vagus nerve stimulation in humans: I. EEG effects. Epilepsia 33(6):1013–1020

Harden CL, Pulver MC, Ravdin LD, Nikolov B, Halper JP, Labar DR (2000) A pilot study of mood in epilepsy patients treated with Vagus nerve stimulation. Epilepsy Behav 1(2):93–99

Henry TR, Bakay RA, Votaw JR, Pennell PB, Epstein CM, Faber TL, Grafton ST, Hoffman JM (1998) Brain blood flow alterations induced by therapeutic vagus nerve stimulation in partial epilepsy: I. Acute effects at high and low levels of stimulation. Epilepsia 39(9):983–990

Holtzheimer PE, Husain MM, Lisanby SH, Taylor SF, Whitworth LA, McClintock S, Slavin KV, Berman J, McKhann GM, Patil PG, Rittberg BR, Abosch A, Pandurangi AK, Holloway KL, Lam RW, Honey CR, Neimat JS, Henderson JM, DeBattista C, Rothschild AJ, Pilitsis JG, Espinoza RT, Petrides G, Mogilner AY, Matthews K, Peichel D, Gross RE, Hamani C, Lozano AM, Mayberg HS (2017) Subcallosal cingulate deep brain stimulation for treatment-resistant depression: a multisite, randomised, sham-controlled trial. Lancet Psychiatry 4(11):839–849

Jiménez F, Velasco F, Salin-Pascual R, Hernández JA, Velasco M, Criales JL, Nicolini H (2005) A patient with a resistant major depression disorder treated with deep brain stimulation in the inferior thalamic peduncle. Neurosurgery 57(3):585–593. discussion 585–93

Kumar A, Bunker MT, Aaronson ST, Conway CR, Rothschild AJ, Mordenti G, Rush AJ (2019) Durability of symptomatic responses obtained with adjunctive vagus nerve stimulation in treatment-resistant depression. Neuropsychiatr Dis Treat 15:457–468

Malone DA Jr, Dougherty DD, Rezai AR, Carpenter LL, Friehs GM, Eskandar EN, Rauch SL, Rasmussen SA, Machado AG, Kubu CS, Tyrka AR, Price LH, Stypulkowski PH, Giftakis JE, Rise MT, Malloy PF, Salloway SP, Greenberg BD (2009) Deep brain stimulation of the ventral capsule/ventral striatum for treatment-resistant depression. Biol Psychiatry 65(4):267–275

Martin JL, Martín-Sánchez E (2012) Systematic review and meta-analysis of vagus nerve stimulation in the treatment of depression: variable results based on study designs. Eur Psychiatry 27(3):147–155

Mayberg HS (1997) Limbic-cortical dysregulation: a proposed model of depression. J Neuropsychiatry Clin Neurosci 9(3):471–481

Mayberg HS (2009) Targeted electrode-based modulation of neural circuits for depression. J Clin Invest 119(4):717–725

Mayberg HS, Brannan SK, Mahurin RK, Jerabek PA, Brickman JS, Tekell JL, Silva JA, McGinnis S, Glass TG, Martin CC, Fox PT (1997) Cingulate function in depression: a potential predictor of treatment response. Neuroreport 8(4):1057–1061

Mayberg HS, Lozano AM, Voon V, McNeely HE, Seminowicz D, Hamani C, Schwalb JM, Kennedy SH (2005) Deep brain stimulation for treatment-resistant depression. Neuron 45(5): 651–660

Mohr P, Rodriguez M, Slavíčková A, Hanka J (2011) The application of vagus nerve stimulation and deep brain stimulation in depression. Neuropsychobiology 64(3):170–181

Nahas Z, Anderson BS, Borckardt J, Arana AB, George MS, Reeves ST, Takacs I (2010) Bilateral epidural prefrontal cortical stimulation for treatment-resistant depression. Biol Psychiatry 67(2): 101–109

Naritoku DK, Terry WJ, Helfert RH (1995) Regional induction of fos immunoreactivity in the brain by anticonvulsant stimulation of the vagus nerve. Epilepsy Res 22(1):53–62

Pezawas L, Meyer-Lindenberg A, Drabant EM, Verchinski BA, Munoz KE, Kolachana BS, Egan MF, Mattay VS, Hariri AR, Weinberger DR (2005) 5-HTTLPR polymorphism impacts human cingulate-amygdala interactions: a genetic susceptibility mechanism for depression. Nat Neurosci 8(6):828–834

Raichle ME, Snyder AZ (2007) A default mode of brain function: a brief history of an evolving idea. NeuroImage 37(4):1083–1090. discussion 1097–9

Ramsay RE, Uthman BM, Augustinsson LE, Upton AR, Naritoku D, Willis J, Treig T, Barolat G, Wernicke JF (1994) Vagus nerve stimulation for treatment of partial seizures: 2. Safety, side effects, and tolerability. First international Vagus nerve stimulation study group. Epilepsia 35(3):627–636

Riva-Posse P, Choi KS, Holtzheimer PE, McIntyre CC, Gross RE, Chaturvedi A, Crowell AL, Garlow SJ, Rajendra JK, Mayberg HS (2014) Defining critical white matter pathways mediating successful subcallosal cingulate deep brain stimulation for treatment-resistant depression. Biol Psychiatry 76(12):963–969

Riva-Posse P, Choi KS, Holtzheimer PE, Crowell AL, Garlow SJ, Rajendra JK, McIntyre CC, Gross RE, Mayberg HS (2018) A connectomic approach for subcallosal cingulate deep brain stimulation surgery: prospective targeting in treatment-resistant depression. Mol Psychiatry 23(4):843–849

Rush AJ, George MS, Sackeim HA, Marangell LB, Husain MM, Giller C, Nahas Z, Haines S, Simpson RK Jr, Goodman R (2000) Vagus nerve stimulation (VNS) for treatment-resistant depressions: a multicenter study. Biol Psychiatry 47(4):276–286

Rush AJ, Marangell LB, Sackeim HA, George MS, Brannan SK, Davis SM, Howland R, Kling MA, Rittberg BR, Burke WJ, Rapaport MH, Zajecka J, Nierenberg AA, Husain MM, Ginsberg D, Cooke RG (2005) Vagus nerve stimulation for treatment-resistant depression: a randomized, controlled acute phase trial. Biol Psychiatry 58(5):347–354

Sackeim HA, Rush AJ, George MS, Marangell LB, Husain MM, Nahas Z, Johnson CR, Seidman S, Giller C, Haines S, Simpson RK Jr, Goodman RR (2001) Vagus nerve stimulation (VNS) for treatment-resistant depression: efficacy, side effects, and predictors of outcome. Neuropsychopharmacology 25(5):713–728

Schlaepfer TE, Bewernick BH, Kayser S, Mädler B, Coenen VA (2013) Rapid effects of deep brain stimulation for treatment-resistant major depression. Biol Psychiatry 73(12):1204–1212

Steele JD, Christmas D, Eljamel MS, Matthews K (2008) Anterior cingulotomy for major depression: clinical outcome and relationship to lesion characteristics. Biol Psychiatry 63(7):670–677

Wang S (2011) Habenula – a new target for treatment of intractable depression. Sheng Li Ke Xue Jin Zhan 42(6):407–412

The manufacturer's authorised representative in the EU is Springer Nature Customer Service Centre GmbH, Europaplatz 3, 69115 Heidelberg, Germany. If you have any concerns regarding our products, please contact ProductSafety@springernature.com

Printed and bound by CPI Group (UK) Ltd, Croydon, CR0 4YY
01/10/2026
02225961-0014